AF598086

METHODS IN MOLECULAR BIOLOGY™

For further volumes:
http://www.springer.com/series/7651

Jasmonate Signaling

Methods and Protocols

Edited by

Alain Goossens and Laurens Pauwels

Department of Plant Systems Biology, VIB, Gent, Belgium;
Department of Plant Biotechnology and Bioinformatics, Ghent University, Gent, Belgium

Editors
Alain Goossens
Department of Plant Systems Biology
VIB, Gent, Belgium

Department of Plant Biotechnology and Bioinformatics
Ghent University
Gent, Belgium

Laurens Pauwels
Department of Plant Systems Biology
VIB, Gent, Belgium

Department of Plant Biotechnology and Bioinformatics
Ghent University
Gent, Belgium

ISSN 1064-3745 ISSN 1940-6029 (electronic)
ISBN 978-1-62703-413-5 ISBN 978-1-62703-414-2 (eBook)
DOI 10.1007/978-1-62703-414-2
Springer New York Heidelberg Dordrecht London

Library of Congress Control Number: 2013935309

Printed on acid-free paper

Humana Press is a brand of Springer
Springer is part of Springer Science+Business Media (www.springer.com)

Preface

It is now well established that jasmonates, originally identified as the major component of jasmine scent, play a universal role in the plant kingdom and are involved in the regulation of diverse aspects of plant biology, including growth, development, metabolism, and interaction with the environment.

The research field of jasmonate signaling has moved forward rapidly over the last years, and the jasmonate research community is growing considerably. The models of the molecular pathways implicated in jasmonate perception and signaling gain in complexity almost weekly, and the size of the crosstalk network with other hormones or signaling pathways increases as well. As a consequence, it became evident that jasmonates affect far more cellular and physiological processes than originally anticipated.

Hence, in our opinion, an overview of the experimental protocols is very timely, not only of those already used in this field but also more general ones that certainly will become useful in the near future. Omics tools provide unprecedented ways to map and explore jasmonate signaling in plants. However, despite the power of the emerging omics platforms, a reductionist approach is often still required to achieve an unambiguous understanding of the role or function of a particular gene or protein in a signaling cascade. The aim of this book is to unite these two streams and to cover both omics and reductionist tools and protocols. Thereby, we attempted to provide a comprehensive picture of the research possibilities on jasmonate signaling. We collected contributions from the top laboratories studying jasmonate-related processes as well as from experts investigating other plant (hormonal) signaling cascades. In general, we tried to gather essential techniques and methods that can be applied with standard equipment in plant (molecular) biology facilities. As such, the chapters in this book embrace physiological, environmental, molecular, omics, and bioinformatics approaches that allow dissecting jasmonate actions in the model species *Arabidopsis thaliana* or in other plants.

Collectively, we hope that these protocols will empower interested researchers to dissect all steps of jasmonate signaling and the processes they modulate. We wish the reader good luck in this undoubtedly spectacular endeavor!

Gent, Belgium

Alain Goossens
Laurens Pauwels

Contents

Contributors

ANNA AMTMANN • *Plant Science Group, Institute of Molecular, Cell, and Systems Biology, College of Medical Veterinary and Life Sciences, University of Glasgow, Glasgow, UK*
PETER A.H.M. BAKKER • *Plant-Microbe Interactions, Department of Biology, Faculty of Science, Institute of Environmental Biology, Utrecht University, Utrecht, The Netherlands*
IAN T. BALDWIN • *Department of Molecular Ecology, Max Planck Institute for Chemical Ecology, Jena, Germany*
SUSANNE BERGER • *Pharmaceutical Biology, Julius-von-Sachs-Institute for Biosciences, University of Wuerzburg, Wuerzburg, Germany*
STEFANIE DE BODT • *Department of Plant Systems Biology, VIB, Gent, Belgium; Department of Plant Biotechnology and Bioinformatics, Ghent University, Gent, Belgium*
JOHN BROWSE • *Institute of Biological Chemistry, Washington State University, Pullman, WA, USA*
VOLKAN ÇEVIK • *The Sainsbury Laboratory, Norwich, UK*
YOUNG HAE CHOI • *Natural Product Laboratory, Institute of Biology, Leiden University, Leiden, The Netherlands*
REBECCA DE CLERCQ • *Department of Plant Systems Biology, VIB, Gent, Belgium; Department of Plant Biotechnology and Bioinformatics, Ghent University, Gent, Belgium*
JANINE COLLING • *Department of Plant Systems Biology, VIB, Gent, Belgium; Department of Plant Biotechnology and Bioinformatics, Ghent University, Gent, Belgium*
AMPARO PÉREZ CUÉLLAR • *Department of Plant Systems Biology, VIB, Gent, Belgium; Department of Plant Biotechnology and Bioinformatics, Ghent University, Gent, Belgium*
BRECHT DEMEDTS • *Department of Plant Systems Biology, VIB, Gent, Belgium; Department of Plant Biotechnology and Bioinformatics, Ghent University, Gent, Belgium*
SANDRA FONSECA • *Departamento de Genética Molecular de Plantas, Centro Nacional de Biotecnología, Consejo Superior de Investigaciones Científicas, Madrid, Spain*
SUSANNE FORNER • *Leibniz Institute of Plant Biochemistry, Halle, Germany*
EMMANUEL GAQUEREL • *Department of Molecular Ecology, Max Planck Institute for Chemical Ecology, Jena, Germany*
GAËTAN GLAUSER • *Chemical Analytical Service of the Swiss Plant Science Web, Institute of Biology, University of Neuchâtel, Neuchâtel, Switzerland*
ALAIN GOOSSENS • *Department of Plant Systems Biology, VIB, Gent, Belgium; Department of Plant Biotechnology and Bioinformatics, Ghent University, Gent, Belgium*
BETTINA HAUSE • *Leibniz Institute of Plant Biochemistry, Halle, Germany*
SHENG YANG HE • *Department of Energy Plant Research Laboratory, Howard Hughes Medical Institute -Gordon and Betty Moore Foundation, Michigan State University, East Lansing MI, USA; Department of Plant Biology, Howard Hughes Medical Institute -Gordon and Betty Moore Foundation, Michigan State University, East Lansing, MI, USA*

MARCO HERDE • *Department of Energy-Plant Research Laboratory, Michigan State University, East Lansing, MI, USA*
GREGG A. HOWE • *Department of Energy-Plant Research Laboratory, Michigan State University, East Lansing, MI, USA; Department of Biochemistry and Molecular Biology, Michigan State University, East Lansing, MI, USA*
DIRK INZÉ • *Department of Plant Systems Biology, VIB, Gent, Belgium; Department of Plant Biotechnology and Bioinformatics, Ghent University, Gent, Belgium*
YUSUKE JIKUMARU • *Life Science Group, Agilent Technologies, Tokyo, Japan*
MARIO KALLENBACH • *Department of Molecular Ecology, Max Planck Institute for Chemical Ecology, Jena, Germany*
YUJI KAMIYA • *RIKEN Plant Science Center, Yokohama, Japan*
KEMAL KAZAN • *Plant Industry Division, Commonwealth Scientific and Industrial Research Organization, Queensland Bioscience Precinct, St Lucia, Queensland, Australia*
FABIAN KELLERMEIER • *Plant Science Group, Institute of Molecular, Cell, and Systems Biology, College of Medical Veterinary and Life Sciences, University of Glasgow, Glasgow, UK*
HYE KYONG KIM • *Natural Product Laboratory, Institute of Biology, Leiden University, Leiden, The Netherlands*
ABRAHAM J.K. KOO • *Department of Energy-Plant Research Laboratory, Michigan State University, East Lansing, MI, USA; Department of Biochemistry and Molecular Biology, Michigan State University, East Lansing, MI, USA*
NAUSICAÄ LANNOO • *Laboratory of Glycobiology and Biochemistry, Department of Molecular Biotechnology, Faculty of Bioscience Engineering, Ghent University, Gent, Belgium*
NOKWANDA P. MAKUNGA • *Institute for Plant Biotechnology, Department of Genetics, Stellenbosch University, Matieland, Stellenbosch, South Africa*
AJIN MANDAOKAR • *DuPont Knowledge Centre, ICICI Knowledge Park, Hyderabad, India*
HIDEYUKI MATSUURA • *Graduate School of Agriculture, Hokkaido University, Hokkaido, Japan*
JOHAN MEMELINK • *Sylvius Laboratory, Institute of Biology, Leiden University, Leiden, The Netherlands*
KATI MIELKE • *Leibniz Institute of Plant Biochemistry, Halle, Germany*
LAURENS PAUWELS • *Department of Plant Systems Biology, VIB, Gent, Belgium; Department of Plant Biotechnology and Bioinformatics, Ghent University, Gent, Belgium*
CORNÉ M.J. PIETERSE • *Plant-Microbe Interactions, Department of Biology, Faculty of Science, Institute of Environmental Biology, Utrecht University, Utrecht, The Netherlands*
JACOB POLLIER • *Department of Plant Systems Biology, VIB, Gent, Belgium; Department of Plant Biotechnology and Bioinformatics, Ghent University, Gent, Belgium*
TIANCONG QI • *School of Life Sciences, Tsinghua University, Beijing, China*
STEPHANE ROMBAUTS • *Department of Plant Systems Biology, VIB, Gent, Belgium; Department of Plant Biotechnology and Bioinformatics, Ghent University, Gent, Belgium*
MARTHA L. ROWE • *Department of Agronomy and Horticulture, University of Nebraska–Lincoln, Lincoln, NE, USA*
KAZUKI SAITO • *RIKEN Plant Science Center, Yokohama, Japan*
MARTIN A. SELTMANN • *Pharmaceutical Biology, Julius-von-Sachs-Institute for Biosciences, University of Wuerzburg, Wuerzburg, Germany*
MITSUNORI SEO • *RIKEN Plant Science Center, Yokohama, Japan*

NOBUKAZU SHITAN • *Laboratory of Natural Medicinal Chemistry, Kobe Pharmaceutical University, Kobe, Japan*
ROBERTO SOLANO • *Departamento de Genética Molecular de Plantas, Centro Nacional de Biotecnología, Consejo Superior de Investigaciones Científicas, Madrid, Spain*
SUSHENG SONG • *School of Life Sciences, Tsinghua University, Beijing, China*
PAUL E. STASWICK • *Department of Agronomy and Horticulture, University of Nebraska–Lincoln, Lincoln, NE, USA*
MICHAEL STITZ • *Department of Molecular Ecology, Max Planck Institute for Chemical Ecology, Jena, Germany*
AKIFUMI SUGIYAMA • *Laboratory of Plant Gene Expression, Research Institute for Sustainable Humanosphere, Kyoto University, Uji, Japan*
BRYAN THINES • *Keck Science Department, Claremont McKenna, Pitzer, and Scripps Colleges, Claremont, CA, USA*
ELS J.M. VAN DAMME • *Laboratory of Glycobiology and Biochemistry, Department of Molecular Biotechnology, Faculty of Bioscience Engineering, Ghent University, Gent, Belgium*
ROBIN VANDEN BOSSCHE • *Department of Plant Systems Biology, VIB, Gent, Belgium; Department of Plant Biotechnology and Bioinformatics, Ghent University, Gent, Belgium*
RUDY VANDERHAEGHEN • *Department of Plant Systems Biology, VIB, Gent, Belgium; Department of Plant Biotechnology and Bioinformatics, Ghent University, Gent, Belgium*
JOHAN A. VAN PELT • *Plant-Microbe Interactions, Department of Biology, Faculty of Science, Institute of Environmental Biology, Utrecht University, Utrecht, The Netherlands*
SASKIA C.M. VAN WEES • *Plant-Microbe Interactions, Department of Biology, Faculty of Science, Institute of Environmental Biology, Utrecht University, Utrecht, The Netherlands*
ROBERT VERPOORTE • *Natural Product Laboratory, Institute of Biology, Leiden University, Leiden, The Netherlands*
JOHN WITHERS • *Plant Research Laboratory, Department of Energy, Michigan State University, East Lansing, MI, USA; Department of Plant Biology, Michigan State University, East Lansing, MI, USA*
JEAN-LUC WOLFENDER • *School of Pharmaceutical Sciences, EPGL, University of Geneva, University of Lausanne, Geneva, Switzerland*
DAOXIN XIE • *School of Life Sciences, Tsinghua University, Beijing, China*
JIAN YAO • *Department of Energy Plant Research Laboratory, Michigan State University, East Lansing, MI, USA*
KAZUFUMI YAZAKI • *Laboratory of Plant Gene Expression, Research Institute for Sustainable Humanosphere, Kyoto University, Uji, Japan*
KEIKO YONEKURA-SAKAKIBARA • *RIKEN Plant Science Center, Yokohama, Japan*

Part I

Physiology

Chapter 1

Phenotyping Jasmonate Regulation of Senescence

Martin A. Seltmann and Susanne Berger

Abstract

Osmotic stress induces several senescence-like processes in leaves, such as specific changes in gene expression and yellowing. These processes are dependent on the accumulation of jasmonates and on intact jasmonate signaling. This chapter describes the treatment of *Arabidopsis thaliana* leaves with sorbitol as an osmotic stress agent and the determination of the elicited phenotypes encompassing chlorophyll loss, degradation of plastidial membrane lipids, and induction of genes regulated by senescence and jasmonate.

Key words Chlorophyll degradation, Osmotic stress, Galactolipid degradation, Gene induction, *Arabidopsis*

1 Introduction

Leaf senescence is characterized by loss of chlorophyll, induction of the expression of senescence-related genes, decrease in photosynthesis, and degradation of macromolecules, such as proteins and lipids [1]. Senescence occurs naturally during aging. Processes similar to senescence can also be induced by exogenous factors, such as prolonged darkness, osmotic stress, and application of plant signaling molecules. Studies on gene expression and metabolites showed that these induced processes share some features with age-related senescence, such as chlorophyll loss. However, there are fundamental differences regarding the molecular processes and the hormonal regulation [2]. Exogenous application of methyl jasmonate on leaves of different plant species, e.g., oat (*Avena sativa*), barley (*Hordeum vulgare*), and *Arabidopsis thaliana*, elicits senescence-like phenotypes, comprising leaf yellowing, down-regulation of photosynthesis and of ribulose-1,5-bisphosphate carboxylase expression, as well as the induction of genes that are also up-regulated during natural senescence [3–5]. This biological activity together with the fact that levels of endogenous jasmonates increase during natural senescence as well as

Alain Goossens and Laurens Pauwels (eds.), *Jasmonate Signaling: Methods and Protocols*, Methods in Molecular Biology, vol. 1011, DOI 10.1007/978-1-62703-414-2_1, © Springer Science+Business Media, LLC 2013

dark-induced and osmotic stress-induced processes has led to the proposal that jasmonates are involved in regulating initiation and progression of leaf senescence. Nevertheless, analysis of *Arabidopsis* mutants and transgenic plants with alterations in jasmonate biosynthesis and signaling revealed that jasmonates are not necessary for age-related and dark-induced senescence [4, 6]. In contrast, accumulation of endogenous jasmonates is involved in senescence-like phenotypes induced by sorbitol treatment that imposes an osmotic stress [7]. Therefore, sorbitol-induced chlorophyll loss, gene induction, and degradation of plastidial membrane lipids represent processes that depend on and are regulated by jasmonates.

The method presented here describes the treatment with sorbitol as an osmotic stress agent to activate the endogenous jasmonate pathway and the analysis of different parameters related to the senescence-like phenotype that can be tested.

2 Materials

2.1 General Equipment

1. Centrifuge for 1.5-mL tubes.
2. Vacuum freeze-drier.

2.2 Plant Material

Arabidopsis thaliana (L.) Heynh. plants grown in soil under short-day conditions (9 h light [100–120 μmol photons/m^2/s], 22 °C, 15 h dark). Fully expanded leaves of 6-week-old plants are used (*see* **Note 1**).

2.3 Sorbitol Treatment

1. Growth chamber with controllable light and temperature.
2. Transparent plastic containers with lids.
3. Sorbitol solution, 500 mM in water.

2.4 Chlorophyll Extraction and Determination

1. Extraction solvent: Acetone/water 4:1 (v/v).
2. VIS spectrophotometer (for wavelengths 647 and 664 nm).

2.5 Determination of Gene Expression

1. Plant RNA extraction kit available from a molecular biology supplier (we used E:Z:N:A Plant RNA Mini Kit; Omega Bio-Tek, Norcross, GA, USA).
2. UV spectrophotometer (for wavelength 260 nm).
3. Reverse transcriptase available from a molecular biology supplier (we used M-MLV reverse transcriptase; Promega, Madison, WI, USA).
4. Oligonucleotide primers (*see* Table 1).
5. Quantitative (q)PCR SYBR Green Master mix available from a molecular biology supplier (we used absolute qRT-PCR

Table 1
Sequence of oligonucleotide primers

Gene	Sequence forward 5′–3′	Sequence reverse 5′–3′	Product length (bp)
LOX2 (At3g45140)	GTACGTCTGACGATACC	TCTGGCGACTCATAGAA	398
AOS (At5g42650)	CCATACATTTAGTCTACCAC	GCTAATCGGTTATGAACTTG	243
VSP1 (At5g24780)	ACAAAGAGGCATATTTTTAC	GGTTCAATCCCGAGTTCAA	241
SAG13 (At2G29350)	TGTCCTTGGTATATCACAACT	TTCATAGATTATGGATGCGG	414
SEN1 (At4g35770)	AACATGTGGATCTTTCAAGTGCC	GTCGTTGCTTTCCTCCATCG	96
ACTIN2/8 (At5g09810/ At1g49240)	GGTGATGGTGTGTCT	ACTGAGCACAATGTTAC	434

SYBR Green Mix from Thermo Fisher Scientific, Waltham, MA, USA).

6. A thermal cycler for qPCR (we used Mastercycler ep realplex S, Eppendorf, Hamburg, Germany).

2.6 Extraction of Digalactosyldiacylglycerols and Monogalactosyldiacylglycerols

1. Water bath or incubator.
2. Ultrasonic bath.
3. 2-Propanol and chloroform/methanol 1:3 (v/v).
4. Internal standards: Monogalactosyldiacylglycerols (MGDG) 18:0–18:0 and digalactosyldiacylglycerols (DGDG) 18:0–18:0 in chloroform/methanol/water 4:1:0.1 (v/v/v) (Matreya LLC, Pleasant Gap, PA, USA).
5. Ammonium acetate in methanol (1 mM).

2.7 Separation and Quantification of MGDG and DGDG

1. Ultra high-performance liquid chromatography (UPLC)/ tandem mass spectrometry (MS/MS) for separation and quantification of galactolipids (we used a Quattro Premier XE triple-quadrupole mass spectrometer (Waters Corporation, Milford, MA, USA), with an electrospray interface (ESI) coupled to an Aquity UPLC [Waters]).
2. Column: Aquity UPLC BEH C8 (2×50 mm, 1.7 μm particle size with a 2.1×5 mm guard column (Waters) or equivalent).

3. Solvents: 1 mM aqueous ammonium acetate and 1 mM ammonium acetate in methanol.
4. Argon as collision gas, nitrogen as desolvation and cone gas.

3 Methods

3.1 Sorbitol Treatment

1. Detach leaves (*see* **Note 2**) and place them upside down on a 500-mM sorbitol solution in a transparent container with transparent lid (*see* **Note 3**).
2. In parallel, place detached leaves in containers with water instead of sorbitol solution as controls.
3. Incubate the leaves floating on the solution under continuous light conditions (approximately 120 μmol photons/m^2/s) at 22 °C. We used 200 mL of solution in containers of 21 cm × 10 cm × 5 cm, keeping approximately 40 cm distance from the illumination source to avoid warming and evaporation of the solution.
4. Harvest before treatment and after 24 and 48 h (*see* **Note 4**). Quickly rinse the leaves twice with water to remove the sorbitol and immediately freeze the material in liquid nitrogen (store at −80 °C).
5. Freeze-dry (*see* **Note 5**) the sample material.
6. Grind material to a fine powder with mortar and pestle and store at −20 °C. Use the same material for all analyses (*see* **Note 6**).

3.2 Analysis of Chlorophyll Loss

The method for determination of chlorophyll levels has been described previously [8]. The amount of chlorophyll is calculated relative to the dry weight; therefore, the exact determination of the weight of the material used for extraction is important.

1. Add 1 mL 80 % acetone (v/v) to 5–10 mg of freeze-dried material.
2. Extract for 3–6 h in the dark at 4 °C and shake vigorously from time to time.
3. Centrifuge for 3 min at 2,700 × *g* to pellet the leaf material.
4. Transfer the supernatant to a new tube and determine the extinction against 80 % acetone in a spectrophotometer at 664 and 647 nm. If extinction is ≥1, the extract has to be diluted with 80 % acetone.
5. Calculate as follows:

   Chlorophyll a (mg/mL) = $(11.78 \times E_{664} - 2.29 \times E_{647}) \times$ dilution factor

   Chlorophyll b (mg/mL) = $(20.05 \times E_{647} - 4.77 \times E_{664}) \times$ dilution factor

   Chlorophyll a + b/dry weight (mg/g)

3.3 Analysis of Gene Expression

For expression analysis, genes are suitable that are induced by jasmonates (e.g., *LOX2*, *AOS*, and *VSP1*) or by senescence and jasmonates (such as *SAG13* and *SEN1*) (*see* **Note 7**). Expression of these genes will always be calculated relative to constitutively expressed genes such as *ACTIN 2* and *ACTIN 8*. Throughout the work with RNA, use RNAse-free water (for instance, treated with diethylpyrocarbonate).

1. Isolate RNA by using a plant RNA extraction kit according to manufacturer's instructions. We used 10 mg of freeze-dried material for each sample.
2. Determine the concentration by measuring the absorption at 260 nm.
   Calculate $c\ (\mu g/mL) = E_{260} \times 40 \times \text{dilution factor}$.
3. Remove DNA by DNAse treatment during the RNA extraction or afterward.
4. Reverse transcribe 1 μg RNA into cDNA and dilute to 25 ng/μL.
5. Perform qPCR at an annealing temperature of 58 °C for all reactions.

   Reaction: Total volume of 20 μL containing cDNA corresponding to 50 ng RNA (2 μL), 12 pmol forward primer, 12 pmol reverse primer, and qPCR-SYBR-Green-mix;

   Cycler conditions: Initial denaturation for 15 min at 95 °C, followed by 45 cycles, comprising denaturation for 15 s at 95 °C, annealing for 20 s at 58 °C, elongation for 20 s at 72 °C, detection for 10 s at 79 °C, and a final denaturation for 10 s at 95 °C.

3.4 Determination of the Decrease in MGDG and DGDG

The amounts of galactolipids are calculated relative to the dry weight; therefore, it is important to determine precisely the weight of the material used for the extraction.

1. Add 1.5 mL of 2-propanol at a temperature of 75 °C to 20 mg freeze-dried material. Conveniently, 2-mL screw cap tubes are used.
2. Add as internal standard 3 μg of MGDG 18:0–18:0 and 3 μg of DGDG 18:0–18:0.
3. Incubate for 15 min at 75 °C to deactivate lipases.
4. Sonicate for 5 min in an ultrasonic bath.
5. Spin for 10 min at 20,500 × *g* at room temperature.
6. Transfer supernatants to test tubes and extract the pellet with 1 mL chloroform/isopropanol (1:3 v/v).
7. Spin, combine the supernatants, and extract the pellets with 1 mL chloroform/methanol (1:3 v/v) again.

8. Spin, combine the supernatants, and dry under a nitrogen stream in a water bath at 60 °C.
9. Reconstitute in 100 μL methanol containing 1 mM ammonium acetate for the liquid chromatography (LC)-MS/MS analysis.
10. Inject into the UPLC system.
11. For the chromatographic separation, from the reversed-phase column elute with a gradient at a flow rate of 0.3 mL/min at 30 °C (a) 25 % 1 mM aqueous ammonium acetate and 75 % 1 mM ammonium acetate in methanol for 1 min; (b) from 25 to 0 % 1 mM aqueous ammonium acetate and from 75 to 100 % 1 mM ammonium acetate in methanol in 10 min; (c) purge with 100 % ammonium acetate in methanol for 1 min; and (d) equilibrate with 25 % 1 mM aqueous ammonium acetate and 75 % ammonium acetate in methanol for 4 min.
12. For detection, operate the ESI source in negative ionization mode and quantify with a multiple reaction-monitoring with a scan time of 0.025 s per transition. Use nitrogen as the desolvation and cone gas with a flow rate of 800 and 100 L/h, respectively, and argon as the collision gas at a pressure of approximately 3.10×10^{-3} bar. For mass-to-charge ratios, *see* Table 2. Conditions are as follows: capillary voltage, 3.0 kV at 120 °C; cone voltage and collision energy, 26 eV (40 and 30); capillary, 3.0 kV; source temperature, 120 °C; and desolvation temperature, 450 °C.

4 Notes

1. Take leaves that have reached their final size and without any sign of senescence. Starting from the first fully expanded leaf, we used the next 5–8 leaves. Older leaves differ in their response and the chlorophyll loss is quicker than that of younger leaves, whereas the gene expression (for instance, *VSP1*) might be less induced.
2. Take most of the petiole with the leaves; place the petiole in sorbitol solution.
3. Freshly prepare a solution with sterile water. As the leaves are not free of contamination, it might become a problem after some days. In our hands, addition of one single antibiotic was not sufficient to prevent microbial growth, whereas antibiotics or their degradation products might affect the experiments. Nevertheless, contamination was observed only at incubation times of 96 h and longer. Therefore, we did not add antibiotics nor extended the incubation times to more than 72 h.
4. Time points are critical for the analysis. Here, we give the time points that gave the best results in our hands, but they might

Table 2
Mass-to-charge ratios of parent and daughter ions from the analyzed molecules

Analyte	*m/z* Parent ion	*m/z* Daughter ion
MGDG 18:0–18:0 (standard)	785.5	283.0
DGDG 18:0–18:0 (standard)	947.5	283.0
DGDG 18:3–16:0	913.0 913.0	277.0 255.0
DGDG 18:3–18:0	941.0 941.0	283.0 277.0
DGDG 18:3–10:1	939.0 939.0	281.0 277.0
DGDG 18:3–16:1	911.0 911.0	277.0 253.0
DGDG 18:3–18:3	935.0	277.0
DGDG 18:3–16:2	909.0 909.0	277.0 251.0
DGDG 18:3–16:3	907.0 907.0	277.0 249.0
DGDG 18:3–18:2	937.0 937.0	279.0 277.0
MGDG 18:3–16:0	751.0 751.0	277.0 255.0
MGDG 18:3–16:1	749.0 749.0	277.0 253.0
MGDG 18:3–18:2	775.0 775.0	279.0 277.0
MGDG 18:3–16:2	747.0 747.0	277.0 251.0
MGDG 18:3–16:3	745.0 745.0	277.0 249.0
MGDG 18:3–18:3	773.0	277.0
MGDG 18:3–18:1	777.0 777.0	281.0 277.0
MGDG 18:3–18:0	779.0 779.0	283.0 277.0

vary with the plant material and the actual incubation conditions. Leaf yellowing was best observed at 48 and 72 h (Fig. 1). Differences in chlorophyll content between wild type and plants with defective jasmonate pathways were most pronounced at

Fig. 1 *Arabidopsis* (Col-0) leaves floating on 500 mM sorbitol. Pictures were taken 24, 48, and 72 h after flotation on 500 mM sorbitol (**a**) or water (**b**)

these time points [7]. However, after 72 h on sorbitol solution, wild-type leaves showed deterioration already. Maximum induction of gene expression of all genes tested was observed at 24 h—correlating with the accumulation of jasmonic acid—and at this time point, differences between the wild type and mutants were the strongest [7, 9]. Decrease in MGDG and DGDG levels and differences between wild type and mutants were best detectable at 48 h [7]. The most prominent galactolipids, MGDG 18:3–16:3 and DGDG 18:3–18:3, showed the strongest decrease and differences.

5. The advantage of the use of freeze-dried material and the calculation based on dry weight is that variation in the water content/remaining solution on the leaves does not affect the results, improving comparability and reproducibility. If data relative to fresh weight are preferred, a calculation factor can be determined by measuring the fresh and dry weights from

the same material. We found that dry weight constituted 6 % of the fresh weight. Consequently, if fresh, instead of dry, material will be used, take approximately 10 times the above-specified amount of freeze-dried material.

6. Between 15 and 20 leaves should give enough material for one replicate for the analysis of chlorophyll content, lipid levels, and gene expression. Use at least three plants for one replicate to obtain an average of different plants.
7. Only a small or no increase in expression was seen for the senescence-regulated gene *SAG12* and for the jasmonate-regulated gene *PDF1.2*.

Acknowledgments

We would like to acknowledge the help of Nadja Stingl and Martin J. Mueller. This work was supported by the SFB567 and the GK1342 of the "Deutsche Forschungsgemeinschaft."

References

1. Lim PO, Kim HJ, Nam HG (2007) Leaf senescence. Annu Rev Plant Biol 58:115–136
2. van der Graaff E, Schwacke R, Schneider A, Desimone M, Flügge U-I, Kunze R (2006) Transcription analysis of Arabidopsis membrane transporters and hormone pathways during developmental and induced leaf senescence. Plant Physiol 141:776–792
3. Ueda J, Kato J (1980) Isolation and identification of a senescence-promoting substance from wormwood (*Artemisia absinthium* L.). Plant Physiol 66:246–249
4. He Y, Fukushige H, Hildebrand DF, Gan S (2002) Evidence supporting a role of jasmonic acid in Arabidopsis leaf senescence. Plant Physiol 128:876–884
5. Reinbothe S, Reinbothe C, Parthier B (1993) Methyl jasmonate represses translation initiation of a specific set of mRNAs in barley. Plant J 4:459–467
6. Schommer C, Palatnik JF, Aggarwal P, Chételat A, Cubas P, Farmer EE, Nath U, Weigel D (2008) Control of jasmonate biosynthesis and senescence by miR319 targets. PLoS Biol 6:e230
7. Seltmann MA, Stingl NE, Lautenschlaeger JK, Krischke M, Mueller MJ, Berger S (2010) Differential impact of lipoxygenase 2 and jasmonates on natural and stress-induced senescence in Arabidopsis. Plant Physiol 152:1940–1950
8. Arnon DI (1949) Copper enzymes in isolated chloroplasts. Polyphenoloxidase in *Beta vulgaris*. Plant Physiol 24:1–15
9. Seltmann MA, Hussels W, Berger S (2010) Jasmonates during senescence: signals or products of metabolism? Plant Signal Behav 5:1493–1496

Chapter 2

Characterizing Jasmonate Regulation of Male Fertility in Arabidopsis

Bryan Thines, Ajin Mandaokar, and John Browse

Abstract

Coordination of events leading to fertilization of *Arabidopsis* flowers is tightly regulated, with an essential developmental cue from jasmonates (JAs). JAs coordinate stamen filament elongation, anther dehiscence, and pollen viability at stage 12 of flower development, the stage immediately prior to flower opening. Characterization of JA-biosynthesis and JA-response mutants of *Arabidopsis*, which usually have a complete male sterility phenotype, has contributed to the understanding of how JAs work in these reproductive processes. These mutants have also been fundamental to the identification of JA-dependent genes acting in male reproductive tissues that accomplish fertilization. The list of JA-dependent genes continues to grow, as does the necessity to characterize novel JA mutant and related transgenic plants. It is therefore instructive to place these genes and mutants in the framework of established JA responses. Here, we describe the phenotypic characterization of flowers that fail to respond to the JA signal. We also measure gene expression in male reproductive tissues of flowers with the aim of identifying their role in JA-dependent male fertility.

Key words Jasmonic acid, Flower development, Stamen, Anther, Pollen, Sterility

1 Introduction

Jasmonoyl-isoleucine, the active form of the jasmonate hormone, is an oxylipin signaling molecule derived from α-linolenic acid (18:3). Characterization of the *fad3-2 fad7-2 fad8* mutant, which is defective in the activity of three fatty acid desaturases required for synthesis of this JA precursor, led to the discovery that JA is required for male reproductive processes in *Arabidopsis thaliana* [1]. Other *Arabidopsis* mutants lacking either the ability to synthesize JA or the capacity to perceive JA are male sterile as well [2–6]. Here, we refer collectively to both JA biosynthesis and response/signaling mutants as "JA mutants." Stamen filaments in these JA mutants do not elongate to place anthers at the stigmatic (female) surface for self-pollination. Furthermore, the anthers do not dehisce and, although pollen develops to the trinucleate stage, it is predominantly nonviable. Whether this nonviability is due to pollen-specific defects or the

Alain Goossens and Laurens Pauwels (eds.), *Jasmonate Signaling: Methods and Protocols*, Methods in Molecular Biology, vol. 1011, DOI 10.1007/978-1-62703-414-2_2, © Springer Science+Business Media, LLC 2013

inability of anther tissues to provide nutritive support in later stages of pollen development is still an open question.

Flowers in JA mutants develop and mature normally until just before opening, at which point they are designated as "stage 12" [7]. In contrast to the wild-type stage 12 that lasts 24–48 h before the flowers open, flower opening in JA mutants is somewhat delayed and flowers are sterile for the reasons outlined above. Notably, exogenous treatment of stage-12 flower buds on JA biosynthesis mutants with methyl jasmonate (MeJA) rescues this male-sterile phenotype, with full silique development and seed set as a consequence. Importantly, only stage-12 flowers respond to the JA signal, which supports the notion that JA acts as a trigger within this narrow developmental window. Based on previous investigations of JA-responsive genes in stamens during stage 12, three key regulators have been identified that serve as marker genes for JA-dependent male reproductive processes. Among these are two genes encoding transcription factors, *MYB21* (At3g27810) and *MYB24* (At5g40350), which are induced by JA within 30 min of treatment [8]. A third transcription factor-encoding transcript, *MYB108* (At3g06490), begins to accumulate 8 h after JA treatment [9].

Characterization of phenotypes and JA-inducible gene expression patterns in reproductive tissues of wild type and JA mutants continues to provide important clues regarding the role of JAs in *Arabidopsis* flower development. In JA mutants, lack of stamen elongation and anther dehiscence, as well as their nonviable pollen, is accepted as a complete absence of the JA signal in stage-12 flowers. Identification of these mutants and their gene responses has yielded quantifiable aspects of this male-sterile phenotype. A critical tool provided by JA biosynthesis mutants is the ability to synchronize gene expression and developmental events after the JA signal is given [8, 10]. As new JA-responsive genes are discovered, characterization of their corresponding overexpression and knockout lines can help elucidate gene function by fitting sometimes incomplete male-sterile phenotypes into the established framework of the complete JA male-sterile phenotype [9]. Below, we outline methods used to recognize and quantify JA-dependent male sterility in *Arabidopsis* as well as methods for synchronizing and studying gene expression in male reproductive tissues of JA-biosynthetic mutants.

2 Materials

2.1 Characterization of Stamen Elongation

1. Soil (Sunshine Mix #1 [Sun Gro Horticulture, Vancouver, Canada], or other potting soil).
2. Wild-type and JA mutant *Arabidopsis thaliana* (L.) Heyhn. seeds.
3. Growth chamber or greenhouse.
4. Fine-tipped forceps.
5. Crossing goggles (jewelers magnifying goggles).

6. Glass slides.
7. Light microscope equipped with digital camera.
8. Computer with ImageJ software (http://rsbweb.nih.gov/ij).

2.2 Characterization of Anther Dehiscence

1. Soil (Sunshine Mix #1, or other potting soil).
2. Wild-type and JA mutant *Arabidopsis* seeds.
3. Growth chamber or greenhouse.
4. Fine-tipped forceps.
5. Crossing goggles.

2.3 Characterization of Pollen Viability

1. Round 5-cm Petri dishes containing fresh pollen-germination medium: 17 % (w/v) sucrose, 1 mM $CaCl_2$, and 102 mg/L boric acid, pH 7.0, solidified with 0.6 % (w/v) agarose.
2. UV/light microscope equipped with digital camera.
3. Stock solution of 2 mg/mL fluorescein diacetate (Sigma-Aldrich, St. Louis, MO, USA) in acetone. For working solution, add stock solution dropwise to 17 % (w/v) sucrose until the solution becomes milky.
4. Stock solution of 1 mg/mL propidium iodide in water. Working solution is diluted to 100 μl/mL with 17 % (w/v) sucrose.
5. Immediately before experiment, equal amounts of fluorescein diacetate and propidium iodide working solutions are mixed.

2.4 Chemical Treatment of Flower Buds with Methyl Jasmonate

1. Soil (Sunshine Mix #1, or other potting soil).
2. Wild-type and JA mutant *Arabidopsis* plants grown in 4-in. (10-cm) pots.
3. Growth chamber or greenhouse.
4. MeJA (Bedoukian Research Inc., Danbury, CT, USA) supplemented to a final concentration of 0.01–0.03 % in a solution of 0.1 % (v/v) Tween-20. As MeJA does not immediately dissolve when initially added to the 0.1 % Tween-20 solution, it must be mixed with a stir bar until the oily bubbles disappear.

2.5 Collection of Stamens for Measuring Gene Induction by JA

1. Styrofoam float.
2. Fine-tipped forceps.
3. Liquid nitrogen.
4. RNase-free 1.5-mL microcentrifuge tubes.

2.6 Total RNA Preparation

1. RNase-free 1.5-mL microcentrifuge tubes and matching pestles.
2. RNase-free water.
3. TRIzol (Life Technologies, Carlsbad, CA, USA).
4. RNeasy kit (Qiagen, Hilden, Germany).
5. UV/Vis spectrophotometer.

2.7 cDNA Synthesis

1. SuperScriptIII First-strand cDNA synthesis kit (Life Technologies).
2. RNase-free 0.2-mL PCR tubes.
3. Thermal cycler.

2.8 Measurement of JA-Dependent Reproductive Marker Genes by Quantitative PCR

1. First-strand cDNA (*see* Subheading 2.7).
2. qPCR master mix (Invitrogen).
3. 1.5 mM $MgCl_2$.
4. 0.2 mM dNTPs.
5. SYBR Green mix (Life Technologies).
6. ROX dye (Life Technologies).
7. *Taq* polymerase.
8. Real-time thermal cycler.
9. Gene-specific primers:
   (a) *MYB21*
   Fw-5′-TAAAACGAACCGGGAAAAGTT-3′
   Rv-5′-GCGGCCGAATAGTTACCATAG-3′
   (b) *MYB24*
   Fw-5′-CAAAATGGGGAAATAGGTGGT-3′
   Rv-5′-TCATC TCATCGACGCTCCAATAGTTT-3′
   (c) *MYB108*
   Fw-5′-AATGGAGAAGGTCGCTGGAACTCT-3′
   Rv-5′-CGTTGTCCGTTCTTCCCGGTAAAT-3′
   (d) *ACTIN2*
   Fw-5′-GGTGATGGTGTGTCTCACACTG-3′
   Rv-5′-GAGGTTTCCATCTCCTGCTCGTAG-3′

3 Methods

3.1 Characterization of Stamen Elongation

1. Grow *Arabidopsis* wild-type and JA mutant plants on soil until plants reach the reproductive phase and produce a primary shoot and flowers.
2. Collect stage-12 flowers (*see* Fig. 1a and Subheading 1) and flowers that are in the process of opening from flower bud clusters on both wild-type and JA mutant plants.
3. With fine-tipped forceps, carefully remove sepal and petal organs from all flowers leaving intact the pistil and all six stamens (two short and four long), including anthers. Use crossing goggles to aid with visualization.
4. Arrange the pistils and stamens (anther along with filament) together with a length marker of known size on a glass slide

Fig. 1 Flower development and anther dehiscence in *Arabidopsis*. (**a**) Time course of flower development. The "0" designates stage 12 of development. Numbers continue through the later stages, including fertilization and flower opening (reproduced from [9] with permission © American Society of Plant Biologists; www.plantphysiol.org). (**b**) Open flowers in the wild type (*left*) and the *opr3* mutant (*right*). In wild-type flowers, stamen filaments elongate to place dehisced anthers on the stigmatic surface where pollen is deposited. In the *opr3* mutant, stamen filaments do not elongate and anthers do not dehisce so that no pollen is deposited and fertilization does not occur

and view with a microscope. Take a series of digital images of these floral organs, including the length marker.

5. Open images in ImageJ on a computer, calibrate length by using the size marker, and calculate the exact length of all six stamens from the base of the filament to the top of the anther. Likewise, measure the length of the carpel (from the base to the stigma surface) of all the flowers and then calculate the carpel-to-stamen length ratio in the flowers.

3.2 Characterization of Anther Dehiscence

1. Grow *Arabidopsis* wild-type and JA mutant plants on soil until plants reach the reproductive phase and produce a primary shoot and flowers.
2. From a single flower bud cluster, select a set of seven flowers to use as a developmental series as follows. (a) Identify the most mature, but unopened, flower bud, which corresponds to stage 12 of floral development, and designate this as flower "0." In wild-type plants, this flower typically does not have any dehisced anthers. Flowers at later stages of development will be farther from the center and/or below the bud cluster and will be in the process of opening. (b) Use flower "0" and the set of six open flowers as a developmental series, starting at stage 12, and designate these flowers "1" to "6" (*see* Fig. 1a). Typically, the first open flowers in the wild type will have six dehisced anthers.
3. Count the number of dehisced anthers in each flower from the wild-type and JA mutant plants. In dehisced anthers, the anther locule folds back and releases pollen. Some of this pollen is deposited on the stigmatic surface. Conversely, anthers that have not dehisced remain closed and, in the mutant, no pollen is observed on the stigmatic surface (*see* Fig. 1b).
4. Continue the analysis for 3–4 days, or until 50 flowers are analyzed.

3.3 Characterization of Pollen Viability

1. Before starting, make sure that the plates containing pollen germination medium are at room temperature. Harvest pollen from mature flowers by gently using forceps to peel back sepals and petals of flowers that are in the process of opening (*see* **Note 1**). Alternatively, open flowers and anther locules manually by using the forceps. Gently touch the anthers to the surface of the plate to distribute the pollen.
2. Incubate plates in the dark for 16–20 h at room temperature (*see* **Note 1**).
3. Calculate the percentage of pollen germination for wild-type control and mutants as follows. In five or more randomly selected microscope fields, count the number of germinated and ungerminated pollen grains. Use these numbers to calculate the average germination (±standard error). Wild-type pollen should have a germination rate greater than 90 %.
4. Pollen viability can also be tested by chemical staining. In this method, place freshly isolated pollen (*see* Subheading 3.3, **step 1**) on a glass microscope slide (*see* **Note 2**).
5. Immediately add fluorescein diacetate/propidium iodide mix dropwise to pollen.
6. Cover pollen with a coverslip and view immediately with the microscope under UV light. Pollen can be viewed immediately after

placing coverslip. This protocol stains viable pollen blue-green and inviable pollen red-brown (*see* **Note 3**). Typically, more than 90 % of the wild-type pollen grains are viable.

3.4 Chemical Treatment of Flower Buds with Methyl Jasmonate

1. Aliquot MeJA solution at room temperature into 2-mL microcentrifuge tubes (*see* **Note 4**).
2. Gently lay plant-containing pots on their side and immerse the primary bud cluster into the MeJA solution for a few seconds (*see* **Note 5**). By laying pots on their side, bending of the shoot is minimized as is handling of the stem and bud cluster. Return pots to their upright position and place back on growth chamber shelf.
3. Observe initial silique elongation 1–2 days after dipping and full elongation up to 3 days after treatment. The rate at which siliques are produced will vary depending on environmental conditions, especially the ambient temperature and photocycle. Healthy flower bud clusters typically contain 2–5 unopened stage-12 flowers, the stage at which this treatment is effective, and, correspondingly, will produce a number of siliques from a single MeJA treatment.

3.5 Stamen Collection for Measuring Gene Induction by JA

1. Grow *Arabidopsis* plants as indicated above (*see* Subheading 3.1, **step 1**). Use a JA biosynthesis mutant, such as *opr3*, treated and untreated for a positive and negative JA control, respectively.
2. Treat flower buds with MeJA solution (*see* Subheading 3.4, **step 1**) and leave in growth chamber for 30 min, or other duration, as required by the experiment.
3. For stamen collection, prepare the 1.5-mL microcentrifuge tube in a float and place it in liquid nitrogen so that the tube is chilled, but the top is open and accessible.
4. Flowers should be removed one at a time from plants kept in the growth chamber. Remove a single MeJA-treated stage-12 flower bud from the flower bud cluster. Harvest stamens (filaments along with anthers) from one flower before taking another.
5. From the harvested flower, peel back or remove sepals and petals with fine-tipped forceps and expose the stamens. Remove the pistil at the base, which helps increase the yield of male tissues. Lift the stamens from the base with forceps. Dip stamen along with forceps in open 1.5-mL microcentrifuge tubes in liquid nitrogen, making sure that all the stamens are released from the tip of the forceps. Minimize the time between harvest of a flower and transfer of stamens to liquid nitrogen to minimize postharvest changes in gene expression.
6. Continue to harvest flowers one at a time, dissect organs, and harvest stamens until at least 10 mg of tissue is collected, which corresponds to approximately 60 flowers.

7. Alternate treatment and harvest of flowers as needed to ensure that treatment time for each flower is within approximately 5 % of the specified time.
8. Isolate RNA immediately or keep the tissue in liquid nitrogen or at −80 °C until further processing.

3.6 Total RNA Isolation

1. Use at least 10 mg of stamen tissue for ease of processing and maximum recovery of RNA. Grind 10–30 mg of stamen (filament + anther) tissue with a plastic pestle in 1.5-mL microcentrifuge tubes in liquid nitrogen.
2. Resuspend the ground tissue in TRIzol according to the manufacturer's instructions for RNA isolation and use the RNeasy kit for RNA purification.
3. Redissolve RNA in RNase-free water.
4. Measure the RNA concentration and purity with an UV/Vis spectrophotometer. It is desirable for both the 260/230 and 260/280 ratios to be above 1.9.
5. Store RNA at −80 °C.

3.7 First-Strand cDNA Synthesis

1. Perform first-strand cDNA synthesis with 2 μg of total RNA with the SuperScript III cDNA synthesis kit according to the manufacturer's instructions.
2. Set up the first-strand synthesis reaction as follows:

Total RNA (2.0 μg)	*n* μL
Oligo dT primer	1.0 μL
10 mM dNTPS	1.0 μL
RNase/DNase-free water to make	10.0 μL

3. Incubate in a thermal cycler at 65 °C for 5 min, and then place immediately on ice for at least 1 min. Move the contents of the tube to the bottom by brief centrifugation for a few seconds at maximum speed.
4. Add the following to the tube on ice:

10× RT buffer	2.0 μL
25 mM $MgCl_2$	4.0 μL
0.1 M DTT	2.0 μL
RNaseOUT™	1.0 μL
SuperScript III™ RT	1.0 μL

5. Incubate reaction mix in thermal cycler at 50 °C for 50 min.
6. Terminate reaction at 85 °C for 5 min and then chill on ice.
7. Store the first-strand cDNA synthesis reaction at −20 °C, or proceed directly to quantitative PCR.

3.8 Measurement of JA-Dependent Reproductive Marker Genes by Quantitative PCR

1. Run each reaction in triplicate and use three biological replicates. Use *ACTIN2* as the normalizing gene, and an untreated JA mutant sample a control.
2. Set up 20-μL reactions for JA-treated and control samples as follows:

   2 μL of diluted first-strand cDNA.

   2.0 μL 1× PCR buffer.

   1.5 mM $MgCl_2$.

   0.2 mM dNTPs.

   0.2 μM of each sequence-specific primer.

   0.6 μL of ROX dye (diluted to 1:500).

   10 μL SYBR Green mix with *Taq* polymerase.

   Reaction conditions are as follows: Denaturation at 95 °C for 2 min, followed by 40 cycles at 95 °C for 15 s, at 55 °C for 30 s, and at 72 °C for 30 s. At the end of the run, perform a melt curve analysis of the PCR products.
3. Collect C_T values for the samples and calculate the relative expression of the JA-inducible genes using $\Delta\Delta C_T$ values (*see* **Note 6**).

4 Notes

1. Pollen germination assays have a reputation for being highly variable and temperature is a major factor in determining viability and reproducibility [11]. Therefore, we suggest that incubation at room temperature is monitored closely, because the building temperature can vary with the time of day. Alternatively, incubate in a temperature-controlled environment.
2. The ideal time to harvest pollen is when stamens have elongated to the point that anthers are placed just level with the stigmatic surface, but have been dehisced for less than 1 day. If grown in a photoperiod, morning is the best time to harvest pollen.
3. Fluorescein diacetate is taken up by living cells and converted to fluorescein, which emits blue–green light under UV irradiation [12]. Propidium iodide is excluded from living cells, but labels dead cells with red-orange fluorescence under UV irradiation [13].
4. Fresh MeJA solution is best, but it can be stored at 4 °C. When stored solution is utilized, let it warm up to growth chamber temperature before use. Depending on the topic of study,

other JAs or JA precursors may be employed as chemical treatment. For example, 12-oxo phytodienoic acid (Cayman Chemical Company, Ann Arbor, MI, USA) and fatty acids (Nuchek, Elysian, MN, USA) have been used to rescue fertility.

5. Alternative methods used to treat flower buds include (a) spraying with an atomizer of spritzer bottle, or (b) painting. Instead of treating large batches of plants, dipping is the most effective and reproducible method. It is critical to consider the presence and location of plants in the same growth chamber meant to stay untreated. MeJA is volatile and, if present at high enough concentrations, can trigger fertility in untreated JA-biosynthetic mutant plants. Therefore, we favor dipping of flower buds over spraying, because application is much more controlled and plants can be kept in the growth chamber during treatment. Furthermore, by dipping specific flower buds (i.e., those on the primary bud cluster), treatment is more uniform. Healthy, approximately 1-month-old plants respond best to treatment, whereas overly mature and stressed plants will yield lower silique/seed production and the floral tissue will become necrotic. Another advantage of dipping over spraying is that the potentially harmful effects of MeJA are kept off the vegetative tissues. Damage may also occur when the MeJA concentration is too high, which is often the case when flower buds appear red around the base (anthocyanins). Although the outcome remains apparently unchanged, it may be a sign that the upper end of the useful concentration range is being approached.

6. This method assumes that both target and reference genes are amplified with near 100 % efficiency. Perform dilution series with cDNA and each primer pair to ensure that reactions have near 100 % amplification efficiency. Relative quantification of experimental samples is compared to the calibrator or the untreated control. Here, the experimental sample and the control are JA-treated and untreated *opr3* stamens, respectively. JA-inducible genes used are *MYB21, MYB24, and MYB108*, and *ACTIN* is used as a reference gene. Calculate the ΔC_T value for both the experimental and calibrator samples by subtracting the reference C_T from the target C_T for each (i.e., $\Delta C_{T(\text{experimental})} = C_{T(\text{MYB21})} - C_{T(\text{ACTIN})}$ and $\Delta C_{T(\text{calibrator})} = C_{T(\text{MYB21})} - C_{T(\text{ACTIN})}$). Then, calculate the $\Delta\Delta C_T$ by normalizing the ΔC_T of the experimental sample to the ΔC_T of the calibrator ($\Delta\Delta C_T = C_{T(\text{experimental})} - C_{T(\text{calibrator})}$). Finally, calculate the fold-change (normalized expression $= 2^{-\Delta\Delta C_T}$).

References

1. McConn M, Browse J (1996) The critical requirement for linolenic acid is pollen development, not photosynthesis, in an Arabidopsis mutant. Plant Cell 8:403–416
2. Feys BJF, Benedetti CE, Penfold CN, Turner JG (1994) Arabidopsis mutants selected for resistance to the phytotoxin coronatine are male sterile, insensitive to methyl jasmonate, resistant to a bacterial pathogen. Plant Cell 6:751–759
3. Stintzi A, Browse J (2000) The *Arabidopsis* male-sterile mutant, *opr3*, lacks the 12-oxophytodienoic acid reductase required for jasmonate synthesis. Proc Natl Acad Sci USA 97: 10625–10630
4. von Malek B, van der Graaff E, Schneitz K, Keller B (2002) The *Arabidopsis* male-sterile mutant *dde2-2* is defective in the *ALLENE OXIDE SYNTHASE* gene encoding one of the key enzymes of the jasmonic acid biosynthesis pathway. Planta 216:187–192
5. Ishiguro S, Kawai-Oda A, Ueda J, Nishida I, Okada K (2001) The *DEFECTIVE IN ANTHER DEHISCENCE1* gene encodes a novel phospholipase A1 catalyzing the initial step of jasmonic acid biosynthesis, which synchronizes pollen maturation, anther dehiscence, flower opening in Arabidopsis. Plant Cell 13:2191–2209
6. Xie D-X, Feys BF, James S, Nieto-Rostro M, Turner JG (1998) *COI1*: an *Arabidopsis* gene required for jasmonate-regulated defense and fertility. Science 280:1091–1094
7. Smyth DR, Bowman JL, Meyerowitz EM (1990) Early flower development in *Arabidopsis*. Plant Cell 2:755–767
8. Mandaokar A, Thines B, Shin B, Lange BM, Choi G, Koo YJ, Yoo YJ, Choi YD, Choi G, Browse J (2006) Transcriptional regulators of stamen development in Arabidopsis identified by transcriptional profiling. Plant J 46:984–1008
9. Mandaokar A, Browse J (2009) MYB108 acts together with MYB24 to regulate jasmonate-mediated stamen maturation in Arabidopsis. Plant Physiol 149:851–862
10. Mandaokar A, Kumar VD, Amway M, Browse J (2003) Microarray and differential display identify genes involved in jasmonate-dependent anther development. Plant Mol Biol 52: 775–786
11. Boavida LC, McCormick S (2007) Temperature as a determinant factor for increased and reproducible in vitro pollen germination in *Arabidopsis thaliana*. Plant J 52:570–582
12. Heslop-Harrison J, Heslop-Harrison Y (1970) Evaluation of pollen viability by enzymatically induced fluorescence; intracellular hydrolysis of fluorescein diacetate. Stain Technol 45:115–120
13. Regan SM, Moffatt BA (1990) Cytochemical analysis of pollen development in wild-type *Arabidopsis* and a male-sterile mutant. Plant Cell 2:877–889

Chapter 3

Phenotyping Jasmonate Regulation of Root Growth

Fabian Kellermeier and Anna Amtmann

Abstract

Root architecture is a complex and highly plastic feature of higher plants. Direct treatments with jasmonates and alterations in jasmonate signaling have been shown to elicit a range of root phenotypes. Here, we describe a fast, noninvasive, and semiautomatic method to monitor root architectural responses to environmental stimuli using plant tissue culture and the software tool EZ-Rhizo.

Key words Jasmonate, Root architecture, Root response, Phenotyping, EZ-Rhizo

1 Introduction

Changes in distinct subparts of the root system occur in response to environmental stimuli, such as nutrients or phytohormones. For example, treatments with oxylipins cause root waving [1] and cross talk of jasmonate with auxin signaling pathways provokes changes in primary and lateral root growth [2, 3]. Moreover, remodeling of root architecture in nutrient deficiency may be linked to changes in jasmonate levels [4, 5]. Therefore, analysis of root growth is a valuable tool to study molecular processes involving jasmonate perception and signaling. Unfortunately, a lot of phenotypic information may be lost when root growth is scored by pen and ruler. Quick and accurate results can be achieved by imaging of seedlings growing on vertical agar plates. Subsequent image analysis with the semiautomatic software tool EZ-Rhizo [6] greatly facilitates the generation of comprehensive phenotypic datasets. The built-in database function enables easy data storage and handling and simplifies further statistical analyses.

Alain Goossens and Laurens Pauwels (eds.), *Jasmonate Signaling: Methods and Protocols*, Methods in Molecular Biology, vol. 1011, DOI 10.1007/978-1-62703-414-2_3, © Springer Science+Business Media, LLC 2013

2 Materials

Unless otherwise stated, prepare all solutions with deionized water and analytical grade reagents. For long-term storage, stock solutions should be autoclaved for 20 min at 121 °C and kept at 4 °C to avoid growth of contaminants.

2.1 Culture Media

1. Macronutrient stock solutions: For each solution, weigh the amount of salt needed into a 1-L beaker and dissolve in approximately 900 mL water using a magnetic stirrer. Transfer the solution to a graded 1-L measuring cylinder and fill up to 1 L. Transfer the solution to a Pyrex glass bottle. Concentrations for stock solutions are as follows: 0.125 M $CaCl_2$ (18.38 g $CaCl_2 \cdot 2H_2O$ or 13.87 g $CaCl_2$ anhydrous per liter), 0.25 M $MgSO_4$ (61.62 g $MgSO_4 \cdot 7H_2O$ per liter), 1 M KNO_3 (101.1 g KNO_3 per liter), 0.2 M NaH_2PO_4 (31.20 g $NaH_2PO_4 \cdot 2H_2O$ per liter), 42.5 mM Fe(III)Na_2-EDTA (15.60 g Fe(III)Na_2-EDTA per liter).
2. Micronutrient stock solution (1,000×): Weigh all substances into a single 1-L measuring cylinder, then add 900 mL water, mix using a magnetic stirrer, and make up to 1 L. Transfer to a Pyrex bottle. Substances and weights: 304.2 mg $MnSO_4 \cdot H_2O$, 2.78 g H_3BO_3, 109.3 mg $ZnSO_4 \cdot 7H_2O$, 18.5 mg $(NH_4)_6Mo_7O_{24} \cdot 4H_2O$, 39.9 mg $CuSO_4 \cdot H_2O$, 2.4 mg $CoCl_2 \cdot 6H_2O$.
3. Buffer solutions: 0.2 M 2-(*N*-morpholino)ethanesulfonic acid (MES) (39.04 g/L), 0.1 M tris(hydroxymethyl)aminomethane (Tris) (12.11 g/L). Prepare solutions as in **step 1**.
4. To prepare 1 L of standard growth medium, add approximately 900 mL deionized water to a 1-L graded measuring cylinder. Add the following volumes of stock solutions: 4 ml of 0.125 M $CaCl_2$, 1 mL of 0.25 M $MgSO_4$, 2 mL of 1 M KNO_3, 2.5 mL of 0.2 M NaH_2PO_4, 1 mL of 42.5 mM Fe(III)Na_2-EDTA, 1 mL of micronutrient stock solution, 5 mL of 0.1 M Tris, and 14 mL of 0.2 M MES (*see* **Note 1**).
5. Mix the solution using a magnetic stirrer, adjust to pH 5.6 using small volumes of either MES or Tris solutions, and fill up to 1 L.
6. Dissolve 5 g of sucrose in the solution.
7. Transfer solution to a 1-L Pyrex bottle and add 10 g of agar or agarose powder.
8. Autoclave at 121 °C for 20 min.
9. Let the medium cool down to approximately 45–55 °C.
10. If media supplements, such as jasmonate or mock, are used, add them now as sterile filtered solutions in a sterile environment

(laminar flow hood). For preparation of solutions of the desired treatment agent refer to respective protocols.

11. Proceed immediately with steps described in Subheading 3.2.

2.2 Tissue Culture

1. Square Petri dishes: 120 mm × 120 mm.
2. Racks or boxes (10 cm height) to hold the plates vertically upright in the growth chamber.
3. Bleach sterilizing solution: Add 10 mL commercial bleach containing 6–14 % of active chlorine (HOCl) and 50 μL of Tween®-20 (Sigma-Aldrich, St. Louis, MO, USA) to 40 mL of deionized water.
4. 3 M Micropore tape: 1.25 cm × 10 m (3M, St. Paul, MN, USA).

2.3 Phenotyping

1. Conventional flatbed scanner.
2. Black cloth as background.
3. EZ-Rhizo software and EasyPHP database: Both software tools are available as a package from http://www.psrg.org.uk/plant-biometrics.html.
4. Software for data analysis of comma-delimited data spreadsheets (*.csv*).

3 Methods

3.1 Seed Sterilization

1. Put a small amount of *Arabidopsis thaliana* seeds into a 1.5-mL tube (*see* **Note 2**). Proceed with all further steps in a sterile environment (laminar flow hood).
2. Add 1 mL of 96–100 % (v/v) ethanol and mix for 1 min by inversion. Sediment seeds, take off the supernatant, and discard it.
3. Incubate for 5 min in 1 mL of bleach sterilizing solution following the procedure described in Subheading 3.1, **step 2**.
4. Rinse five times with 1 mL distilled, sterilized water as in Subheading 3.1, **step 2**.
5. Add 1 mL distilled, sterilized water and incubate for 2 days in the dark at 4 °C (stratification) before seeds can be sown on agar plates.

3.2 Preparation of Agar Plates

All steps must be carried out in a sterile environment.

1. Pour 35 mL of medium into the square Petri dish (*see* **Note 3**).
2. Let the medium solidify for 1–2 h.
3. With a sterile knife and a spatula, remove 2 cm of agar from one side of the dish. This will be the top end of the vertically positioned plate (*see* **Note 4**).

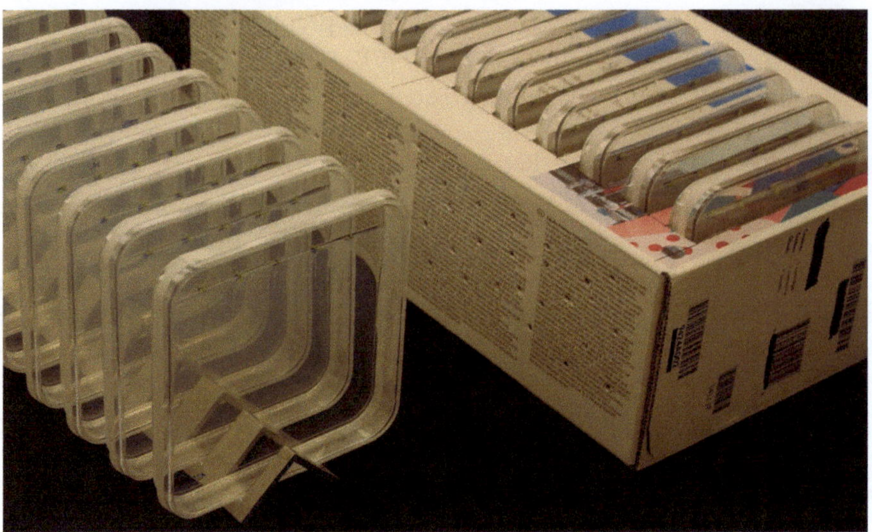

Fig. 1 Petri dishes placed vertically in the growth chamber using metal racks (*left*) or cardboard boxes (*right*). Height of boxes should be smaller than the plates so that the top 2 cm are in full light

4. Sow 4–6 stratified seeds on the surface of the agar close to the cut edge using a small pipette (*see* **Note 5**). If seeds get stuck in the pipette tip, cut off the tip with a sterile knife or sterile scissors. To avoid effects of jasmonate treatment on germination, plants can also be precultured on standard (control) medium without jasmonates and transferred onto jasmonate-containing plates on day 3 after germination (*see* Subheading 3.3, **step 4**).
5. Seal the Petri dishes with micropore tape.

3.3 Cultivation in Growth Chamber

1. Place Petri dishes vertically upright with a slight angle toward the back in racks or growth boxes (Fig. 1; *see* also **Note 6**).
2. Cultivate the seedlings for up to 2 weeks in a controlled environment (*see* **Note** 7).
3. On day 3 after sowing (=day of germination), score germination and mark seeds that have not germinated. Nongerminated seeds will not be taken into account for further analysis.
4. When plants are precultured on control media, transfer seedlings to Petri dishes containing growth medium supplemented with jasmonate on day 3 after germination using sterile forceps in a sterile environment (laminar flow hood). After the transfer, mark the position of the root tip on the plate, seal it with micropore tape, and put it back in the growth chamber.
5. Randomize the plate position within the growth chamber at regular intervals to avoid position effects (*see* **Note 8**).

3.4 Image Acquisition and Analysis of Root Architecture with EZ-Rhizo

1. Scan plates on a flatbed scanner at regular intervals or once at the endpoint of growth. To ensure a high image quality, avoid water vapor condensation (*see* **Note 9**) and scan plates from the backside (through the agar). Putting a black cloth on top of the plate will greatly increase the contrast of the image. Save files in bitmap format (*.bmp*) with a resolution of 200 dpi. Leave some space around all sides of the plate when scanning.
2. Follow the EZ-Rhizo (*see* **Note 10**) analysis procedure as described [6] and summarized below.
3. Load the image and set the result folder path ("options—set result folder path").
4. "Make black and white": Choose a threshold that reduces noise, but does not create large gaps in the root structure.
5. "Remove box."
6. "Remove noise": Choose an algorithm that reduces noise, but does not reduce the root structure.
7. "Dilate."
8. "Skeletonize."
9. "Re-touch": All crossovers of roots need to be avoided because the algorithm is designed to find an end point for the main root and each lateral root (Fig. 2; *see* also **Note 11**). Use the "skeletonize" function again after any changes have been made in this step.
10. "Find roots."
11. "Confirm roots."
12. "Save results" (*see* **Note 12**).
13. Optionally: "Add to database" (*see* **Note 13**) and "query database" (*see* **Notes 14** and **15**).

4 Notes

1. Although widely used in plant tissue culture, we prefer not to use Murashige and Skoog (MS) salts for the analysis of root phenotypes. The nutrient concentrations are rather high in MS media, possibly repressing some root architectural parameters; for instance, high nitrate levels decrease lateral root length [7]. Therefore, we have developed the standard growth medium that contains moderate nutrient levels. These chosen nutrient concentrations avoid repression due to excess, but are high enough to ensure healthy growth of the whole plant.
2. This surface sterilization method works well for small seeds. For large seeds, please refer to the respective protocols.

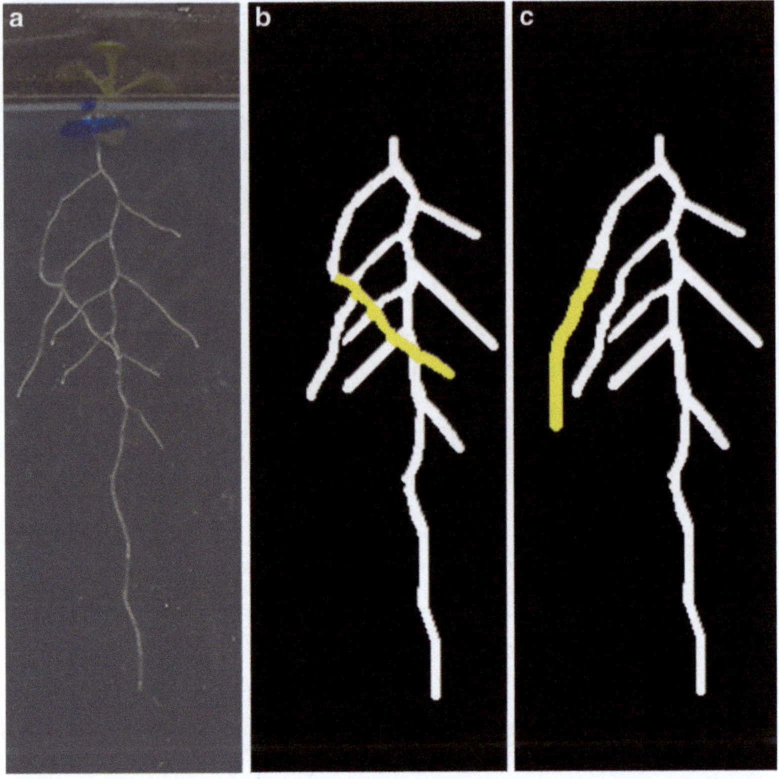

Fig. 2 Detachment of lateral roots crossing over other parts of the root system in the "re-touch" function of EZ-Rhizo (*see* also **Note 11**). The original image (**a**) and the black and white picture (**b**) generated during the EZ-Rhizo analysis show one lateral root crossing from *left* to *right* (highlighted in *yellow*). Crosses can be eliminated, for instance, by redrawing the root toward the *left* (**c**)

3. As the volume is rather low, the surface of the flow hood must be flat to prevent substance gradients due to an uneven distribution of the media.
4. Avoiding contact of shoots with the growth medium is physiologically relevant. For example, sucrose taken up via the shoots in tissue-cultured seedlings can generate an abscisic acid signal that affects root architecture [8].
5. Make sure that seeds are on the surface of the agar, and not inside it, because only roots growing along the surface can be used for analysis. Scratching of the agar surface can already cause a small cut through which the roots grow into the medium.
6. Growing seedlings in boxes has two advantages. First, shading of the root system generates a light gradient from above that leads to elongation of the hypocotyl, facilitating separation of shoots and roots in the latter picture analysis and preventing uptake via shoots (*see* **Note 4**). Second, less condensation occurs on the lid, because both the airflow and water evaporation from the media are reduced.

7. We use either long-day (16/8-h light/dark cycle) or short-day (9/15 h) conditions with temperatures at 22 °C in the light and 18 °C in the dark cycle. The relative humidity should be kept constant at 60–70 %.
8. When different genotypes are used in the study, they should be mixed on plates to ensure full randomization. To avoid subjectivity, a random number calculator can be used for genotype position on plates and plate location in the growth chamber.
9. Condensation mainly occurs on the plate lid in the presence of a temperature gradient. As different environmental conditions can create this gradient, immediate scanning of the plates is recommended once they have been taken out from the growth chamber. In case of high condensation amounts, gently tap one side of the plate onto a hard surface to combine small water droplets into larger ones, turn the plate upside down, and swirl it slowly to "collect" the remaining smaller droplets with the bigger ones.
10. Currently EZ-Rhizo runs on Windows XP only. A new platform-independent version of EZ-Rhizo will be released later in 2013.
11. Some image manipulation may become necessary to enable computer analysis. Whereas it is acceptable within certain limits, any manipulations made need to be documented and taken into account during data analysis.
12. It is advised to enter metadata in the fields provided to allow tracking of each individual plant back to the original picture. The only compulsory field, however, is the resolution for correct conversion of pixels into cm. For example, type "200," if you scanned your plates at 200 dpi. If you want to use the database function for subsequent analyses, we recommend stringent use of codes and abbreviations (e.g., for media and genotypes), because it will help handling large datasets and increase the effectiveness of database usage.
13. "*EasyPHP*" has to be loaded before this function can be used. Not each individual result file has to be loaded immediately after the picture analysis, but can be added at any time in any quantity. All result *.txt* files within a specified folder will be added to the database. Be careful not to add the same files multiple times because they will appear again as multiple entries when the database is queried. We usually copy the result files from their storage folders into a temporary folder before adding them to the database. Afterward, we delete any unwanted files again.
14. The whole database or subsets of data can be queried according to the options given. Each specified field will be considered for the query, i.e., the more filled in, the fewer data will be in

the output. We also recommend to separate the queries for the main and lateral root parameters, making data analysis in Excel (averages, standard deviations, etc.) more convenient.

15. Phenotypic analysis of roots with EZ-Rhizo provides a large set of root architectural parameters. The database function also enables quick and easy handling of these large datasets. Especially when root phenotypes related to jasmonate signaling are studied, the whole range of root parameters should be taken into consideration. For example, root waving elicited by oxylipin treatment [1] could be quantified with the "*straightness*" parameter (*see* also ref. 6 for a full description of analyzed traits).

References

1. Vellosillo T, Martínez M, López MA, Vicente T, Dolan L, Hamberg M, Castresana C (2007) Oxylipins produced by the 9-lipoxygenase pathway in *Arabidopsis* regulate lateral root development and defense responses through a specific signaling cascade. Plant Cell 19: 831–846
2. Sun J, Xu Y, Ye S, Jiang H, Chen Q, Liu F, Zhou W, Chen R, Li X, Tietz O, Wu X, Cohen JD, Palme K, Li C (2009) *Arabidopsis ASA1* is important for jasmonate-mediated regulation of auxin biosynthesis and transport during lateral root formation. Plant Cell 21:1495–1511
3. Monzón GC, Pinedo M, Lamattina L, de la Canal L (2012) Sunflower root growth regulation: the role of jasmonic acid and its relation with auxins. Plant Growth Regul 66:129–136
4. Armengaud P, Breitling R, Amtmann A (2004) The potassium-dependent transcriptome of Arabidopsis reveals a prominent role of jasmonic acid in nutrient signaling. Plant Physiol 136: 2556–2576
5. Troufflard S, Mullen W, Larson TR, Graham IA, Crozier A, Amtmann A, Armengaud P (2010) Potassium deficiency induces the biosynthesis of oxylipins and glucosinolates in *Arabidopsis thaliana*. BMC Plant Biol 10:172
6. Armengaud P, Zambaux K, Hills A, Sulpice R, Pattison RJ, Blatt MR, Amtmann A (2009) EZ-Rhizo: integrated software for the fast and accurate measurement of root system architecture. Plant J 57:945–956
7. Deak KI, Malamy J (2005) Osmotic regulation of root system architecture. Plant J 43:17–28
8. MacGregor DR, Deak KI, Ingram PA, Malamy JE (2008) Root system architecture in *Arabidopsis* grown in culture is regulated by sucrose uptake in the aerial tissues. Plant Cell 20:2643–2660

Part II

Environment

Chapter 4

Bioassays for Assessing Jasmonate-Dependent Defenses Triggered by Pathogens, Herbivorous Insects, or Beneficial Rhizobacteria

Saskia C.M. Van Wees, Johan A. Van Pelt, Peter A.H.M. Bakker, and Corné M.J. Pieterse

Abstract

Jasmonates, together with other plant hormones, are important orchestrators of the plant immune system. The different hormone-controlled signaling pathways cross-communicate in an antagonistic or a synergistic manner, providing the plant with a powerful capacity to finely regulate its immune response. Jasmonic acid (JA) signaling is required for plant resistance to harmful organisms, such as necrotrophic pathogens and herbivorous insects. Furthermore, JA signaling is essential in interactions of plants with beneficial microbes that induce systemic resistance to pathogens and insects. The role of JA signaling components in plant immunity can be studied by performing bioassays with different interacting organisms. Determination of the level of resistance and the induction of defense responses in plants with altered JA components, through mutation or ectopic expression, will unveil novel mechanisms of JA signaling. We provide detailed protocols of bioassays with the model plant *Arabidopsis thaliana* challenged with the pathogens *Botrytis cinerea* and *Pseudomonas syringae*, the insect herbivore *Pieris rapae*, and the beneficial microbe *Pseudomonas fluorescens*. In addition, we describe pharmacological assays to study the modulation of JA-regulated responses by exogenous application of combinations of hormones, because a simultaneous rise in hormone levels occurs during interaction of plants with other organisms.

Key words *Arabidopsis thaliana*, SA, JA, Plant hormones, Bioassay, ISR, Herbivorous insect, Plant immunity

1 Introduction

The use of bioassays to study the effects of treatments on the resistance level of plants against an attacker has been recorded for the first time at the beginning of the twentieth century [1, 2]. Numerous examples were described in which plants were protected against pathogen infection after pretreatment with (attenuated) pathogens or extracts obtained from pathogens [1, 2]. In nature, plants encounter a plethora of harmful and beneficial organisms, including

Alain Goossens and Laurens Pauwels (eds.), *Jasmonate Signaling: Methods and Protocols*, Methods in Molecular Biology, vol. 1011, DOI 10.1007/978-1-62703-414-2_4,

bacteria, fungi, oomycetes, viruses, nematodes, and insects. Each of these interacting organisms exploits highly specialized features to establish an intimate relationship with its host plant.

The plant responds differently to various types of ingression by interacting organisms through changes in levels of and sensitivity to plant hormones. Plant hormones play an important role in the organization of the immune signaling network that induces defense responses. The hormones jasmonic acid (JA) and salicylic acid (SA) are recognized as major players in plant immune signaling, whereas other hormones have modulating roles in the JA- and SA-controlled responses [3]. JA-regulated defenses triggered by wounding control resistance to insect herbivores [4] and also to pathogens with a necrotrophic lifestyle. These pathogens first kill the cells and then live on the contents [5]. Here, we describe bioassays with *Arabidopsis thaliana* and the JA-controlled necrotrophic pathogen *Botrytis cinerea* and the herbivorous insect *Pieris rapae*. Biotrophic pathogens, such as *Hyaloperonospora arabidopsidis*, that keep the host cells alive and retrieve nutrients by forming specialized feeding structures (haustoria), are controlled by SA-regulated defense responses [5]. Some plant pathogens display both necrotrophic and biotrophic lifestyles, depending on the stage of their life cycle, and are called hemi-biotrophs. The chapter also provides a description of a bioassay with the hemi-biotrophic bacterial pathogen *Pseudomonas syringae*.

Beneficial soil-borne microorganisms, such as mycorrhizal fungi and plant growth-promoting rhizobacteria, can cause induced systemic resistance (ISR) in distant plant parts [6, 7]. During ISR, a mild, but effective, immune response is activated in systemic tissues that in many cases is regulated by JA-dependent signaling pathways. ISR is associated with priming for accelerated JA-dependent defense gene expression rather than with direct activation of defense responses, and is predominantly effective against a broad spectrum of pathogens and insects that are sensitive to JA-controlled defenses [8, 9]. In *Arabidopsis,* ISR triggered by the rhizosphere-colonizing bacterium *Pseudomonas fluorescens* WCS417 is well studied and bioassays to assess WCS417-ISR are described in this chapter.

In recent years, molecular, genetic, and genomic tools have been used to uncover the complexity of the hormone-regulated induction of the defense signaling network. Besides balancing of the relative abundance of different hormones, intensive interplay between hormone signaling pathways has emerged as an important regulatory mechanism by which the plant is able to tailor its immune response to the type of invader encountered [10, 11]. For example, resistance of *Arabidopsis* to *P.* was shown to depend on activation of SA signaling, and was associated with suppression of JA signaling [12, 13]. JA-dependent resistance to the necrotrophic fungal pathogen *B. cinerea* was found to be synergized by ethylene, but antagonized by abscisic acid (ABA) [14, 15]; opposite effects of ethylene (antagonistic) and ABA (synergistic) on JA-dependent resistance against insect herbivores were reported [16, 17]. Pharmacological assays in which hormones are applied to the plant have further elucidated some of the molecular mechanisms involved in the communication between different hormone signaling pathways [3, 18–22]. Modulation of JA signaling by other hormones has been reported to occur by interfering with the function of certain JA signaling components, such as the transcription factor MYC2 in the case of ethylene and ABA, but their exact influence on MYC2 is still not clear [14, 23]. For the antagonistic effect of SA on JA signaling, the JA-regulated transcription factor ORA59 has been suggested as target [20]. Pharmacological assays with combinations of defense-related hormones are described in this chapter.

Despite its unquestioned role in the plant's immunity, many aspects of JA signaling are still unresolved. The use of proper bioassays and pharmacological assays, as described here, will help us to piece the JA puzzle together.

2 Materials

2.1 Equipment

1. Growth chambers set at 21 °C, 70 % relative humidity, and 10-h/14-h day/night regime with a light intensity during the day of 200 μE/m^2/s provided by bulb HPI lamps (Philips, Eindhoven, The Netherlands) or LuxLine plus F58W/840 cool white tube lamps (Havells Sylvania, London, UK).
2. Large autoclave (50 L) and autoclavable plastic bags (40 × 60 cm).
3. Containers (30–50 L) for mixing water and solutions through soil.
4. Sieved potting soil mixed with river sand (12:5 v/v).
5. Small trays (100–500 mL; 4 cm high) for seedling cultivation.
6. Pots (60 mL) with holes in the bottom for plant cultivation after the seedling stage.
7. Small 5-cm Petri dishes.
8. Trays (approximately 45 × 30 × 8 cm) to contain small trays or pots that can be covered with transparent lids to achieve 100 % relative humidity.
9. Tweezers with curved beak tip.
10. Table centrifuge.
11. Spectrophotometer.
12. Incubator set at 22 °C, 10-h day/14-h night, Philips TL-D 36 W/33 lamps for fungus growth or at 28 °C for bacterial growth.
13. Hemocytometer.

14. Light microscope.
15. Empty pipette tip box.
16. Needleless 1-mL syringe.
17. One-hole puncher (diameter 6 mm from an office supplier).
18. 96-Deep-well microplate (96-well format boxes containing 12 disposable 8-strip tubes and caps (Greiner Bio-one, Frickenhausen, Germany)).
19. Stainless steel beads (diameter 2.3 mm).
20. Orbital shaker at 28 °C.
21. Plate shaker MM301 (Retsch, Haan, Germany) or a regular paint shaker.
22. 8-Channel pipette (10, 20, 180 μL).
23. 96-Well dilution plates (≥200 μL).
24. Fine paintbrush.
25. A desiccator or other device that can be air-tightly closed.

2.2 Buffers, Media, and Solutions

2.2.1 Arabidopsis thaliana Cultivation

1. Seeds of *Arabidopsis thaliana* (L.) Heynh.
2. 0.1 % (w/v) agar.
3. Half-strength, modified Hoagland nutrient solution: 2 mM KNO_3, 5 mM $Ca(NO_3)_2$, 1 mM KH_2PO_4, 1 mM $MgSO_4$, trace elements, pH 7 [24], 10 μM Fe-ethylenediamine-di[*o*-hydroxyphenylacetic acid] (Sequestreen; Ciba-Geigy, Basel, Switzerland) (*see* **Note 1**).
4. Plant labels.

2.2.2 Pseudomonas fluorescens ISR Bioassay

1. *P. fluorescens* strain WCS417 [25] or any other biocontrol pseudomonad strain (stocks stored in 25 % glycerol at −80 °C).
2. King's B (KB) medium agar [26]: 20 g proteose peptone no. 3 (Difco™ BD Diagnostics, Franklin Lakes, NJ, USA), 10 g glycerol, 1.5 g $MgSO_4$, 1.2 g KH_2PO_4 per liter demineralized water supplemented with 13 g of granulated agar (Difco™) for the solid medium in Petri dishes (*see* **Note 2**).
3. Sterilized 10 mM $MgSO_4$.

2.2.3 Botrytis cinerea Bioassay

1. Pathogen *B. cinerea* isolate B0510 (stocks stored in 25 % glycerol at −80 °C).
2. Half-strength potato dextrose broth (PDB; Difco™).
3. Half-strength potato dextrose agar (PDA; Difco™), supplemented with 0.75 % granulated agar (Difco™) to obtain a final concentration of 1.5 % agar.

2.2.4 Pseudomonas syringae Bioassay

1. *P. syringae* pv. *tomato* DC3000 [27] or another virulent *P. syringae* strain (stocks stored in 25 % glycerol at −80 °C).
2. KB liquid medium and KB agar supplemented with 25 mg/mL rifampicin to select DC3000 (*see* Subheading 2.2.2 and **Note 2**).
3. Sterilized 100-mL Erlenmeyer flasks with cotton plugs containing 25 mL of liquid KB.
4. Sterilized 10 mM $MgSO_4$.
5. Silwet L-77 (Van Meeuwen Chemicals, Weesp, The Netherlands).

2.2.5 Pieris rapae Two-Choice Bioassay

1. First-instar (L1) larvae of *P. rapae*. Request the caterpillars from a collaborator or use caterpillars of your own collection (*see* **Note 3**).
2. *Brassica oleracea* (white cabbage) or *Brassica campestris* (Chinese cabbage) as food sources for the caterpillars.
3. *Lantana* sp. (shrub verbena) plants that supply nectar to the butterflies.

2.2.6 Combinatorial Hormone Application Pharmacological Assay

1. SA (Mallinckrodt Baker, Deventer, The Netherlands) or sodium salt SA (Na-SA; Sigma-Aldrich, St. Louis, MO, USA) (*see* **Note 4**).
2. Methyl jasmonate (MeJA; Brunschwig Chemie, Amsterdam, The Netherlands) (*see* **Note 5**).
3. 96 % Ethanol.
4. Silwet L-77 (Van Meeuwen).
5. Optionally, 1-aminocyclopropane-1-carboxylic acid (ACC; Sigma-Aldrich).
6. For plate assays with seedlings, Murashige and Skoog (MS) medium supplemented with vitamins (pH 5.7; Duchefa, Haarlem, The Netherlands), 5 % sucrose, and plant agar (0.85 %; Duchefa) in 10 × 10 cm square plates.
7. For liquid assays with seedlings, MES buffer (5 mM 2-(*N*-morpholino)ethanesulfonic acid monohydrate (MES), 1 mM KCl, pH 5.7) in 24-well plates.
8. For seed surface sterilization: HCl (37 %), household chlorine (original Glorix; Unilever, London, UK).

3 Methods

The introduction of microbes and insects by plant pathologists and entomologists in the plant growth facilities is harmless. The described experiments, with the exception of those with caterpillars, can be done in close proximity to other plant experiments without the risk for cross-contamination.

3.1 *Arabidopsis* Cultivation

1. Suspend *Arabidopsis* seeds (3× more than the number of plants needed; 100 seeds weigh approximately 1.5 mg) in 0.1 % agar in 1.5- or 15-mL tubes and imbibe at 4 °C for 2–4 days (*see* **Note 6**).
2. Autoclave (moist) river sand in (double) plastic bags (with 5–10 kg sand) for 20 min at 121 °C.
3. Autoclave (moist) potting soil:river sand mixture (12:5) in (double) plastic bags (with 5–10 kg mix) for 1 h at 121 °C. Repeat the next day.
4. Add half-strength modified Hoagland nutrient solution to the sand (250 mL/kg) and supply water until sand is nearly saturated with fluid.
5. Fill up the 4 cm high small trays (100–500 mL) with the sand.
6. With a Pasteur pipette, distribute the seeds (in 0.1 % agar) evenly onto the sand (60 seeds/25 cm^2).
7. Place the sown trays in a large tray covered with a transparent lid (to achieve 100 % relative humidity) and place in a growth chamber for 12 days.
8. In a large container, mix the autoclaved soil mixture with Hoagland nutrient solution (50 mL/kg).
9. Supply water if needed: a filled 60-mL pot should weigh 75 g.
10. Fill 60-mL pots with holes in the bottom with the soil mix, push slightly on the soil top for firmness, and make one hole in the middle of the soil with the conical end of a 15-mL tube.
11. Place the pots on small Petri dishes that function as saucers to allow individual water/nutrient supply and to prevent cross-contamination between different treatments (*see* **Note 7**).
12. Flood the small trays containing 12-day-old seedlings in sand with water and use tweezers to gently transfer single seedlings from the sand to the planting holes in the potting soil.
13. Close the planting hole lightly by pushing the soil back around the root, leaving the above-ground plant parts free of soil.
14. Stick a color plant label in every pot for genotype/treatment indication.
15. Place the seedling-containing pots in a randomized order in large plant trays (30 plants/tray).
16. Cover the trays with transparent lids for 2 days, after which they are removed.
17. Every other day, water the plants with approximately 10 mL per pot during the first 10 days, and up to 20 mL at later growth stages.
18. Once a week, give the plants 10 mL of Hoagland solution (*see* **Notes 7** and **8**).

3.2 *Pseudomonas fluorescens* ISR Bioassay

1. Start a culture of *P. fluorescens* strain WCS417 or another biocontrol pseudomonad strain by inoculating bacteria from a glycerol stock on two KB agar plates and incubating them for 1 day at 28 °C (*see* **Note 9**).
2. Harvest the bacteria by scraping them off the plates in 10 mM $MgSO_4$.
3. Wash the bacterial cells by spinning down in Eppendorf tubes at $1,500 \times g$ for 5 min in a table centrifuge and resuspend in 10 mM $MgSO_4$.
4. Measure the density of the bacterial suspension in a spectrophotometer at the optical density (OD) 660 nm ($1 = 10^9$ cells/mL).
5. Mix 50 mL of 10^9 colony-forming units (cfu)/mL per kg of soil to obtain 5×10^7 cfu/kg, whereas the control treatment receives 50 mL of 10 mM $MgSO_4$ per kg of soil.
6. Proceed with the plant cultivation (*see* Subheading 3.1) and treat with pathogens/insects/hormones as described below.

3.3 *Botrytis cinerea* Bioassay

1. To determine the level of disease resistance to *B. cinerea*, use 20 plants per genotype/treatment (*see* **Note 10**).
2. For gene expression analysis, harvest ten inoculated leaves in triplicate of a total of ten plants per time point (e.g., $t = 0$, 1, and 2 days after inoculation) (*see* **Note 10**).
3. Start a culture of *B. cinerea* by inoculating conidia from a glycerol stock on half-strength PDA plates and incubate them for 2 weeks at 22 °C under a 10-h day/14-h night regime.
4. Around 1:00 p.m., harvest conidia by scraping them off the plates in half-strength PDB.
5. Filter the suspension through glass wool.
6. Measure the conidial density in a hemocytometer with a light microscope.
7. Dilute the suspension with PDB to a final concentration of 5×10^5 conidia/mL.
8. Leave the conidia in PDB for 2 h at room temperature.
9. Around 3:00 p.m., inoculate the plants by pipetting a 5-μL droplet of the conidial suspension on approximately five fully grown leaves per plant (*see* **Note 11**).
10. Place two wet towels in the plant trays and tape-shut transparent lids to the trays to create 100 % relative humidity.
11. Record disease symptoms at 3–7 days after inoculation and categorize them in different disease severity classes depending on the size and appearance of the lesions (Fig. 1) (*see* **Note 12**).
12. Determine the percentage of leaves per plant falling in each disease class and by means of the Chi-square test, whether the

Fig. 1 Classification of disease symptoms caused by infection with *B. cinerea*. From *left* to *right*: Stage I, lesion 2 mm; stage II, lesion 2 mm + chlorosis; stage III, lesion 2–4 mm + chlorosis; stage IV, lesion > 4 mm + chlorosis

distribution between the different classes differs between genotypes/treatments.

13. Determine the number of *in planta*-formed spores on *B. cinerea*-infected leaves in three pools of 16 inoculated leaves of four plants per genotype/treatment.
14. Shake the leaves vigorously in a test tube containing 10 mL of water to release the spores from the leaf surface.
15. Use tweezers to remove the leaves, centrifuge the remaining spore suspension at 200 × *g* for 10 min, and resuspend the spores in 500 μL of water.
16. Count the spores in a hemocytometer with a light microscope.
17. Log-transform the data and perform a Tukey's honestly significant difference test to analyze the differences between genotypes/treatments.

3.4 Pseudomonas syringae Bioassay

Basically, the resistance level against *P. syringae* can be determined by two different inoculation methods: (a) dipping and (b) pressure infiltration of the leaves with the bacterial suspension. By dipping, the bacteria enter through the stomata and start colonizing the leaves from there, whereas by infiltration the bacteria are immediately present everywhere in the apoplast of the infiltrated area. The dipping method is commonly used in ISR bioassays, whereas the infiltration method is used in most other experiments with *P. syringae*.

1. Use 20 plants per genotype/treatment for the dipping bioassay and 10–20 plants per treatment for the infiltration bioassay (*see* **Notes 10** and **13**).

2. For gene expression analysis, harvest ten inoculated leaves in triplicate of a total of ten plants per time point (e.g., $t=0$, 6, and 24 h after inoculation) (*see* **Note 10**).
3. At around 4:00 p.m., start a culture of *P. syringae* by inoculating bacteria from a glycerol stock in an Erlenmeyer flask containing liquid KB and incubate overnight at 28 °C in an orbital shaker (225 rpm).
4. The next morning, wash the bacterial cells by spinning them down in Eppendorf tubes at $1{,}500 \times g$ for 5 min in a table centrifuge and resuspend them in 10 mM $MgSO_4$.
5. Measure the density of the bacterial suspension in a spectrophotometer at OD_{660} ($1 = 10^9$ cells/mL).
6. For the dipping bioassay, dilute the bacteria in $MgSO_4$ until 2.5×10^7 cfu/mL and amend with Silwet L-77 to 0.02 % (v/v) to facilitate entry of the bacteria into the leaves. For the infiltration assay, dilute the bacteria to $OD_{660} = 0.0005$ for bioassays and to $OD_{660} = 0.005$ (thus tenfold higher) for gene expression analyses (*see* **Note 14**).
7. Proceed with **steps 8** and **12** for the dipping and infiltration assay, respectively.
8. For dipping, turn the plant in the pot upside down in the bacterial suspension, so that all the leaves are immersed, for 3 s (*see* **Note 15**).
9. Refresh the inoculum at least once every 30 plants and use separate boxes for differently pretreated plants to prevent cross-contamination.
10. After inoculation, place the transparent lids on the plant trays.
11. After 4 days, score the percentage of leaves with disease symptoms (presence of water-soaked lesions and chlorosis) per plant and analyze the differences between genotypes/treatments with the Tukey's honestly significant difference test.
12. For pressure infiltration, gently turn the leaf so that its adaxial side is pressed on the index finger and gently press the plunger of a needleless 1-mL syringe firmly placed on the abaxial side to release the bacterial suspension into the leaf.
13. First, indicate with a marker pen on the petioles which leaves will be infiltrated (*see* **Note 11**).
14. After 3 days, determine the disease symptoms (*see* Subheading 3.4, **step 11**).
15. Determine the bacterial growth *in planta* by analyzing eight samples containing two leaf discs of two leaves of one plant, which are collected in 96-deep-well plates containing two beads per well (*see* **Notes 16** and **17**).

16. After all the samples for a time point are collected, add 400 μL of 10 mM $MgSO_4$ to each sample with a multichannel pipette and homogenize the tissue in a plate shaker.
17. Make dilution series in 96-well dilution plates by pipetting 20 μL of homogenate into 180 μL of 10 mM $MgSO_4$ (*see* **Note 18**).
18. Plate the serial dilutions on KB agar containing 25 mg/mL rifampicin to select for *P. syringae* pv. *tomato* DC3000.
19. For high-throughput plating, split the plate into two with a stripe on the back of the plate and streak 2.5-cm lines of 10 μL of a dilution of 8 samples with an 8-channel pipette (one treatment) on one half and repeat on the other half of the plate (*see* **Note 19**).
20. Incubate for 2 days at 28 °C and count the cfu.
21. From these data, calculate the ^{10}log-transformed cfu/cm^2 leaf surface area and subject to the Tukey's honestly significant difference test to analyze differences between genotypes/treatments.

3.5 *Pieris rapae* Two-Choice Bioassay

The caterpillars of *P. rapae* (small cabbage white butterfly) are specialists on cabbage plants and because *Arabidopsis* is also a member of the Cruciferaceae, they can also feed on *Arabidopsis*. As specialists, their performance is hardly influenced by activation of JA-dependent responses, but when given a choice, they prefer to feed on plants that express the ERF branch of the JA signaling pathway that is controlled by the ERF transcription factor ORA59 rather than be deterred by induction of the MYC branch [17]. In case of two-choice assays, the preference of the caterpillars for either one of two genotypes or treatments is tested.

1. For the two-choice bioassay, place four 6-week-old plants, two of each genotype/treatment, close together so that the leaves overlap and the caterpillars can move from one plant to the other (*see* **Note 20**).
2. Create an empty space of at least 30 cm between each plant arena to prevent crossing-over of the caterpillars.
3. To get reliable data, test the choice of the caterpillars in at least 20 arenas.
4. For gene expression analysis, plants can grow in the usual (no-choice) setup.
5. Harvest ten infested leaves in triplicate per genotype/treatment of a total of ten plants per time point (e.g., $t=0$, 6, 24, 48 h).
6. Collect L1 larvae from the insect-rearing facility by cutting leaves from cabbage plants harboring caterpillars that are 1–2 days old (*see* **Note 21**).
7. Using a fine paintbrush, place two caterpillars on each plant so that in each plant arena eight caterpillars are released.

8. For two-choice assays, allow the caterpillars to feed for 4 days.
9. For gene expression analyses, remove the caterpillars from the plants with a paintbrush, after 24 h of feeding.
10. Cut through the hypocotyl and inspect the rosette carefully to monitor the presence of the caterpillars on the different plant genotypes/treatments in each arena (*see* **Note 22**).
11. Calculate the frequency distribution of the caterpillars over the different genotypes/treatments per two-choice arena and test for statistical difference from a 50 % distribution (equal choice) using the Student's *t*-test.

3.6 Combinatorial Hormone Application Pharmacological Assay

Preparation and application of the hormonal solutions is the same for combinatorial pharmacological assays as for hormonal induction treatments in bioassays with interacting organisms. In most of the hormone combination assays with SA and JA, we use 5-week-old soil-grown plants that are dipped in combinatorial hormonal solutions, but sterile, plate-grown or liquid medium-grown seedlings can be assayed for SA/JA cross talk as well. Usually, in hormone dipping assays, 1 mM SA and 100 μM MeJA are applied to study cross-communication between hormone signaling pathways by means of their effect on gene expression 24 h after treatment. However, other experimental scenarios are suitable as well, because the antagonistic effect of SA on JA signaling is apparent when SA is supplied up to 30 h before the MeJA application and the SA/JA cross talk effects last for at least 96 h [19]. Moreover, SA concentrations as low as 0.1 μM suffice to antagonize the JA-induced signaling.

3.6.1 Soil-Grown Plants

1. To determine the effect of SA and MeJA on each other's action (such as induction of gene expression), use 30 plants per treatment in a dipping assay that allows for sampling at $t=0$ and $t=24$ h (*see* **Note 10**).
2. Prepare SA and MeJA solutions (*see* **Notes 4** and **5**). For dipping, add Silwet L-77 to a final concentration of 0.015 % to facilitate entry into the leaves.
3. For the dipping assay, follow instructions as described in Subheading 3.4, **step 8**, except that lids on the trays are not fully closed, but cracked (*see* **Notes 14** and **23**).

3.6.2 Sterile-Grown Seedlings

1. Put <200 seeds in an open Eppendorf tube and place in a desiccator together with a 200-mL beaker containing 97 mL of HCl.
2. Add briefly 3 mL of chlorine to the HCl and mix with a pipette, immediately followed by closure of the desiccator with its lid (*see* **Note 24**).
3. Take out the seeds after 3 h and transfer the seeds to MS plates.

4. Imbibe the seeds for 2 days at 4 °C, after which the sown plates are placed vertically in a growth chamber for 12 days (*see* **Note 25**).
5. Transfer the seedlings either to fresh MS agar medium supplemented with 0.5 mM SA, 20 μM MeJA, or a combination of both chemicals or to 1.5 mL of liquid MES buffer medium in 24-well plates (5 seedlings per well), where they are left to acclimatize for 1 day before addition of SA and MeJA at the final concentrations of 0.5 mM and 100 μM, respectively (*see* **Note 26**).

4 Notes

1. We make 20× concentrated stock solutions and store them at room temperature. Prepare a working solution by filling a 50-L tap can with 25 L tap water; add 25 mL of the stock solutions; fill the can up to 50 L with tap water. Mix solution well. To avoid algae growth, place a black bin over the can.
2. To make 1 L of KB, place a 2-L Erlenmeyer flask containing 500 mL demineralized water and a magnet on a magnetic stirrer. To avoid precipitation, add all the ingredients one by one and dissolve completely before adding the next compound. Adjust with water to obtain 1 L and pour into bottles for autoclaving. Agar has to be added to the bottle and not to the Erlenmeyer because it will not dissolve without heating.
3. Keeping an in-house colony of *P. rapae* is laborious and demands large temperature-controlled growth facilities, as one chamber is used for rearing of *P. rapae* and another to cultivate the plants needed for the rearing of *P. rapae* (see Subheading 2.2.5).
4. SA is acidic and should be buffered to neutral pH. As at a (common) 1 mM SA concentration, the buffering capacity of tap water suffices, we usually prepare SA solutions in tap water. Whereas Na-SA readily dissolves in water, SA does not and has to be boiled. Stock solutions of 100 mM SA can be stored at room temperature, but boiling is required to dissolve the precipitated SA. SA (stock) solutions of a low concentration appear to lose their defense-inducing activity when stored.
5. MeJA is available as a 4.46 M solution. Make a 1,000-fold concentrated stock in 96 % ethanol by adding 10 μL MeJA to 436 μL ethanol, resulting in a 100 mM MeJA solution. To all the solutions without MeJA, a similar volume of ethanol is added (0.1 %).
6. The 0.1 % agar prevents the seeds from sinking to the tube bottom and allows an equal distribution of the seeds in the solution.

7. Water and nutrients are supplied via the saucers by using a bottle dispenser fused to rubber tubing. This prevents contamination between pots and allows the fluids to stream from the bottom up and not from the top down.
8. Assays as described in this chapter can usually be performed when the plants are 5 weeks old.
9. They grow also at room temperature, albeit more slowly.
10. Plants are 5 weeks old at the time of inoculation.
11. Be careful to select leaves that are younger than leaf four, because the round-shaped older leaves tend to be very susceptible, irrespective of the genetic background or treatment of the plant.
12. The rate of disease progression and also the symptom appearance can differ between experiments and, thus, the day of symptom scoring and the criteria of the disease classes might have to be adjusted accordingly.
13. Ten plants are needed when only one time point is harvested and more plants are needed for multiple time points to determine bacterial titer.
14. Inoculation with *P. syringae* and treatment with SA preferably take place in the morning, because then SA-dependent signaling is activated stronger and the difference in disease level between resistant and susceptible plants is greater.
15. A pipette tip box can be used to contain the suspension.
16. The infiltration method is usually coupled with assessment of *in planta* bacterial growth, which is a highly valuable but also laborious method. Therefore, we have tried to automate it as much as possible. A one-hole puncher significantly speeds up the process of cutting leaf discs compared to the classical cork borer.
17. When samples are taken at $t=0$ to determine the number of bacteria that entered the leaves (in practice 1 h after inoculation), then the leaves need to be washed briefly (3 s) in 70 % ethanol and subsequently rinsed with water and dried with a tissue, before leaf discs can be sampled.
18. The $t=0$ samples are diluted 10×; the $t=3$ samples are diluted 10,000×, but when plants are very susceptible 1,000,000× dilutions can be needed.
19. This way, if you plate three dilutions, you use three plates per genotype/treatment.
20. Plants of the same genotype/treatment are placed diagonally to each other.
21. Seven days prior to the experiment, a fresh cabbage plant is introduced into the insect-rearing room on which butterflies are allowed to deposit eggs for 1 day, after which the plants are

placed in a closed cage; 5 days later, the caterpillars hatch from the eggs.

22. The caterpillars are hard to track down: they are small and green and in addition, tend to crawl on the abaxial side of the leaves. Chances of finding them back increase if the rosette is held against the light, which shines through the leaves except where the caterpillars are.
23. Treat plants with hormones in the morning (before 12:00 p.m.) and sample tissue to analyze for JA-induced gene expression around 2:00 p.m., because then the plants show high sensitivity to JA while the basal expression level of genes like *PDF1.2* is low. The basal *PDF1.2* transcript levels are high at the end of the day due to the circadian rhythm.
24. The HCl–chlorine mixture needs to be freshly prepared.
25. There is still chlorine gas in the seed coat after surface-sterilization and this will eventually kill the seeds if it cannot be released. Therefore, the tubes with seeds should be left open in a sterile hood for at least half an hour before they are transferred to plates or stored at 4 °C.
26. To enhance the induction of JA-sensitive genes that are co-regulated by ethylene, like those under control of the ERF branch of the JA signaling pathway, 0.002 mM ACC can be added to the medium. Be careful with increasing the ethylene concentration in the assay, because it is known to suppress the antagonism by SA on JA signaling [20].

Acknowledgments

The authors would like to thank other (previous) members of the laboratory who have contributed to developing the foregoing protocols. The authors are supported by the Netherlands Organization of Scientific Research (VICI grant no. 865.04.002 and VIDI grant no. 11281) and a European Research Council Advanced Grant (no. 269072).

References

1. Chester KS (1933) The problem of acquired physiological immunity in plants. Q Rev Biol 8:129–154
2. Chester KS (1933) The problem of acquired physiological immunity in plants (Continued). Q Rev Biol 8:275–324
3. Pieterse CMJ, Van der Does D, Zamioudis C, Leon-Reyes A, Van Wees SCM (2012) Hormonal modulation of plant immunity. Annu Rev Cell Dev Biol 28:489–521. doi: 101146/annurev-cellbio-092910-154055
4. Howe GA, Jander G (2008) Plant immunity to insect herbivores. Annu Rev Plant Biol 59:41–66
5. Glazebrook J (2005) Contrasting mechanisms of defense against biotrophic and necrotrophic pathogens. Annu Rev Phytopathol 43: 205–227
6. Van Wees SCM, Van der Ent S, Pieterse CMJ (2008) Plant immune responses triggered by beneficial microbes. Curr Opin Plant Biol 11:443–448

7. Van Loon LC, Bakker PAHM, Pieterse CMJ (1998) Systemic resistance induced by rhizosphere bacteria. Annu Rev Phytopathol 36: 453–483
8. Van der Ent S, Van Wees SCM, Pieterse CMJ (2009) Jasmonate signaling in plant interactions with resistance-inducing beneficial microbes. Phytochemistry 70:1581–1588
9. Zamioudis C, Pieterse CMJ (2012) Modulation of host immunity by beneficial microbes. Mol Plant-Microbe Interact 25:139–150
10. De Vos M, Van Oosten VR, Van Poecke RMP, Van Pelt JA, Pozo MJ, Mueller MJ, Buchala AJ, Métraux J-P, Van Loon LC, Dicke M, Pieterse CMJ (2005) Signal signature and transcriptome changes of *Arabidopsis* during pathogen and insect attack. Mol Plant-Microbe Interact 18:923–937
11. Pieterse CMJ, Leon-Reyes A, Van der Ent S, Van Wees SCM (2009) Networking by small-molecule hormones in plant immunity. Nat Chem Biol 5:308–316
12. Van Wees SCM, Luijendijk M, Smoorenburg I, Van Loon LC, Pieterse CMJ (1999) Rhizobacteria-mediated induced systemic resistance (ISR) in *Arabidopsis* is not associated with a direct effect on expression of known defense-related genes but stimulates the expression of the jasmonate-inducible gene *Atvsp* upon challenge. Plant Mol Biol 41: 537–549
13. Spoel SH, Koornneef A, Claessens SMC, Korzelius JP, Van Pelt JA, Mueller MJ, Buchala AJ, Métraux J-P, Brown R, Kazan K, Van Loon LC, Dong X, Pieterse CMJ (2003) NPR1 modulates cross-talk between salicylate- and jasmonate-dependent defense pathways through a novel function in the cytosol. Plant Cell 15:760–770
14. Anderson JP, Badruzsaufari E, Schenk PM, Manners JM, Desmond OJ, Ehlert C, Maclean DJ, Ebert PR, Kazan K (2004) Antagonistic interaction between abscisic acid and jasmonate-ethylene signaling pathways modulates defense gene expression and disease resistance in Arabidopsis. Plant Cell 16:3460–3479
15. Penninckx IAMA, Thomma BPHJ, Buchala A, Métraux J-P, Broekaert WF (1998) Concomitant activation of jasmonate and ethylene response pathways is required for induction of a plant defensin gene in Arabidopsis. Plant Cell 10:2103–2113
16. Bodenhausen N, Reymond P (2007) Signaling pathways controlling induced resistance to insect herbivores in *Arabidopsis*. Mol Plant-Microbe Interact 20:1406–1420
17. Verhage A, Vlaardingerbroek I, Raaymakers C, Van Dam NM, Dicke M, Van Wees SCM, Pieterse CMJ (2011) Rewiring of the jasmonate signaling pathway in Arabidopsis during insect herbivory. Front Plant Sci 2:47
18. Robert-Seilaniantz A, Grant M, Jones JDG (2011) Hormone crosstalk in plant disease and defense: more than just jasmonate-salicylate antagonism. Annu Rev Phytopathol 49: 317–343
19. Koornneef A, Leon-Reyes A, Ritsema T, Verhage A, Den Otter FC, Van Loon LC, Pieterse CMJ (2008) Kinetics of salicylate-mediated suppression of jasmonate signaling reveal a role for redox modulation. Plant Physiol 147:1358–1368
20. Leon-Reyes A, Du Y, Koornneef A, Proietti S, Körbes AP, Memelink J, Pieterse CMJ, Ritsema T (2010) Ethylene signaling renders the jasmonate response of *Arabidopsis* insensitive to future suppression by salicylic acid. Mol Plant-Microbe Interact 23:187–197
21. Leon-Reyes A, Spoel SH, De Lange ES, Abe H, Kobayashi M, Tsuda S, Millenaar FF, Welschen RAM, Ritsema T, Pieterse CMJ (2009) Ethylene modulates the role of nonexpressor of pathogenesis-related genes1 in cross talk between salicylate and jasmonate signaling. Plant Physiol 149:1797–1809
22. Leon-Reyes A, Van der Does D, De Lange ES, Delker C, Wasternack C, Van Wees SCM, Ritsema T, Pieterse CMJ (2010) Salicylate-mediated suppression of jasmonate-responsive gene expression in Arabidopsis is targeted downstream of the jasmonate biosynthesis pathway. Planta 232:1423–1432
23. Lorenzo O, Chico JM, Sánchez-Serrano JJ, Solano R (2004) Jasmonate-insensitive1 encodes a MYC transcription factor essential to discriminate between different jasmonate-regulated defense responses in Arabidopsis. Plant Cell 16:1938–1950
24. Hoagland DR, Arnon DI (1938) The water-culture method for growing plants without soil. Calif Agric Exp Stn Circ 347:1–39
25. Lamers JG, Schippers B, Geels FP (1988) Soil-borne diseases of wheat in the Netherlands and results of seed bacterization with pseudomonads against *Gaeumannomyces graminis* var. *tritici*. In: Jorna ML, Slootmaker LAJ (eds) Cereal breeding related to integrated cereal production. Pudoc, Wageningen, The Netherlands, pp 134–139
26. King EO, Ward MK, Raney DE (1954) Two simple media for the demonstration of pyocyanin and fluorescin. J Lab Clin Med 44: 301–307
27. Whalen MC, Innes RW, Bent AF, Staskawicz BJ (1991) Identification of *Pseudomonas syringae* pathogens of *Arabidopsis* and a bacterial locus determining avirulence on both *Arabidopsis* and soybean. Plant Cell 3:49–59

Chapter 5

Elicitation of Jasmonate-Mediated Defense Responses by Mechanical Wounding and Insect Herbivory

Marco Herde, Abraham J.K. Koo, and Gregg A. Howe

Abstract

Many plant immune responses to biotic stress are mediated by the wound hormone jasmonate (JA). Functional and mechanistic studies of the JA signaling pathway often involve plant manipulations that elicit JA production and subsequent changes in gene expression in local and systemic tissues. Here, we describe a simple mechanical wounding procedure to effectively trigger JA responses in the *Arabidopsis thaliana* rosette. For comparison, we also present a plant–insect bioassay to elicit defense responses with the chewing insect *Trichoplusia ni*. This latter procedure can be used to determine the effect of JA-regulated defenses on growth and development of insect herbivores.

Key words Jasmonate, Plant–insect interaction, Mechanical wounding, Wound response, Herbivory, Plant defense, Arabidopsis, *Trichoplusia ni*

1 Introduction

A pioneering study by C.A. Ryan and colleagues demonstrated that leaf damage inflicted by insect herbivory or mechanical wounding results in rapid expression of defensive proteinase inhibitors [1]. Furthermore, tissue damage had been reported to elicit defense responses not only in wounded leaves of the plant but also in undamaged ones. This work [1] set the stage for decades of intensive research to elucidate the ecophysiological relevance and underlying molecular mechanism of induced resistance to arthropod herbivores. A major conclusion from a large body of research is that plant defenses induced by wounding and insects are regulated largely by the plant stress hormone jasmonate (JA) [2–4].

JA-regulated plant defense responses are modulated by many environmental inputs [5, 6]. The use of defined experimental conditions to induce this form of plant immunity is therefore paramount to achieving reproducible results within a study, as well as for comparing results generated in different laboratories. Exogenous JA, mechanical wounding (or simulated herbivory), and herbivory

Alain Goossens and Laurens Pauwels (eds.), *Jasmonate Signaling: Methods and Protocols*, Methods in Molecular Biology, vol. 1011, DOI 10.1007/978-1-62703-414-2_5, © Springer Science+Business Media, LLC 2013

are among the most common experimental treatments used to induce resistance. Each of these approaches has its own merits and shortcomings. Although exogenous JA is widely utilized as a potent elicitor of defense responses [7], typical compounds (such as MeJA) are now recognized to be inactive per se at the level of the COI1-JAZ receptor and must be metabolized *in planta* to the bioactive form of the hormone, jasmonoyl-isoleucine (JA-Ile) [8–10]. Such treatments may generate nonphysiological concentrations of the hormone and override the cell- and tissue-specific control of the pathway.

Mechanical wounding provides a simple alternative approach to elicit JA/JA-Ile production and associated responses in damaged (local) and undamaged (systemic) tissues. Among the implements used for mechanical wounding are hole punchers, razor blades or scalpels to cut the leaf surface, pattern wheels to create multiple small wounds, and hemostatic forceps (hemostats) to crush the leaf lamina. These treatments are typically administered at a single time point to activate the JA pathway in a reproducible manner. Relatively severe wounds inflicted, for example, by a hemostat have been used extensively to study the timing of wound-induced changes in JA content in local and systemic tissues, the dynamic behavior of JA signaling components [11], and genome-wide alterations in transcript profiles [12, 13]. This approach is applicable to monocots and dicots, as well as gymnosperms [14]. The specific method of mechanical wounding should be tailored to the particular plant species under study.

Herbivore challenge is arguably the best approach for eliciting defense responses in the context of the natural plant–herbivore interaction. Although herbivory and severe mechanical wounding often elicit similar responses [1, 2], these two treatments differ in important ways. For example, arthropod herbivores (such as chewing insects) use highly specialized mouthparts to continuously remove small pieces of leaf tissue. Specialized robotic devices have been developed to more accurately simulate the spatial and temporal aspects of mechanical damage resulting from insect herbivory [15]. Moreover, oral secretions of herbivores contain chemical factors that modulate the timing and amplitude of the host defense response [2]. For these reasons, caution should be exercised when drawing conclusions about the underlying causes of differential plant responses to mechanical wounding versus herbivory.

Below, we describe the use of hemostat wounding to elicit JA responses in the *Arabidopsis thaliana* rosette as well as a simple protocol to challenge *Arabidopsis* plants with the lepidopteran herbivore *Trichoplusia ni* (cabbage looper). In addition to its utility for studying host plant responses to herbivory, this procedure is also useful for determining the effectiveness of host defenses on the herbivore.

2 Materials

2.1 Elicitation of JA Responses by Mechanical Wounding

1. Commercial bleach 40 % (v/v).
2. Sterile water.
3. Autoclavable bags.
4. Pots (at least 3.5-in. diameter).
5. Scale with 1-mg accuracy.
6. Weigh paper or weigh boats.
7. Prelabeled collection tubes or aluminum foil.
8. Liquid nitrogen in portable container.
9. Hemostats with straight 1.5–2 mm serrated tip and locking mechanism disabled by wrapping tape around the interlocking teeth (Fig. 1a).
10. Razor blade or mini-bud trimmer.
11. Tissue storage box (precooled in −80 °C freezer).
12. Timer.

2.2 Plant–Insect Bioassay

2.2.1 Insects

1. *Trichoplusia ni* (Benzon Research, or obtained from an in-house rearing facility).
2. Featherweight forceps.
3. Paintbrush.
4. Clear plastic cups (9 oz. or 250-mL tumblers).
5. Glue gun.

Fig. 1 Eliciting JA-mediated defense responses in *Arabidopsis*. (**a**) Use of hemostats to mechanically wound *Arabidopsis* leaves. (**b**) Experimental setup for caging insect larvae on *Arabidopsis*

2.2.2 Plants

1. Pots (3.5 in. in diameter).
2. Standard *Arabidopsis* soil mix.
3. *Arabidopsis thaliana* (L.) Heyhn. plants not yet at the reproductive state.
4. Parafilm.

2.2.3 Plant–Insect Arena

1. Remove the bottom of an inverted clear plastic cup with a heated metal piece.
2. Cover the opening with Miracloth attached with a hot glue gun.
3. Affix to the pot top (Fig. 1b).
4. Prepare a sufficient number of cup cages prior to initiating the feeding assay.

3 Methods

3.1 Preparation of Plants for Elicitation

1. Surface sterilize *Arabidopsis* seeds with 40 % (v/v) commercial bleach for 5 min.
2. Rinse seeds thoroughly with sterile water.
3. Imbibe in water for 2–3 days at 4 °C in the dark (*see* **Note 1**).
4. Sterilize soil prior to sowing (*see* **Note 2**) by placing the desired amount of moist soil in an autoclavable bag. Autoclave for an appropriate amount of time (such as 60-min dry cycle for a 45-L bag).
5. Allow soil to cool after autoclaving, but keep the bag unopened until seeds are sown to prevent recontamination.
6. Add soil to the pots and immediately sow seeds. Taking the seed germination efficiency into account, sow a sufficient number of seeds to ensure growth of two to four plants per pot (*see* **Note 3**).
7. Grow plants in a growth chamber maintained at 22 °C with a photoperiod of 16-h light (100 μE/m^2/s) and 8-h darkness. In our standard wound assay we use 30-day-old plants with fully expanded rosette leaves but that have not bolted yet.

3.2 Elicitation of JA Responses by Mechanical Wounding

1. Plan the location and timing of your experiment, with particular attention to environmental factors that affect JA-induced responses (*see* **Notes 4** and **5**).
2. Select specific plants to be included in the experiment and arrange pots so as to facilitate wounding and collection of samples in a coordinated manner. It is critical to use plants that have similarly sized leaves so that the relative proportion of damaged and undamaged areas on the leaf is consistent between plants (*see* **Notes 6** and 7).

3. Administer the wound treatment with the device of choice (*see* **Note 8**) and immediately start the timer. If using a hemostat, briefly clamp the device firmly across the leaf (perpendicular to the midvein; Fig. 1a) and then release (*see* **Note 9**). Serrated wounds should be clearly visible on the leaf.
4. Make one or two such wounds per leaf (*see* **Note 10**), damaging at least five leaves in less than 1 min. Continue to wound a sufficient number of leaves for one biological replicate (*see* **Note 11**).
5. With the timer still running, wound additional plants for replicates two and three for that particular time point. Typically, we do not pause the wound treatment between replicates or between different plant genotypes.
6. Start wounding independent plants for additional time points (*see* **Note 12**).
7. At the appropriate time after wounding, harvest the damaged leaves by cutting the petiole with a razor blade or mini-bud trimmers.
8. Work quickly to measure the fresh weight of the excised leaves.
9. Transfer the tissue to a suitable container (such as aluminum foil) and immediately freeze in liquid nitrogen (*see* **Note 13**). Frozen tissue should be stored at −80 °C until use.
10. Harvest leaf tissue from undamaged control plants at the beginning of the time-course experiment or, alternatively, in parallel with the damaged leaves from a matched set of wounded plants.

3.3 Plant–Insect Bioassay

Reproducibility in plant–insect bioassays designed to measure plant and insect performance (or fitness) is of the utmost importance. One critical factor is the ability to grow healthy plants under controlled conditions. To achieve meaningful results, the number of plants, insects, and replicates required should also be taken into account. Experimental results are influenced by many factors, including larval mortality, the amount of plant tissue consumed during the feeding assay, and the extent to which larval growth on a particular host genotype varies between insects (*see* **Note 14**). If the larval mortality is high, plants can be challenged with multiple larvae. The use of many larvae per plant might result in overconsumption of plant tissue without sufficient larval weight gain during the feeding trial. For statistical analyses, we typically use the average of all larval weights from one plant as a single data point.

The plant and insect species used for bioassays will influence the conclusions drawn from the experiment. For example, the performance of an insect species (such as *Pieris rapae*) that is specialized for feeding on *Arabidopsis* may be unaffected by the presence of glucosinolates, which are the major class of chemical defenses in

Arabidopsis. In contrast, growth and development of crucifer nonspecialists (such as *Spodoptera exigua*) will be negatively affected by glucosinolates. The ecological relevance of the plant–insect interaction (i.e., whether a particular interaction actually occurs in nature) may also be relevant for the experimental design.

Moreover, the defense status of a plant is strongly affected by the plant's developmental age. When assessing the effect of JA-inducible defenses expressed in vegetative tissue (for instance, leaves), it is desirable to use plants that have not yet entered the reproductive phase of development. Prolonged vegetative growth (i.e., in the absence of bolting) of *Arabidopsis* can be accomplished by the use of short-day growth conditions.

Finally, the genetic variation between individual insects is a potential source of experimental variation in insect growth on a given diet. This variation could be minimized by the use of an insect colony that has been inbred for many generations. However, excessive inbreeding within the population may result in the loss of traits (such as host range) exhibited by populations in the wild.

The following protocol describes a bioassay with *Trichoplusia ni* reared on *Arabidopsis* (accession Col-0). The procedure can readily be adapted for use with other lepidopteran herbivores, such as *Spodoptera exigua* and *Pieris rapae*.

1. Grow one plant per pot with soil under short-day conditions to maximize vegetative growth. Our standard growth conditions are 22 °C with a photoperiod of 10 h light (100 μE/m^2/s) and 14 h darkness. Plants should approximately be 5 weeks old at the time of insect challenge (*see* **Note 15**).
2. Place insect eggs (such as commercially obtained *T. ni* eggs) in a Petri dish together with a small piece of moist filter paper to maintain humidity and prevent desiccation of larvae.
3. Seal the dish with parafilm and incubate eggs at room temperature (*see* **Note 16**). Eggs typically begin to hatch within 48 h. Please keep in mind the day/night light cycle under which plants and insects are grown (*see* **Note 17**).
4. Water plants well the evening before placing insects on plants.
5. Use a small paintbrush to transfer larvae to leaves near the center of the rosette (*see* **Notes 18** and **19**). Remain consistent with the location of the insect placement on the plant (*see* **Note 20**).
6. Use cup cages to secure the insect on the plant and tightly seal the junction between the pot and cage with parafilm.
7. Return the plants to the growth chamber in which the plants had been grown previously.
8. Remove the cup cage at different time points (i.e., various days after infestation) to determine the weight of insects (*see* **Notes 21** and **22**).

9. Use a scale with appropriate accuracy for the particular larvae measured (*see* **Note 22**). A setup for automated transfer of weight measurements from the scale to an appropriate computer program (such as Excel) facilitates data recording.
10. Handle heavier larvae carefully with featherweight forceps without damaging or stressing the insects. For this reason, we return individually weighed larvae to their plant of origin prior to processing to the next insect.
11. Thoroughly search the host plant, soil surface, and cage to identify larvae. The number of viable larvae identified provides a measure of insect mortality.
12. Analyze the host plant as dictated by the needs of the experiment, which might include photographing leaves to evaluate the consumed leaf area with an appropriate imaging software (such as ImageJ; http://rsbweb.nih.gov/ij/) and harvesting leaf tissue for RNA isolation or chemical analysis.

4 Notes

1. Alternatively, seeds can be sown and the pots placed in a cold room for 3 days. The drawback of this procedure is that sufficient space is needed in a cold room and plants may inadvertently be exposed to fungal spores. Growth of healthy, unstressed plants is a key factor in obtaining reliable results.
2. We use autoclaved soil for growing *Arabidopsis* because it reduces the risk of unintended interactions with soil-borne fungi.
3. Sowing too many seeds per pot will necessitate the additional manual removal of plants to avoid overcrowding. Extra handling of plants at this stage may cause inadvertent stress to the plants.
4. We typically perform short-term (i.e., <12 h) time-course experiments with plants that have been moved from the growth chamber to a work area within the laboratory with similar light intensity and temperature. Care should be taken to minimize unintentional stress to the plants during transport and handling. For example, mechanical agitation may activate a basal response in unwounded control plants, or may prime plants for a stronger wound response. As an alternative experimental setup, wound elicitation and tissue harvesting may be performed without removing plants from the growth chamber. In this case, all materials should be readily available in the growth chamber.
5. Recent studies indicate that JA-regulated defense responses to herbivore attack are influenced by circadian and/or diurnal inputs [16–18]. Therefore, it is important to plan the timing

of the experiments with respect to the light/dark cycle and to be as consistent as possible from experiment to experiment. In our hands, wound treatments administered in the "morning" (i.e., within a few hours of the dark-to-light transition) result in more robust JA-mediated responses in *Arabidopsis* and *Solanum lycopersicum* (tomato).

6. Generally, the correlation is positive between the area of damaged tissue (as a percentage of the total leaf area) and the accumulated levels of JA and JA-responsive transcripts. If a particular mutant under study exhibits a morphological phenotype (such as a semidwarf stature), the damage (relative to the total leaf area) inflicted to the mutant will be greater than that administered to the wild type. Accordingly, the responses in the mutant might appear stronger than those in the wild type.
7. To analyze wound-induced systemic responses in the *Arabidopsis* rosette, the extent of the vascular connectivity between damaged (local response) and undamaged (systemic response) leaves of the same plant should be considered. In *Arabidopsis* and nonrosette plants as well, parastichous leaves with a strong interconnective vasculature exhibit stronger systemic responses than nonparastichous leaves [19]. In the case of *Arabidopsis*, we typically use three or four developmentally older leaves at the base of the rosette and younger undamaged (parastichous) leaves on the same rosette to assess the local and systemic responses, respectively [11].
8. The amplitude of the wound-induced JA accumulation and JA-dependent gene expression may vary greatly depending on the particular wounding method. In general, the severity of tissue damage correlates with the strength of the response. Crushing-type wounds inflicted with a hemostat are relatively severe and seemingly activate maximal responses. Accordingly, this simple procedure is especially useful for eliciting rapid and uniform responses to be monitored in time-course studies.
9. Wounds inflicted either parallel or perpendicular to the midvein are effective in activating JA responses. However, for the ease and consistency of the treatment, we typically administer wounds perpendicular to the midvein (Fig. 1a).
10. Although multiple wounds inflicted to a single leaf intensify the JA responses, it also increases the time required for the treatment. We found that two wounds per leaf is a good compromise for most applications. If administering more than one wound per leaf, make the initial wound at the distal end of the leaf and the subsequent wounds at the proximal end.
11. We use three biological replicates (i.e., different plants) for each data point. Statistical variation between replicates may be reduced by pooling leaves from multiple plants. In general, we

pool three to five fully expanded rosette leaves (approximately 200–300 mg fresh weight) as a single replicate in hormone measurements and RNA preparations.

12. In short time-course experiments (for instance, tissue harvested 5 min post wounding) or experiments with multiple sampling points in which wounding and harvesting procedures temporally overlap, it is convenient to use multiple timers to keep track of the different samples.

13. Recent studies with *Arabidopsis* [11, 20] indicate that it is crucial to minimize the time between harvesting (i.e., excising) and freezing of tissue in liquid nitrogen because JA levels increase very rapidly in excised leaves. Therefore, it is important to perform the various steps of the protocol (leaf excision, weighing and recording, securing tissue in an appropriate container, and flash freezing) as fast as possible without compromising the accuracy. Wrapping tissue in prelabeled aluminum foil is useful for this purpose. In general, the total time elapsed between excising and freezing leaves in liquid nitrogen should be shorter than 1 min.

14. As for the wound assays, host responses and insect performance on plant genotypes that differ significantly in morphology or development should be compared with caution.

15. The physical position of individual plants within a growth chamber (and watering tray, if used) may have a significant effect on the plant physiology, which, in turn, can influence insect performance. To correct this potential bias, the position of all plants within the growth chamber and watering tray should be randomized at the onset of the experiment.

16. The time required to hatch eggs can be modified by changing the temperature. Higher temperatures (such as 28 °C) accelerate the process, whereas cooler temperatures are useful to slow the hatch rate. Although some insect eggs can be stored temporarily at 4 °C to delay hatching, we recommend that eggs should be hatched as soon as possible to maximize hatching viability and uniformity.

17. The circadian regulation defense responses of *Arabidopsis* are coordinated with the circadian cycle of the *T. ni* larvae [18]. Thus, consistency in the timing (relative to the light/dark cycle) of the plant and insect manipulations should be maintained. In general, we perform these procedures during the second half of the light period, prior to the beginning of the dark period.

18. Neonate (newly hatched) larvae are a suitable choice for feeding trials. Typically, an excess number of eggs are hatched and, for plant infestation, neonates are selected that hatch at a similar time. In the event of unacceptably high neonate mortality

on a particular host genotype, neonates may be reared on an artificial diet for a brief period (for instance, 1 day) prior to transfer to the plant. Although this method tends to reduce larval mortality and facilitates more effective damage to the host plant, bypassing the initial neonate–host interaction may be unsuitable for all purposes.

19. As neonate larvae are very delicate, physical damage during the transfer process should be avoided.
20. Because of inter- and intra-leaf variation in the expression of host defenses [21], every effort should be made to standardize the position of the larvae on the leaves at the onset of the experiment.
21. *T. ni* larvae usually empty their gut content prior to molting and pupation with a significant effect on larval weight as a consequence.
22. Specific time points chosen for the determination of larval weight depend on the objective of the experiment. In general, differences in larval performance on well-defended (such as wild type) versus defense-compromised (such as a JA mutant) host genotypes become more apparent at the later stages of the larval development. A scale with a 0.1-mg accuracy will facilitate weight measurements of early-instar larvae.

Acknowledgments

This work was supported by the National Institutes of Health (grant R01GM57795) and the Chemical Sciences, Geosciences and Biosciences Division, Office of Basic Energy Sciences, Office of Science, US Department of Energy (grant DE–FG02–91ER20021) for partial support of M.H. M.H. was also the recipient of a fellowship (HE 5949/1-1) from the German Research Foundation (DFG).

References

1. Green TR, Ryan CA (1972) Wound-induced proteinase inhibitor in plant leaves: a possible defense mechanism against insects. Science 175: 776–777
2. Howe GA, Jander G (2008) Plant immunity to insect herbivores. Annu Rev Plant Biol 59: 41–66
3. Koo AJK, Howe GA (2009) The wound hormone jasmonate. Phytochemistry 70: 1571–1580
4. Wu J, Baldwin IT (2010) New insights in plant responses to attack from insect herbivores. Annu Rev Genet 44:1–24
5. Kazan K, Manners JM (2012) JAZ repressors and the orchestration of phytohormone crosstalk. Trends Plant Sci 17:22–31
6. Erb M, Meldau S, Howe GA (2012) Role of phytohormones in insect-specific plant reactions. Trends Plant Sci 17:250–259
7. Farmer EE, Ryan CA (1990) Interplant communication: airborne methyl jasmonate induces synthesis of proteinase inhibitors in plant leaves. Proc Natl Acad Sci USA 87:7713–7716
8. Thines B, Katsir L, Melotto M, Niu Y, Mandaokar A, Liu G, Nomura K, He SY, Howe GA, Browse J (2007) JAZ repressor proteins

are targets of the SCF^{COI1} complex during jasmonate signalling. Nature 448:661–665
9. Staswick PE, Tiryaki I (2004) The oxylipin signal jasmonic acid is activated by an enzyme that conjugates it to isoleucine in Arabidopsis. Plant Cell 16:2117–2127
10. Katsir L, Chung HS, Koo AJK, Howe GA (2008) Jasmonate signaling: a conserved mechanism of hormone sensing. Curr Opin Plant Biol 11:428–435
11. Koo AJK, Gao X, Jones AD, Howe GA (2009) A rapid wound signal activates the systemic synthesis of bioactive jasmonates in Arabidopsis. Plant J 59:974–986
12. Reymond P, Weber H, Damond M, Farmer EE (2000) Differential gene expression in response to mechanical wounding and insect feeding in Arabidopsis. Plant Cell 12:707–719
13. Weber H, Vick BA, Farmer EE (1997) Dinor-oxo-phytodienoic acid: a new hexadecanoid signal in the jasmonate family. Proc Natl Acad Sci USA 94:10473–10478
14. Lippert D, Chowrira S, Ralph SG, Zhuang J, Aeschliman D, Ritland C, Ritland K, Bohlmann J (2007) Conifer defense against insects: proteome analysis of Sitka spruce (*Picea sitchensis*) bark induced by mechanical wounding or feeding by white pine weevils (*Pissodes strobi*). Proteomics 7:248–270
15. Mithöfer A, Wanner G, Boland W (2005) Effects of feeding *Spodoptera littoralis* on lima bean leaves. II. Continuous mechanical wounding resembling insect feeding is sufficient to elicit herbivory-related volatile emission. Plant Physiol 137:1160–1168
16. Kessler D, Diezel C, Baldwin IT (2010) Changing pollinators as a means of escaping herbivores. Curr Biol 20:237–242
17. Radhika V, Kost C, Mithöfer A, Boland W (2010) Regulation of extrafloral nectar secretion by jasmonates in lima bean is light dependent. Proc Natl Acad Sci USA 107:17228–17233
18. Goodspeed D, Chehab EW, Min-Venditti A, Braam J, Covington MF (2012) *Arabidopsis* synchronizes jasmonate-mediated defense with insect circadian behavior. Proc Natl Acad Sci USA 109:4674–4677
19. Glauser G, Dubugnon L, Mousavi SAR, Rudaz S, Wolfender J-L, Farmer EE (2009) Velocity estimates for signal propagation leading to systemic jasmonic acid accumulation in wounded *Arabidopsis*. J Biol Chem 284:34506–34513
20. Glauser G, Grata E, Dubugnon L, Rudaz S, Farmer EE, Wolfender J-L (2008) Spatial and temporal dynamics of jasmonate synthesis and accumulation in *Arabidopsis* in response to wounding. J Biol Chem 283:16400–16407
21. Shroff R, Vergara F, Muck A, Svatoš A, Gershenzon J (2008) Nonuniform distribution of glucosinolates in *Arabidopsis thaliana* leaves has important consequences for plant defense. Proc Natl Acad Sci USA 105:6196–6201

Chapter 6

Pseudomonas syringae Infection Assays in Arabidopsis

Jian Yao, John Withers, and Sheng Yang He

Abstract

Pseudomonas syringae pv. *tomato* DC30000 (*Pst* DC3000) infection of *Arabidopsis thaliana* has been widely used to elucidate many of the general principles underlying the plant immune response and bacterial pathogenesis. Study of *Pst* DC3000 virulence factors has also proven useful in the discovery and elucidation of fundamental mechanisms in plant biology. In particular, *Pst* DC3000 produces a phytotoxin, coronatine, that is a remarkable molecular mimic of the active form of the plant hormone jasmonate. Here we illustrate several common methods used for *Pst* DC3000-based assays, including preparation of *Pst* DC3000 inocula, inoculation of soil-grown *Arabidopsis* plants, and subsequent bacterial quantification *in planta*. We also describe how *Pst* DC3000 infection can be applied to study gene expression and protein degradation associated with jasmonate signaling.

Key words Plant defense, Plant pathogen, Type III secretion, Salicylic acid

1 Introduction

Pseudomonas syringae pv. *tomato* DC3000 (*Pst* DC3000) is a Gram-negative bacterial pathogen that causes bacterial speck disease of tomato (*Solanum lycopersicum*) and *Arabidopsis thaliana* [1] and has become an integral part of molecular and biochemical studies of host–pathogen interactions. *P. syringae* is considered a hemi-biotrophic plant pathogen whose natural infection cycle begins with inocula from infected seeds, plants, or debris. Many *P. syringae* strains exhibit epiphytic growth on the plant surface [2] and can enter the plant through surface wounds and natural openings, such as stomata. Aggressive endophytic growth within the host plant eventually leads to disease. The use of this host–microbe interaction as a model system to study bacterial pathogenesis and host response to infection has provided great insight into plant hormone signal transduction mechanisms involved in defense responses. In particular, the plant hormone jasmonic acid (JA) has been of great interest

Jian Yao and John Withers have contributed equally to this work.

Alain Goossens and Laurens Pauwels (eds.), *Jasmonate Signaling: Methods and Protocols*, Methods in Molecular Biology, vol. 1011, DOI 10.1007/978-1-62703-414-2_6, © Springer Science+Business Media, LLC 2013

in this pathosystem because JA signaling components, most notably a JA receptor protein, CORONATINE INSENSITIVE 1 (COI1), and a major JA transcription factor, MYC2, are necessary for *P. syringae* to infect *Arabidopsis* and tomato [3–6].

The importance of JA signaling in plant pathogenesis is supported by the fact that a number of *P. syringae* strains belonging to pathovars *glycinea*, *atropurpurea*, *morsprunorum*, *maculicola*, and *tomato* all produce the non-host-specific phytotoxin coronatine (COR) [7]. COR consists of two molecular moieties, coronafacic acid (CFA) and coronamic acid (CMA). CFA is a modified polyketide and is linked by an amide bond to CMA, a cyclized derivative of isoleucine [8–10]. COR has been shown to be a potent activator of JA-mediated signal transduction in *Arabidopsis* and tomato [11–13] and to be required for bacterial suppression of host defenses during both bacterial entry through stomata and bacterial multiplication in the apoplast [7, 11, 14, 15]. At the molecular level, COR is a structural mimic of the active form of jasmonate, jasmonoyl-l-isoleucine (JA-Ile) [16] and played an important role in the discovery and characterization of the JASMONATE ZIM-domain (JAZ) family of transcriptional repressors and ligand-dependent formation of the JA receptor complex [17–20].

Although the well-established role of COR/JA signaling in plant–*P. syringae* interactions makes *Pst* DC3000 pathogenesis an excellent functional assay to study JA signaling and response, the assay involves some specialized techniques that are routine in plant pathology laboratories, but may appear alien to plant molecular biologists. In addition, environmental conditions and the developmental stages of plants used for infection studies have significant effects on the responses [21]. The conditions under which *P. syringae* and the host plants are grown must be monitored and kept consistent to maximize reproducibility of phenotypic effects [21]. In this chapter we present detailed protocols for preparation and growth of *Arabidopsis* and *Pst* DC3000. We illustrate how to infect plants using three independent methods of inoculation: syringe injection of individual leaves and either dipping or spraying of whole plants grown in pots. Methods for monitoring host symptom development and bacterial multiplication within the infected plants are presented as well. Finally, we provide examples of methods that have been useful for the elucidation of the role of COR/JA signaling in disease including assays for JAZ protein degradation and expression of JA response genes.

2 Materials

2.1 Growth of Arabidopsis Plants

1. Potting soil (Redi-Earth and perlite [2:1]; W.R. Grace and Co., Elmwood Park, NJ, USA).
2. Square 3.5-in. or 9-cm pots.

3. Standard mesh Phiferglass insect screen (Phifer Inc., Tuscaloosa, AL, USA).
4. Rubber bands.
5. Growth chamber (22 °C; 12-h/12-h day/night photoperiod; light intensity 80–100 μmol/m^2/s).

2.2 Preparation of Bacterial Inoculum

1. Glycerol stocks of *Pst* DC3000 and other desired bacterial strains (for instance, *Pst* DC3118) stored at −80 °C.
2. Modified Luria-Bertani liquid medium (LM): 10.0 g Bacto tryptone, 6.0 g Bacto yeast extract, 1.5 g K_2HPO_4, 0.6 g NaCl, 0.4 g $MgSO_4 \cdot 7H_2O$ per liter H_2O.
3. LM agar plates: LM medium plus 18.0 g Bacto agar per liter and appropriate antibiotics.
4. Sterilized culture tubes and glassware for liquid bacterial cultures.
5. Shaking and static incubators set to 28 °C.

2.3 Bacterial Inoculation

1. Spectrophotometer to determine the optical densities of bacterial cultures.
2. Disposable 1-mL needleless syringes for leaf infiltration.
3. Sterile 500-mL spray bottle for spray inoculation.
4. Sterile glass bowl (approximately 500 mL) for dip inoculation.
5. Silwet L-77 surfactant (OSi Specialties, Friendly, WV, USA).

2.4 Bacterial Quantification

1. Glass beakers.
2. Sterile 1.5-mL microcentrifuge tubes.
3. Sterile plastic micropestles.
4. Cordless drill.
5. 70 % (v/v) ethanol:H_2O.
6. Sterile distilled water.
7. 0.5-cm^2 cork borer for excising leaf discs.
8. LM agar plates with appropriate antibiotics.
9. Parafilm.
10. Forceps and fine-tipped scissors or razor blade.

2.5 Plant Sample Preparation and RNA Extraction

1. Leaves of *Arabidopsis thaliana* inoculated with bacteria.
2. Liquid nitrogen.
3. 2.0-mL impact-resistant screw cap tubes (USA Scientific, Orlando, FL, USA).
4. Metal beads (diameter 2.38 mm) (MO BIO laboratories, Carlsbad, CA, USA).

5. Tissue grinding device (TissueLyser II, Qiagen, Valencia, CA, USA) (*see* **Note 1**).
6. TRIzol Reagent (Life Technologies, Grand Island, NY, USA).
7. RNase-free, diethylpyrocarbonate (DEPC)-treated water (*see* **Note 2**).
8. Chloroform.
9. 2-Propanol.
10. 75 % (v/v) ethanol:H_2O made with DEPC-treated water.

2.6 RNA Purification

1. RNase-free DNase (Roche Diagnostics, Indianapolis, IN, USA).
2. Protector RNase Inhibitor (Roche Diagnostics).
3. 2-Propanol and ethanol.
4. RNeasy Mini Kit (Qiagen).

2.7 First-Strand cDNA Synthesis

1. Purified RNA.
2. 10 mM dNTP mix.
3. 500 μg/μL of oligo$(dT)_{12-18}$ (Life Technologies, Grand Island, NY, USA).
4. Nuclease-free water.
5. M-MLV Reverse Transcriptase (Life Technologies).
6. Protector RNase Inhibitor (Roche Diagnostics).
7. PCR machine.

2.8 Quantitative Real-Time PCR

1. Real Time PCR machine (7500 Fast Real-Time PCR system; Applied Biosystems, Foster City, CA, USA).
2. 96-well PCR plates and sealing films.
3. cDNA.
4. Fast SYBR Green Master Mix (Applied Biosystems).
5. Primers for reference genes and gene(s) of interest (*see* **Note 3**).

2.9 Plant Material and Protein Extraction

1. *Arabidopsis* leaves inoculated with bacteria.
2. Liquid nitrogen.
3. 50 % bleach.
4. 70 % (v/v) ethanol:H_2O.
5. Sterile pestles and matching microfuge tubes (RPI, Mount Prospect, IL, USA).
6. Plant protein extraction buffer (PPEB): 50 mM Tris–HCl (pH 7.5), 150 mM NaCl, 1 % Triton X-100, 0.1 % sodium dodecyl sulfate (SDS), 1 mM ethylenediaminetetraacetic acid (EDTA), and 1 mM dithiothreitol (DTT); stored at −20 °C. Immediately prior to use, add protease inhibitor cocktail (Sigma-Aldrich,

St. Louis, MO, USA) to 1 % (v/v) and 26S proteasome inhibitor MG132 (10 mM stock in dimethylsulfoxide (DMSO)) (Cayman Chemicals, Ann Arbor, MI, USA) to 100 μM.

7. *DC/RC* Protein Assay Kit (Bio-Rad, Hercules, CA, USA).
8. Laemmli SDS sample buffer (4×): 250 mM Tris–HCl (pH 6.8), 8 % (w/v) SDS, 40 % glycerol, 0.04 % bromophenol blue; stored at −20 °C. Immediately prior to use, add β-mercapto-ethanol to 10 % (v/v).

2.10 SDS-Polyacrylamide Gel Electrophoresis Analysis

1. Separating gel buffer (5×): 1.875 M Tris–HCl (pH 8.9) and 0.5 % SDS.
2. Stacking gel buffer (5×): 0.5 M Tris–HCl (pH 6.8) and 0.5 % SDS.
3. 30 % acrylamide/0.8 % bis-acrylamide solution (Bio-Rad).
4. 10 % ammonium persulfate.
5. *N*,*N*,*N*,*N*-tetramethyl-ethylenediamine (TEMED).
6. Running buffer: 25 mM Tris base, 192 mM glycine, and 0.1 % SDS.
7. PageRuler Plus Prestained protein ladder (Fermentas, Glen Burnie, MD, USA).
8. Power supply.
9. Mini-PROTEAN 3 gel system (Bio-Rad) with Bio-Ice cooling system.

2.11 Immunoblotting

1. Tank blotting system (Mini Trans-Blot Cell) (Bio-Rad).
2. Tank transfer buffer: 25 mM Tris base, 150 mM glycine, and 10 % methanol.
3. Gel blotting filter paper, slightly larger than the gel.
4. Polyvinylidene fluoride (PVDF) membrane of the same size as the gel.
5. Methanol.
6. Black Mini-Blotting Containers (RPI).
7. Tris-buffered saline (TBS): 10 mM Tris–HCl (pH 7.5) and 150 mM NaCl.
8. TBS with Tween 20 (TBST): TBS with 0.1 % Tween-20.
9. Primary antibody dilution buffer: TBS with 3 % bovine serum albumin (BSA) and 0.05 % NaN_3; stored at 4 °C.
10. Blocking buffer/secondary antibody dilution buffer: TBS with 5 % nonfat dry milk.
11. Primary antibody and horseradish peroxidase (HRP)-conjugated secondary antibody.

12. SuperSignal West Pico Chemiluminescent Substrate (Thermo Scientific, Rockford, IL, USA) kit containing stable peroxidase and luminol/enhancer solutions.
13. X-ray film and exposure cassette.
14. X-ray film developer.
15. Coomassie Brilliant Blue (CBB) staining solution: 0.1 % CBB R-250, 50 % methanol, and 10 % glacial acetic acid.
16. Destaining solution: 30 % methanol and 10 % glacial acetic acid.

3 Methods

3.1 Inoculation of *Arabidopsis* with *P. syringae*

3.1.1 Growing Arabidopsis Plants for Inoculation

1. Fill pots with moist soil and lightly pack each pot by hand to ensure an evenly firm surface for sowing the *Arabidopsis* seeds (*see* **Note 4**).
2. Place pots in flats, fill flat bottom with water to soak soil thoroughly, and drain off excess water.
3. Sow cold-stratified *Arabidopsis* seeds (1–4 seeds) in each corner of the pot.
4. Cover flats with a transparent plastic dome to maintain humidity for seed germination.
5. Move flats to a growth chamber set to the conditions specified above (*see* **Note 5**).
6. After germination, thin seedlings to four evenly spaced plants per pot.
7. Grow plants for 4–6 weeks before inoculation (*see* **Note 6**).

3.1.2 Preparation of Bacterial Inoculum

1. Streak out bacteria from a −80 °C freezer stock onto a fresh LM agar plate containing appropriate antibiotics.
2. Incubate the plate at 28 °C for 2 days (*see* **Note 7**).
3. The evening before, inoculate 5 mL liquid LM medium (with appropriate antibiotics) in a sterile culture tube with a single bacterial colony.
4. Incubate the culture tube at 28 °C for 12 h (*see* **Note 8**).
5. Harvest bacteria by centrifugation at 2,500 × *g* for 10 min at room temperature.
6. Remove supernatant and resuspend bacteria in sterile water to a desired concentration depending on the inoculation technique to be used (*see* **Notes 9** and **10**).

3.1.3 Syringe Inoculation

1. For each bacterial strain, select four fully developed leaves of each plant.
2. Mark the petioles with permanent ink for identification.

Fig. 1 Syringe inoculation of *Arabidopsis* leaves. (**a**) Supplies needed for preparations of dilutions of bacterial cultures and leaf infiltrations prior to beginning the procedure. (**b**) By means of a 1-mL needleless syringe, enough pressure is applied gently on the abaxial leaf surface of a fully expanded, healthy leaf to infiltrate the leaf tissue. (**c**) During infiltration, water-soaking of the leaf is apparent. (**d**) After the infiltrated leaves have been allowed to dry, plants are placed in a covered container to maintain humidity and returned to previous growth conditions

3. Using a 1-mL needleless syringe filled with bacterial suspension (OD_{600} = 0.002 or 1×10^6 colony-forming units (CFU)/mL), gently infiltrate marked leaves on the abaxial surface (Fig. 1) (*see* **Note 11**).
4. Allow leaves to dry and cover plants with a plastic wrap or a plastic dome to maintain high humidity.
5. Return plants to previous growth conditions for approximately 3 days while the disease progresses (Fig. 1).

3.1.4 Spray or Dip Inoculation

1. Grow plants and bacteria as described (*see* Subheadings 3.1.1 and 3.1.2) except starting a larger volume of bacterial culture to produce sufficient inoculum.
2. For dipping, grow plants on mounded pots covered with mesh (Fig. 2a) (*see* **Note 12**).
3. For spray inoculation, prepare bacterial suspension to 5×10^8 CFU/mL in water supplemented with 0.02 % Silwet L-77 in an ethanol-sterilized spray bottle set to release a fine mist.
4. Spray the surface of the rosettes until saturation, when leaves appear evenly wet.
5. For dip inoculation, prepare a bacterial suspension at the same concentration as for spray inoculation (1×10^8 CFU/mL in water supplemented with 0.02 % Silwet L-77).

Fig. 2 Spray and dip inoculation of *Arabidopsis* rosettes. (**a**) Supplies needed for the preparation of bacterial culture dilutions, including a sterile container for dipping plants or a spray bottle and 4- to 6-week-old plants grown in soil. (**b**) The entire rosette is submerged into the bacterial suspension and plants are swirled gently in the suspension to ensure a uniform leaf exposure to the bacteria. (**c**) The rosettes should be evenly coated with bacterial suspension. (**d**) Inoculated plants are placed immediately into a covered flat to avoid drying of the dipped rosettes and to maintain humidity

6. Pour the suspension into a sterile 500-mL glass beaker.
7. Dip the entire rosette into the bacterial suspension by inverting the pot, submerging the plants, and gently swirling the rosette in the liquid for 5 s (Fig. 2b).
8. Immediately place the plants in a flat and cover with a plastic dome to maintain high humidity.
9. Return plants to previous growth conditions for 3 days (*see* **Note 13**).

3.1.5 Bacterial Quantification

1. Harvest inoculated leaves 3 days post inoculation (dpi) with forceps and fine-tipped scissors or a razor blade.
2. Surface sterilize leaves by submerging them in 70 % ethanol for 10–15 s under gentle shaking.
3. Remove leaves; rinse briefly by submerging them in sterile, distilled water for 10–15 s; and blot on paper towels to dry.
4. Excise one 0.5-cm^2 leaf disc per leaf sample using a cork borer.
5. Place leaf disc in a 1.5-mL Eppendorf tube containing 100 μL sterile distilled water (*see* **Notes 14** and **15**).
6. Repeat **steps 4** and **5** for each plant until all samples have been collected.

7. Grind each leaf disc in the 1.5-mL tube of sterile distilled water using a cordless drill and plastic micropestle until tissue is completely macerated and no intact leaf pieces are visible (*see* **Note 16**).
8. Change micropestle and repeat **step 7** for each sample.
9. Add 900 μL of sterile distilled water to each sample and vortex vigorously.
10. Proceed with serial dilutions by removing 100 μL of the suspension from the tube and add to a new tube containing 900 μL of sterile distilled water (*see* **Note 17**).
11. Spot 10-μL samples in triplet from each dilution on LM agar plates with appropriate antibiotics.
12. Allow spotted samples to dry, cover the plates, and wrap them with parafilm.
13. Incubate plates at 28 °C for approximately 24–30 h or until the colonies are clearly visible under a dissecting microscope.
14. Using a dissecting microscope, count colonies from each dilution, record results, and calculate CFU/cm^2 of leaf tissue (*see* **Note 18**).

3.2 qRT-PCR Analysis of Gene Expression in Arabidopsis During Pst DC3000 Infection

3.2.1 Plant Sample Preparation and RNA Extraction

1. Inoculate 5-week-old *Arabidopsis* wild-type Col-0 and *coi1-30* mutant plants with water and 1×10^8 CFU/mL *Pst* DC3000 and DC3118 by syringe infiltration. Use two leaves per plant and at least three plants per treatment (*see* **Note 19**).
2. Put inoculated plants back under original growth condition for 6 h.
3. During this period, prechill two TissueLyser adaptors in the −80 °C freezer.
4. Label collection tubes and place 12 metal beads into each one.
5. After 6 h cut two inoculated leaves at the base of the leaf with a pair of sharp scissors, put them into a collection tube, and freeze immediately in liquid nitrogen (*see* **Note 20**).
6. Use the TissueLyser to grind the frozen leaves to fine powder at 30 s^{-1} for 1 min.
7. Add 1 mL of TRIzol reagent into the tube and vortex thoroughly (*see* **Note 21**).
8. Incubate the sample at room temperature for 5 min for complete dissociation of the nucleoprotein complex.
9. Add 0.2 mL chloroform into the sample and vortex for 15 s.
10. Incubate the sample at room temperature for 3 min.
11. Centrifuge the sample at 12,000 × *g* for 15 min at 4 °C (*see* **Note 22**).
12. Remove 0.5 mL of the upper aqueous phase and place in a new 1.5-mL RNase-free microfuge tube by pipetting.

13. Add 0.5 mL 2-propanol and incubate at room temperature for 10 min.
14. Centrifuge at 12,000 × *g* for 10 min at 4 °C and discard the supernatant.
15. Wash the pellet with 1 mL 75 % ethanol by brief vortexing.
16. Centrifuge at 7,500 × *g* for 5 min at 4 °C and discard the supernatant.
17. Centrifuge at 7,500 × *g* for 1 min at 4 °C and remove any residual ethanol by pipetting.
18. Air-dry the pellet for 5 min (*see* **Note 23**) and resuspend it in 88.5 μL of RNase-free water by incubating at 60 °C for 10 min.

3.2.2 RNA Purification

1. Prepare the following mixture:

Component	Total volume (100 μL)
Total RNA	88.5 μL (<100 μg)
10× incubation buffer	10 μL
RNase-free DNaseI	1 μL (10 units)
Protector RNase inhibitor	0.5 μL (20 units)

2. Incubate at 37 °C for 20 min.
3. Add 350 μL of buffer RLT from the RNeasy Mini Kit into the DNase-treated RNA sample and mix well.
4. Add 250 μL ethanol to the RLT-diluted sample and mix well by pipetting.
5. Transfer the sample to an RNeasy Mini Spin Column, close lid gently, centrifuge at 12,000 × *g* for 15 s, and discard the flow-through.
6. Add 500 μL of buffer RPE to the column, close lid gently, centrifuge for 15 s to wash the column, and discard the flow-through.
7. Add 500 μL of buffer RPE to the column, close lid gently, and centrifuge for 2 min to wash the column.
8. Place the column in a new 2-mL collection tube, close lid gently, and centrifuge for 1 min to eliminate any carryover of buffer RPE.
9. Place the column in a new 1.5-mL RNase-free tube, add 50 μL RNase-free water directly to the column membrane, close lid gently, and centrifuge for 1 min to elute RNA (*see* **Note 24**).
10. Measure the RNA concentration by spectrophotometry.
11. Adjust the RNA to the same concentration using RNase-free water and store the RNA sample at −80 °C.

3.2.3 First-Strand cDNA Synthesis

1. Add the following components into a nuclease-free PCR tube: 1 μL of 500 μg/μL oligo(dT)$_{12-18}$ (*see* **Note 25**), 1 μL of 10 mM dNTPs, 1 ng to 5 μg total RNA, and nuclease-free H_2O to a final volume of 12 μL.
2. Mix and incubate at 65 °C for 5 min and quickly chill on ice.
3. Briefly spin down the contents in the tube and add the following components: 4 μL of 5× first-strand buffer, 2 μL of 0.1 M DTT, 1 μL of Protector RNase Inhibitor (Promega or Roche), and 1 μL of M-MLV RT (200 units).
4. Mix gently and incubate at 37 °C for 50 min.
5. Inactivate the reaction by incubating at 70 °C for 15 min.
6. Make a 1:10 dilution of the original cDNA in nuclease-free H_2O, aliquot in small volumes in PCR strip tubes, and store at −20 °C.

3.2.4 Quantitative RT-PCR

1. Prepare a PCR reaction premix, consisting for a single reaction of 5 μL of Fast SYBR Green Master Mix, 0.25 μL of each 10 μM primer, 2 μL of diluted cDNA, and nuclease-free H_2O to 10 μL (*see* **Note 26**).
2. Mix and briefly spin to collect all the contents.
3. Run PCR by using Fast Mode following a dissociation curve analysis to check for possible formation of nonspecific side products, such as PCR dimers.
4. Determine the threshold cycle (C_t) according to the manufacturer's program.

3.2.5 Data Analysis

1. For each primer set, test a dilution series of cDNA: undiluted, 1:10, 1:100, 1:1,000, and 1:10,000.
2. Plot C_t against $\log_{10}$ (concentration) and calculate the slope by linear regression.
3. Calculate the PCR efficiency as $E = 10^{(-1/\text{slope})}$.
4. Calculate the relative expression ratio (Ratio) of the gene(s) of interest (GOI) against the reference (REF) gene with the equation [22]

$$\text{Ratio} = \frac{E_{\text{GOI}}^{\Delta C_t(\text{Control}-\text{Treatment})}}{E_{\text{REF}}^{\Delta C_t(\text{Control}-\text{Treatment})}}$$

with E_{GOI} and E_{REF} the PCR efficiencies for GOI and REF, respectively.

Results from an actual experiment are shown in Fig. 3. *Controls* were samples from water-infiltrated Col-0 plants and *Treatments* were wild-type Col-0 or *coi1-30* mutant plants treated with DC3000 or DC3118.

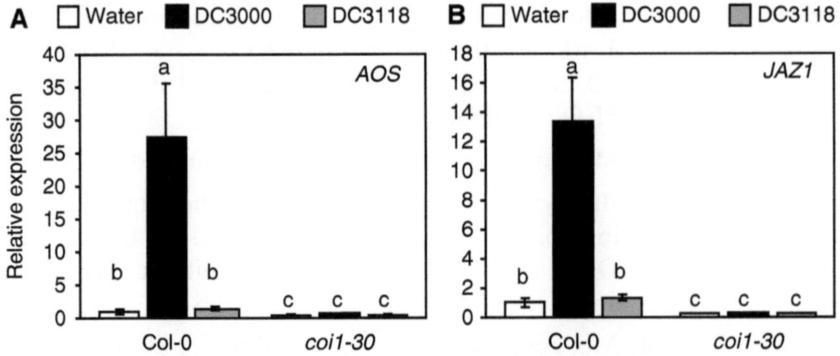

Fig. 3 JA-responsive gene expression induced in *Arabidopsis* during *Pst* DC3000 infection. (**a**) Expression of *ALLENE OXIDE SYNTHASE* (*AOS*, *At5G42650*) and (**b**) *JASMONATE-ZIM-DOMAIN PROTEIN 1* (*JAZ1*, *At1G19180*) 6 h after inoculation. In contrast to *Pst* DC3000, *Pst* DC3118 does not produce coronatine (COR) nor does it induce the expression of *AOS* and *JAZ1*. Induction of *AOS* and *JAZ1* by *Pst* DC3000 infection was compromised in the *coi1* JA receptor mutant. The *Y*-axis shows the expression ratio relative to the transcript levels in water-treated control wild-type Col-0 plants. Data shown are the means of three biological replicates. Error bars represent the standard deviations. Different letters on the columns indicate significant differences that are determined by the Tukey–Kramer multiple comparison test ($P < 0.05$). *PROTEIN PHOSPHATASE 2A SUBUNIT A3* (*PP2AA3*, *AT1G13320*) was used as reference gene to normalize expression levels

3.3 Analysis of JAZ Protein Stability During Pst DC3000 Infection

3.3.1 Plant Material and Protein Extraction

1. Inoculate two leaves of at least four 5-week-old P_{35S}::HA-JAZ2 and P_{35S}::HA-JAZ6 *Arabidopsis* plants with water or 1×10^8 CFU/mL *Pst* DC3000 or DC3118 by syringe infiltration.
2. After 6 h, cut inoculated leaves with a pair of sharp scissors and collect leaf discs using a 0.5 cm^2 cork borer.
3. Collect four leaf discs into a microfuge tube and freeze immediately in liquid nitrogen. Two tubes are collected for one treatment (*see* **Note 27**).
4. Chill a pestle in liquid nitrogen and grind leaf discs into fine powder (*see* **Note 28**).
5. Add 100 μL ice-cold complete PPEB into the pulverized tissue and continue to grind until no tissue clumps are visible.
6. Briefly vortex the homogenized sample and put on ice.
7. Continue to process the remaining samples.
8. Centrifuge samples at 20,000 × *g* for 15 min at 4 °C.
9. Transfer 64 μL supernatant into a new microfuge tube (*see* **Note 29**).
10. Prepare a working Reagent A′ solution by adding 20 μL Bio-Rad *DC/RC* Reagent S to each mL of Reagent A.
11. Prepare 3–5 dilutions of a protein standard containing from 0.2 mg/mL to 1.5 mg/mL of BSA in PPEB and dilute 4 μL of plant protein extract in 16 μL of PPEB (i.e., 1:5 dilution).

12. Pipette 5 μL standard and samples into two wells of a clean microtiter plate.
13. Add 25 μL of Reagent A′ into each well.
14. Add 200 μL of Reagent B into each well and mix the content by gentle agitation.
15. Incubate the plate at room temperature for 15 min.
16. Measure the absorbance at 750 nm using a microplate reader and calculate the protein concentration.
17. Adjust the concentrations of the protein extracts to 4 μg/μL by using 4× SDS sampling buffer and sterile H_2O.
18. Boil the extracts for 10 min and store at −20 °C.

3.3.2 SDS-PAGE Analysis

1. Assemble the mini-PROTEAN 3 gel casting system (*see* **Note 30**).
2. For the preparation of a 15 % separating gel, mix 1.2 mL of water, 0.8 mL of separating buffer, 2 mL of 30 % acrylamide solution, 2 μL of TEMED, and 20 μL of 10 % ammonium persulfate.
3. Briefly mix the solution by swirling.
4. Pour the gel immediately, leaving a 1.5-cm space for the stacking gel.
5. Gently overlay the gel with 2-propanol.
6. After 20 min, remove the 2-propanol overlay as completely as possible, including any remaining alcohol by wicking onto a paper towel.
7. Prepare the stacking gel by mixing 1.5 mL of water, 0.4 mL of stacking buffer, 0.3 mL of 30 % acrylamide solution, 3 μL of TEMED, and 22.5 μL of 10 % ammonium persulfate.
8. Briefly mix the solution by swirling, pour into the gel cast, and insert the comb immediately.
9. After 20 min, carefully remove the comb and rinse wells with sterile water and running buffer (*see* **Note 31**).
10. Boil samples in SDS loading buffer for 5 min.
11. Briefly spin to collect all the liquid.
12. Briefly vortex the samples and load 10 μL of each sample to the gel.
13. Run gel at 100 V (constant voltage) until loading dye reaches the bottom of the gel.

3.3.3 Immunoblotting

1. Mark a corner of a PVDF membrane with a pencil to orient the blot.
2. Immerse the membrane in methanol for 20 s, then in water for 2 min, and finally incubate in ice-cold transfer buffer.

3. Soak precut pieces of 3MM filter paper, fiber pads for the Mini Trans-Blot system, and the gel (*see* Subheading 3.3.2) in ice-cold transfer buffer for 15 min.
4. Assemble the gel transfer sandwich as follows: transparent side of gel holder cassette—fiber pad—filter paper—PVDF membrane—gel—filter paper—fiber pad—grey side of gel holder cassette.
5. Remove air bubbles between layers by rolling a glass rod over the sandwich.
6. Insert the assembled gel cassette in the electrode module with the grey side facing the black side of the module.
7. Put the electrode module in the buffer tank together with a Bio-Ice cooling unit with ice.
8. Fill buffer tank with transfer buffer (*see* **Note 32**) and add a stir bar to help maintain even buffer temperature and ion distribution.
9. Blot the gel in a cold room at a constant voltage of 90 V for a 1-h transfer or at 25 V for an overnight transfer.
10. Wash the membrane twice in 20 mL of TBS at room temperature in the black Mini-Blotting container, each time for 10 min (*see* **Note 33**).
11. Incubate the membrane in 10 mL of blocking solution at room temperature for 1 h.
12. Wash the membrane twice in 20 mL of TBST, each time for 10 min, followed by one wash for 10 min in 20 mL of TBS.
13. Dilute a primary antibody in 5–10 mL of dilution buffer.
14. Incubate the membrane in the primary antibody solution by gently shaking overnight in a cold room (*see* **Note 34**).
15. Wash the membrane twice in 20 mL of TBST, each time for 10 min, followed by one wash for 10 min in 20 mL of TBS.
16. Dilute the HRP-conjugated secondary antibody in 10 mL of blocking buffer according to the manufacturer's instructions.
17. Incubate the membrane in the secondary antibody solution by gently shaking for 1 h at room temperature.
18. Wash four times in 20 mL of TBST, each time for 10 min.
19. Prepare a SuperSignal working solution by mixing 2 mL of stable peroxide solution with 2 mL luminol/enhancer solution.
20. Incubate the membrane in the working solution for 5 min.
21. Remove the membrane and allow excess liquid to drip off.
22. Place the membrane between two sheets of plastic wrap, making sure that no air bubbles are trapped between the membrane and the plastic wrap.
23. Place the wrapped membrane in a film cassette and place an X-ray film on top of the membrane in a darkroom.

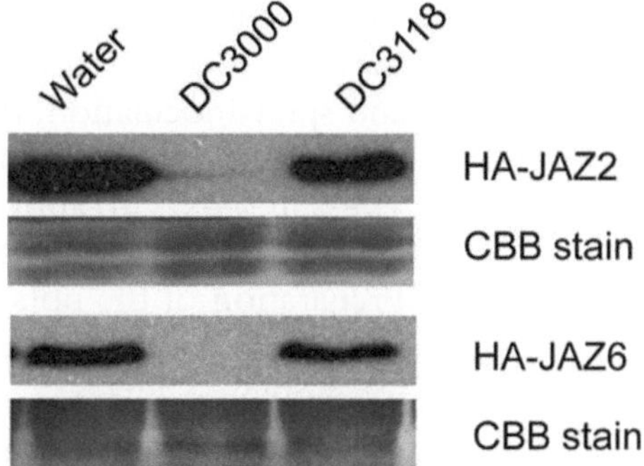

Fig. 4 JAZ protein degradation in *Arabidopsis* induced during *Pst* DC3000 but not *Pst* DC3118 infection. HA-tagged JAZ2 and JAZ6 were degraded 8 h after inoculation. JAZ2 and JAZ6 were detected by an anti-HA antibody. CBB staining was used to check for equal loading of proteins

24. Expose the film for a certain period of time, depending on the signal strength.
25. Incubate the membrane in 10 mL of CBB staining solution for 5 min (*see* **Note 35**).
26. Destain the membrane in 50 mL of destaining solution for 3 h with one change of solution. Results from an actual experiment are shown in Fig. 4.

4 Notes

1. Tissue-grinding devices, such as TissueLyser used in this protocol, can process many samples at the same time. If not available, traditional mortar and pestle work just fine.
2. For qRT-PCR analysis of gene expression, intact and pure RNA is extremely important for success. To prevent RNase contamination, all solutions must be made in DEPC-treated RNase-free water. Water with 0.1 % DEPC added is incubated overnight at room temperature with stirring and then autoclaved to destroy DEPC. All reusable vessels need to be treated with 0.1 % DEPC and autoclaved or baked at 220 °C overnight. Always wear gloves and use filter-fitted pipetting tips to prevent RNase contamination.
3. All primers should have a similar melting temperature (T_m) between 58 and 62 °C. The length of the amplicon is usually designed to be between 70 and 140 bp. The primers must be highly specific for their target sequences to avoid formation of side products, such as primer dimers. Furthermore, one primer is often designed to span an exon–exon junction to prevent amplification of products from genomic DNA which may still be present in the cDNA samples.

4. The specific inoculation methods should be determined before the pots are prepared and the seeds sown. For syringe infiltration and spray inoculation, the seeds can be sown directly onto the soil surface in pots. For dip inoculation, pots with soil mounded 1.5–2 in. (4–5 cm) above the top are covered with insect screen held in place with rubber bands just below the pot lips. Preparation of the pots in this manner will greatly reduce the amount of soil contamination in the bacterial suspension used for dipping plants.
5. Cold stratification of *Arabidopsis* seeds will help ensure synchronized germination. Seeds can be imbibed in sterile water in a 1.5-mL Eppendorf tube or sown on soil, and then placed at 4 °C for 2 days before being moved to a growth chamber.
6. During the 4–6 weeks of growth, plants should be watered by filling the flat bottoms to evenly soak the pots followed by draining off excess water. Approximately one to two times per week is sufficient, but care should be taken not to overwater. When 4–6 weeks old, plants grown under the specified short-day conditions should have established a fully developed rosette of leaves and should not have transitioned to flowering. We do not recommend using plants that have begun developing an inflorescence for disease assays.
7. Once colonies are well developed, they can be used immediately for inoculation or they can be stored temporarily at 4 °C.
8. Bacteria should reach late log phase growth (OD_{600} = 0.8–1.0). If the bacterial culture is overgrown, the estimation of viable cells will become less accurate and bacterial counts and/or symptom development in the plant may become more variable. As an alternative approach, an overnight culture can be diluted in the morning and grown to the desired optical density. In either case, inoculation times should be kept consistent across experimental replications.
9. Adjust the optical density of the cell suspension using a spectrophotometer set to 600 nm. For *Pst* DC3000, an OD_{600} = 0.2 should approximate 1×10^8 CFU/mL. For dip or spray inoculations, bacterial suspensions at an OD_{600} = 0.2 are typically used. For syringe inoculations, bacterial suspensions at an OD_{600} = 0.002 are used. With either inoculation technique, at the recommended bacterial concentration *Pst* DC3000 should cause robust disease symptoms (chlorosis and necrosis) in wild-type Col-0 *Arabidopsis* plants within 3 days.
10. Subtle changes in environmental conditions during growth of plants and bacteria or during the infection and pathogenesis processes can cause significant differences in the host responses. Experimental conditions, such as culture density, may have to be optimized in each laboratory setting.

11. Avoid applying pressure with the syringe over the vascular tissue. Damage to the leaf could adversely affect the outcome of the disease assay. If the infiltration technique is done efficiently, expect only a small volume (e.g., 20 μL) to infiltrate an entire leaf. During infiltration, water soaking of the leaf will be visible (Fig. 1).
12. Spray and dip inoculation are the closest laboratory approximation to natural infection by *P. syringae*. With these methods bacteria must circumvent initial defense responses at the surface of the leaf and enter the apoplastic space in the leaf through stomata or wounds.
13. Maintaining high humidity (80–90 %) is critical for proper disease symptom development. On the other hand, it is equally important to ensure that the humidity does not get too high (e.g., 100 %), because the leaves will become completely water-soaked and disease development will be compromised.
14. Rinse the cork borer in 70 % ethanol and then sterile distilled water after collecting each sample.
15. Typically samples should be collected from three or more leaves per plant to generate statistically significant bacterial colony counts.
16. Grind in short pulses to avoid generation of heat from friction.
17. Generally a dilution series reaching 10^{-5} is sufficient for counting even the highest possible CFUs at 3 dpi for *Pst* DC 3000.
18. For 10-μL spots, the dilution that results in more than 10 but fewer than 80 colonies is ideal for determining the bacterial count within each leaf disc. The CFU/cm^2 infected leaf tissue is calculated with the formula (Average CFU/10 μL spot × Dilution Factor × 10^2)/(area of leaf disc sample [cm^2]). Results from bacterial quantification are typically plotted on a $\log_{10}$ scale.
19. Be careful not to infiltrate major veins or to create excessive wounds. Samples from different plants within the same treatment serve as biological replicates. The whole experiment needs to be repeated at least twice to obtain conclusive results.
20. The sampling process needs to be as rapid as possible. Frozen tissue can be stored at −80 °C for several months.
21. The homogenized samples can be stored at −80 °C for at least 1 month.
22. The mixture separates into a bottom red phenol–chloroform phase, an interphase, and a colorless aqueous top phase. RNA remains exclusively in the aqueous phase. The upper aqueous phase is ~50 % of the total volume.
23. Do not overdry the RNA pellet as this may complicate dissolution. DO NOT use a SpeedVac to dry the RNA.

24. For downstream of qRT-PCR analysis, it is crucial to obtain DNA-free RNA samples. After this step, pure RNA is almost guaranteed.
25. Random primers can be used for plastidic mRNAs that transiently carry polyA tails.
26. Add fast SYBR Green Master Mix, forward and reverse primers and H_2O according to the number of reactions needed.
27. To prevent protein degradation in response to wounding, the sampling process needs to be completed as rapidly as possible. To avoid cross-contamination between samples, sequentially dip the cork borer in 50 % bleach, 70 % ethanol, and sterile water, and then wipe dry by using a clean tissue paper. Samples may be stored at −80 °C for several weeks.
28. Pestles can be used together with a regular hand drill. Keep tissue frozen to prevent degradation of proteins.
29. The extracts can be stored at −20 °C for several weeks.
30. Glass plates for PAGE must be cleaned thoroughly and rinsed with 100 % ethanol before use.
31. Gels can be stored in running buffer at 4 °C for several days.
32. For proteins larger than 100 kDa, add SDS into the transfer buffer to a final concentration of 0.025 % to facilitate transfer. Excessive SDS in the system may hinder the binding of proteins to the PVDF membrane.
33. All incubation and wash steps are done on a rocking platform or orbital shaker set at an appropriate speed.
34. Primary antibodies are usually expensive and/or precious. In most cases, they can be reused three to four times. Collect the diluted primary antibodies in a 15-mL conical tube and store at 4 °C for up to 2 months.
35. Membranes can be stained by CBB to check for equal loading and for the degree of protein transfer.

Acknowledgments

This work was supported by funding from the National Institutes of Health R01AI068718; the Chemical Sciences, Geosciences Division, Office of Basic Energy Sciences, Office of Science, Department of Energy DE–FG02–91ER20021 (support of research infrastructure); and the Gordon and Betty Moore Foundation GBMF3037. SYH is a Howard Hughes Medical Institute–Gordon and Betty Moore Foundation Investigator.

References

1. Whalen MC, Innes RW, Bent AF, Staskawicz BJ (1991) Identification of *Pseudomonas syringae* pathogens of *Arabidopsis* and a bacterial locus determining avirulence on both *Arabidopsis* and soybean. Plant Cell 3:49–59
2. Hirano SS, Upper CD (2000) Bacteria in the leaf ecosystem with emphasis on Pseudomonas syringae—a pathogen, ice nucleus, and epiphyte. Microbiol Mol Biol Rev 64:624–653
3. Feys BJF, Benedetti CE, Penfold CN, Turner JG (1994) Arabidopsis mutants selected for resistance to the phytotoxin coronatine are male sterile, insensitive to methyl jasmonate, and resistant to a bacterial pathogen. Plant Cell 6:751–759
4. Kloek AP, Verbsky ML, Sharma SB, Schoelz JE, Vogel J, Klessig DF, Kunkel BN (2001) Resistance to *Pseudomonas syringae* conferred by an *Arabidopsis thaliana* coronatine-insensitive (*coi1*) mutation occurs through two distinct mechanisms. Plant J 26:509–522
5. Laurie-Berry N, Joardar V, Street IH, Kunkel BN (2006) The *Arabidopsis thaliana JASMONATE INSENSITIVE 1* gene is required for suppression of salicylic acid-dependent defenses during infection by *Pseudomonas syringae*. Mol Plant-Microbe Interact 19:789–800
6. Zhao Y, Thilmony R, Bender CL, Schaller A, He SY, Howe GA (2003) Virulence systems of *Pseudomonas syringae* pv. *tomato* promote bacterial speck disease in tomato by targeting the jasmonate signaling pathway. Plant J 36:485–499
7. Bender CL, Alarcón-Chaidez F, Gross DC (1999) *Pseudomonas syringae* phytotoxins: mode of action, regulation, and biosynthesis by peptide and polyketide synthetases. Microbiol Mol Biol Rev 63:266–292
8. Ichihara A, Shiraishi K, Sato H, Sakamura S, Nishiyama K, Sakai R, Furusaki A, Matsumoto T (1977) The structure of coronatine. J Am Chem Soc 99:636–637
9. Mitchell RE, Young SA, Bender CL (1994) Coronamic acid, an intermediate in coronatine biosynthesis by *Pseudomonas syringae*. Phytochemistry 35:343–348
10. Parry RJ, Mafoti R (1986) Biosynthesis of coronatine, a novel polyketide. J Am Chem Soc 108:4681–4682
11. Brooks DM, Bender CL, Kunkel BN (2005) The *Pseudomonas syringae* phytotoxin coronatine promotes virulence by overcoming salicylic acid-dependent defences in *Arabidopsis thaliana*. Mol Plant Pathol 6:629–639
12. Thilmony R, Underwood W, He SY (2006) Genome-wide transcriptional analysis of the *Arabidopsis thaliana* interaction with the plant pathogen *Pseudomonas syringae* pv. *tomato* DC3000 and the human pathogen *Escherichia coli* O157:H7. Plant J 46:34–53
13. Uppalapati SR, Ayoubi P, Weng H, Palmer DA, Mitchell RE, Jones W, Bender CL (2005) The phytotoxin coronatine and methyl jasmonate impact multiple phytohormone pathways in tomato. Plant J 42:201–217
14. Melotto M, Underwood W, He SY (2008) Role of stomata in plant innate immunity and foliar bacterial diseases. Annu Rev Phytopathol 46:101–122
15. Zeng W, Brutus A, Kremer JM, Withers JC, Gao X, Jones AD, He SY (2012) A genetic screen reveals Arabidopsis stomatal and/or apoplastic defenses against *Pseudomonas syringae* pv. *tomato* DC3000. PLoS Pathog 7:e1002291
16. Staswick PE (2008) JAZing up jasmonate signaling. Trends Plant Sci 13:66–71
17. Fonseca S, Chini A, Hamberg M, Adie B, Porzel A, Kramell R, Miersch O, Wasternack C, Solano R (2009) (+)-7-*iso*-Jasmonoyl-l-isoleucine is the endogenous bioactive jasmonate. Nat Chem Biol 5:344–350
18. Katsir L, Schilmiller AL, Staswick PE, He SY, Howe GA (2008) COI1 is a critical component of a receptor for jasmonate and the bacterial virulence factor coronatine. Proc Natl Acad Sci USA 105:7100–7105
19. Melotto M, Mecey C, Niu Y, Chung HS, Katsir L, Yao J, Zeng W, Thines B, Staswick P, Browse J, Howe GA, He SY (2008) A critical role of two positively charged amino acids in the Jas motif of Arabidopsis JAZ proteins in mediating coronatine- and jasmonoyl isoleucine-dependent interactions with the COI1 F-box protein. Plant J 55:979–988
20. Sheard LB, Tan X, Mao H, Withers J, Ben-Nissan G, Hinds TR, Kobayashi Y, Hsu F-F, Sharon M, Browse J, He SY, Rizo J, Howe GA, Zheng N (2010) Jasmonate perception by inositol-phosphate-potentiated COI1—JAZ co-receptor. Nature 468:400–405
21. Katagiri F, Thilmony R, He SY (2002) The Arabidopsis thaliana–Pseudomonas syringae interaction. The Arabidopsis Book 1:e0039
22. Pfaffl MW (2001) A new mathematical model for relative quantification in real-time RT-PCR. Nucleic Acids Res 29:e45

Chapter 7

Jasmonate Signaling in the Field, Part I: Elicited Changes in Jasmonate Pools of Transgenic *Nicotiana attenuata* Populations

Emmanuel Gaquerel, Michael Stitz, Mario Kallenbach, and Ian T. Baldwin

Abstract

Nicotiana attenuata, a wild tobacco species native of the southwestern USA that grows in the immediate postfire environment, is one of the important host plants for herbivore populations recolonizing recently burned habitats in the Great Basin Desert. Based on more than 20 years of field research on this eco-genomics model system established in our group, we have developed a genetic and analytical toolbox that allows us to assess the importance of particular genes and metabolites for the survival of this plant in its native habitat. This toolbox has been extensively applied to study the activation of jasmonate signaling after the attack of different herbivore species. Here, we provide detailed guidelines for the analysis, under field conditions, of induced changes in jasmonate pools during insect herbivory. The procedures range from selection and field release of well-characterized transgenic lines for testing the physiological consequences of manipulating jasmonate biogenesis, metabolism, or perception to the metabolic elicitation of chewing herbivore attack and the quantification of the resulting changes in jasmonate fluxes.

Key words Jasmonate, Plant–insect interaction, Field work, Metabolic sink

1 Introduction

Over the last two decades of research on a wild tobacco species, *Nicotiana attenuata* (coyote tobacco), our group has developed a set of molecular, chemical, and ecological procedures to analyze in the field the ecological consequences of silencing or overexpressing a particular gene. In this approach, field trials are placed early in the gene function discovery process, allowing us to describe the essential role of jasmonate signaling in almost every aspect of this plant's defenses to insects [1], to discover the importance of herbivory-specific shifts in volatile emissions in the attraction of the herbivores' natural enemies [2], and to identify unexpected functions of differently regulated genes, such as the role of RNA-dependent RNA polymerases in mediating resistance to herbivores [3]

Alain Goossens and Laurens Pauwels (eds.), *Jasmonate Signaling: Methods and Protocols*, Methods in Molecular Biology, vol. 1011, DOI 10.1007/978-1-62703-414-2_7, © Springer Science+Business Media, LLC 2013

or to ultraviolet-B [4] and in the plant's intraspecific competitive ability in nature [5].

N. attenuata germinates in the first two growing seasons after fires, from long-lived seed banks to produce ephemeral populations. Burned habitats provide nutrient-rich environments, but these soil resources rapidly become immobilized and unavailable due to microbial activity and competition from other early colonizing plants. *N. attenuata* plants are important host plants for herbivores that colonize the postfire environment every year and have evolved efficiently induced resistance strategies to cope with insect attack from different feeding guilds. Although ethylene and other small signaling molecules (e.g., NO and H_2O_2) fulfill regulatory functions in this process [6, 7], their contribution is relatively minor in comparison to that of jasmonic acid (JA) and its cyclic precursors and derivatives, collectively referred to as jasmonates [1, 8, 9].

JA is synthesized de novo upon activation of lipases that release fatty acids from membrane lipids in insect-damaged leaves [10]. Free linolenic acid is oxygenated by lipoxygenase enzymes and subsequently converted to 12-oxo-phytodienoic acid (OPDA) through the combined actions of allene oxide synthase (AOS) and allene oxide cyclase (AOC). OPDA is subsequently transformed to JA by reduction and three cycles of β-oxidation. After enzyme-dependent conjugation of JA to isoleucine, JA-Ile is the bioactive jasmonate interacting with the F-box protein CORONATINE INSENSITIVE1 (COI1) [11, 12]. The amplitude of the jasmonate bursts activated by different insect herbivores is not only conditioned by the severity of the mechanical stress inflicted to the attacked tissues but also highly influenced by the presence of specific elicitors contained in the oral secretions (OS) of the insects. Fatty acid amino acid conjugates contained in the OS from the tobacco hornworm (*Manduca sexta*) strongly amplify jasmonate accumulation and signaling in *N. attenuata* and thereby, trigger many of the transcriptional and metabolic reconfigurations required for elevating the plant's defensive status [13, 14].

Although our attention is frequently biased toward transcription-mediated controls, substrate availability and competing homeostatic pathways within the jasmonate enzymatic cascade also impose strong regulatory control over the signaling outcome of this pathway. Conversely, to uncover new signaling outputs, manipulating the rate of certain enzymatic reactions controlling jasmonate metabolism and characterizing its consequences at the organismic level is a valuable approach, in addition to traditional loss-of-function analyses. The expression "metabolic sink" has previously been used to define transgenic manipulations in which the metabolic and/or signaling outputs of specific biosynthetic pathways are diverted or inactivated [15]. Methylation is one of the catalytic reactions used by plants to adjust their pools of active hormones to

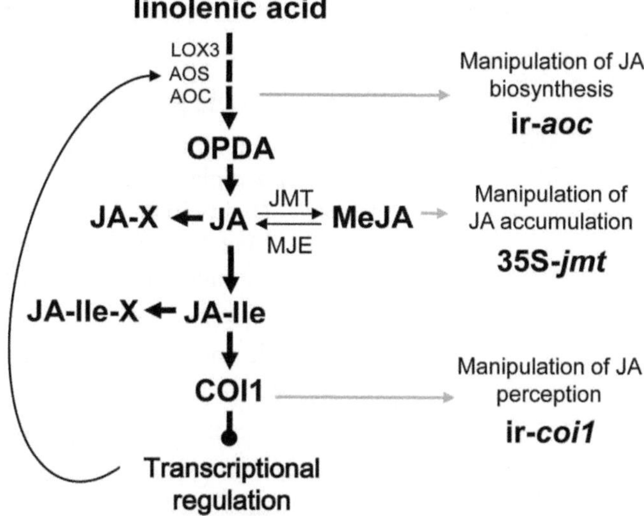

Fig. 1 Creation of *Nicotiana attenuata* plants with altered jasmonate biosynthesis, accumulation, or perception. *Grey arrows* localize the different nodes within the jasmonate enzymatic and signaling cascade that have been manipulated by gene silencing and ectopic expression approaches. Total jasmonate biosynthesis is strongly repressed by silencing *N. attenuata allene oxide cyclase* using an inverted-repeat construct (ir-*aoc*). Herbivory-induced JA and JA-Ile levels are depleted by ectopically expressing, under a constitutive promoter, an *Arabidopsis thaliana* jasmonic acid-specific methyltransferase (35S-*jmt*) that redirects the JA flux toward the inactive methyl jasmonate (MeJA). JA signaling and the transcriptional activation of dependent defense responses are interrupted by silencing the JA-Ile receptor, the *coronatine insensitive1*, with an inverted-repeat construct (ir-*coi1*)

environmental conditions and therefore represents an ideal target for the genetic manipulation of bioactive hormonal levels [16, 17]. In *N. attenuata*, overexpressing *Arabidopsis thaliana JASMONIC ACID METHYLTRANSFERASE* (*JMT*) plants redirects the jasmonate flux toward the formation of inactive methyl jasmonate (MeJA) (Fig. 1) [18]. 35S-*JMT* transgenic plants have depleted herbivory-induced pools of JA and JA-Ile and are more susceptible to native herbivores [19].

In this chapter, we provide extensive guidelines for the selection of adequate transgenic lines and their release into the field and for the measurement of herbivory-induced changes in jasmonate pools. Altogether, this chapter and its companion chapter illustrate the value of combining chemical profiling and field observations for assessing the defensive function of jasmonates in nontransformed *N. attenuata* as well as in *AOC*-silenced, *COI1*-silenced, and *Arabidopsis JMT*-overexpressing transformants (Fig. 1).

2 Materials

2.1 Growing N. attenuata Plants in Field Conditions

1. Field plot in the plant's native habitat.
2. Seeds from untransformed and jasmonate-deficient transgenic *N. attenuata* plants can be obtained upon request from Ian T. Baldwin (baldwin@ice.mpg.de), Department of Molecular Ecology, Max Planck Institute for Chemical Ecology in Jena (http://www.ice.mpg.de/ext/).
3. Sterilization solution: 2 % (w/v) of dichloroisocyanuric acid sodium salt and 0.05 % (v/v) polysorbate 20 (Tween-20) in deionized water.
4. 1 mM gibberellic acid A3 (GA3).
5. Liquid smoke (House of Herbs, Passiac, NJ, USA; but other brands from your local grocer also work).
6. Gamborg's B5 medium (Duchefa, Haarlem, The Netherlands).
7. 7.5 mg/L $Na_2[B_4O_5(OH)_4]\cdot 8H_2O$ (borax) in water. Approval from legal authorities to work with genetically modified plants in native habitats (*see* **Note 1**).
8. Automatic watering system to grow plants used in common-garden experiments.
9. Shade house.
10. Jiffy 703 pots (Jiffy International AS, Stange, Norway).
11. Hand tools (shovel, rake, hoe, etc.).

2.2 Jasmonate Analysis

1. Liquid nitrogen for freezing and grinding of samples.
2. Mortars and pestles or mechanical grinder (ball mill).
3. Steel balls (ASK Kugellagerfabrik Artur Seyfert GmbH, Korntal-Münchingen, Germany).
4. Geno/Grinder (Spex SamplePrep, Metuchen, NJ, USA).
5. Extraction solvent: Ethyl acetate (HPLC grade).
6. 70/30 (v/v) methanol/deionized water (HPLC grade).
7. High-performance liquid chromatography-electron spray ionization (HPLC-ESI)/triple-quadrupole-MS (Varian, Palo Alto, CA, USA) instrument with an analytical reverse-phase column.
8. ProntoSil colum (C18; 5 μm; 50 × 2 mm) (Bischoff, Leonberg, Germany) attached to a precolumn (C18; 4 × 2 mm) (Phenomenex, Terrace, CA, USA).
9. Isotopically labeled phytohormone standards (*see* **Note 2**).
10. Vacuum rotary evaporator (Speed-Vac) with trap for organic solvents.

11. Refrigerated microcentrifuge.
12. LC-MS/MS instrument equipped with an analytical column.
13. LC-MS/MS solvents: 0.05 % formic acid in water (solvent A), methanol HPLC grade (solvent B).

2.3 Plant Elicitations

1. Preparation of *M. sexta* OS is described in Subheading 3.2.
2. *M. sexta* larvae for collecting herbivore OS (*see* **Note 3**).
3. Deionized water.
4. Fabric pattern tracing wheel with serrated edge (available everywhere where sewing supplies are sold).
5. Support for leaf to be treated (such as plastic discs with a diameter of 5–10 cm).
6. Dry ice blocks for sample collection.

3 Methods

The most commonly utilized method to analyze *in planta* the signaling functions of jasmonates entails silencing or knockout mutations in known jasmonate biosynthetic or perception genes. However, these loss-of-function approaches cannot per se unravel the importance of flux regulations in tuning jasmonate-dependent responses. We recently developed a novel strategy to analyze molecular and phenotypic responses controlled by jasmonate signaling. This strategy is based on the creation of jasmonate sinks by ectopically expressing the *Arabidopsis JMT* to divert JA metabolism toward inactive MeJA formation [18], allowing JA signaling to be silenced without disrupting the feedback/forward regulatory circuits that adjust JA synthesis to its signaling output [18]. Mutant 35S-*JMT* plants have depleted herbivory-induced pools of JA and JA-Ile and are more susceptible to native herbivores [19].

Transgenic plants (overexpression and silenced lines) produced by *Agrobacterium*-mediated transformation are screened and characterized first under laboratory conditions 12–18 months prior to field release. These screening procedures are fully described in a recent article from our group [20].

3.1 Release of Genetically Modified Plants into the Field

The release of genetically modified plants is subject to country-specific legal requirements (*see* **Note 1**).

1. Prepare a fresh sterilization solution containing 2 % (w/v) of dichloroisocyanuric acid sodium salt and 0.005 % Tween-20 in deionized water.
2. Incubate seeds in 5 mL of the sterilization solution for 5 min in 15-mL conical tubes; shake occasionally. The following steps should be carried out under a sterile hood.

3. Decant solution and wash seeds in 5–10 mL sterile deionized water.
4. Repeat decanting and washing steps at least three times.
5. Treat seeds with 1 mM GA_3 in 1:50 (v/v) diluted liquid smoke (*see* **Note 4**).
6. Germinate seeds on agar plates containing Gamborg's B5 medium under fluorescent lights.
7. When seedlings produce their first true leaves (7–10 days), transplant young seedlings into Jiffy 703 pots.
8. Gradually adapt the plants to the high-light environment of the field by first growing them in covered and shaded waterproof plastic shoe boxes in a shade house, floated on water for temperature regulation, and later moving them to shaded tents.
9. Fertilize seedlings with approximately 300 μg borax (*see* **Note 5**).
10. Transplant the transgenic plants, each paired with a size-matched empty vector (EV) control plant (*see* **Note 6**) with the appropriate labeling system required by the APHIS regulatory authorities into the field plot (*see* **Notes 7** and **8**).
11. To meet the regulatory requirements of the release, we monitor all transgenic plants daily during reproductive growth to remove all flowers before their corollas open and can release pollen.
12. After the experiments, harvest all plants completely (including roots and rhizosphere) and incinerate them.

3.2 Simulation of *M. sexta* Herbivory on *N. attenuata* Leaves

The responses of *N. attenuata* to *M. sexta* feeding can be simulated by mechanically wounding the leaf lamina and immediately applying *M. sexta* OS to freshly produced puncture wounds [21]. This procedure provides a convenient means of accurately standardizing herbivore elicitation of *N. attenuata* leaves, allows detailed kinetic analyses of the elicitation process, and mimics a large proportion of the changes activated during direct *M. sexta* feeding at the levels of the leaf transcriptome [22], proteome [23], and endogenous [24] and volatile metabolomes [25].

1. Collect OS from *M. sexta* larvae on ice with Teflon tubing into a GC vial with a septum lid connected to a vacuum and store under argon at −20 °C (*see* **Notes 9** and **10**).
2. Prior to plant treatment, centrifuge OS for 10 min at 16,000 × *g* at 4 °C and transfer the supernatant into a new tube.
3. Select a specific leaf on each plant to be used for elicitation (*see* **Note 11**).
4. For the native *N. attenuata* plant screening, select the least damaged, nonsenescing leaf available on each plant.

5. Wound this leaf by rolling a fabric pattern wheel three times on each side of the midvein.
6. Immediately after wounding, treat each leaf side with either 20 μL of water or 1:5 (v/v) *M. sexta* OS/deionized water (v/v) (*see* **Note 10**).
7. At time points of interest, excise the leaf below the leaf base with a sharp, clean razor.
8. Wrap the leaf in aluminum foil and immediately freeze between blocks of dry ice to prevent degradation of leaf metabolites.

3.3 Jasmonate Extraction with Ethyl Acetate

1. Transfer and crush frozen leaf material in a mortar cooled with liquid nitrogen.
2. Aliquot 100–150 mg of frozen leaf tissue into 2-mL microcentrifuge tubes containing two metal balls.
3. Record fresh mass of the samples. Prevent defrosting of the samples at all times.
4. Add 1 mL of cooled ethyl acetate, containing 2 μL of internal standard solution.
5. Extract jasmonates for 5 min at 1,000 strokes/min with a Geno/Grinder (*see* **Note 12**).
6. Centrifuge tubes at 4 °C for 20 min at 16,000 × *g*.
7. Transfer supernatants into fresh 2-mL microcentrifuge tubes.
8. Repeat **steps 4–7** with 0.5 mL of cooled ethyl acetate and combine supernatants.
9. Evaporate the solvent to dryness in a Speed-Vac.
10. Reconstitute the dry residue in 0.5 mL 70 % (v/v) methanol/water.

3.4 HPLC-ESI/Triple-Quadrupole-MS Analysis of Jasmonate Levels

Tandem mass spectrometry coupled to high-pressure liquid chromatography (HPLC-MS/MS) is the ideal analytical platform for the highly reproducible analysis of minute changes in phytohormone levels in complex biological matrices, such as plant tissues. The high precision in jasmonate quantification by means of HPLC-ESI/triple-quadrupole-MS measurements is achieved by the high selectivity in the integration of precursor ion-to-fragment ion specific to each natural jasmonate and the use of isotopically labeled internal standards.

1. Inject 10 μL of the sample onto a ProntoSIL column attached to a precolumn.
2. Use as mobile phases 0.05 % (v/v) formic acid/water (solvent A) and methanol (solvent B) in a gradient mode under the following conditions: time/concentration (min/%) for B: 0.0/5; 2.5/5; 5.5/98; 11.0/98; 11.5/5; 15.0/5;

time/flow (mL/min): 0.0/0.4; 0.4/0.4; 1.0/0.2; 10.0/0.2; 10.5/0.4; 15.0/0.4. Additional parameters include the following: collision gas (2.1 mTorr); atmospheric-pressure ionization (API) drying gas (19 psi; 300 °C); API nebulizing gas (60 psi); needle (4,500 V); shield (600 V); detector (1,800 V).

3. Detect OPDA, JA, 12-OH-JA, JA-Ile, and 12-OH-JA-Ile in the ESI-negative mode and MeJA in the ESI-positive mode, and multiple reaction monitoring. Precursor ion/fragment ion transitions used for natural and isotopically labeled jasmonate detection are presented in Fig. 2.
4. Integrate peak areas for natural and isotopically labeled jasmonate standards.
5. Calculate jasmonate concentrations in ng/g fresh weight (Fig. 3).
6. Use correction factors (response factors) if no labeled standards are available and the calculation is done according to another structurally labeled standard (*see* **Note 2**).

4 Notes

1. For field experiments in the USA, all transgenic lines require the permission from the Animal and Plant Health Inspection Service (APHIS) for import and release. For more information for import and plantation of transgenic plants in the USA, please go to http://www.aphis.usda.gov/contact_us/. The minimum requirement for such permissions is the demonstration that transgenes will not be released into the environment. Other requirements, such as those pertaining to the potential impact of the transformed plants on endangered species, may also apply. This process can be lengthy and should be started at least one year in advance of the desired release date.
2. Internal standards allow quantification. Standards should be structurally as closely related as possible to the analytes to be quantified because they should mimic the behavior of the analyte throughout all of the steps in the extraction and analysis process. To avoid misleading quantifications, internal standards should not occur naturally in the plant tissues. For all these reasons, identical molecules containing one or more heavy isotopes, such as deuterium, ^{13}C or ^{18}O, are preferred. Isotopically labeled JA and JA-Ile can be purchased or synthesized. When no isotopically labeled internal standard is available for a natural jasmonate, a response factor can be used for correction during quantification versus a structurally related isotopically labeled internal standard. The response factor is defined as the ratio between the jasmonate to be analyzed and the internal standard

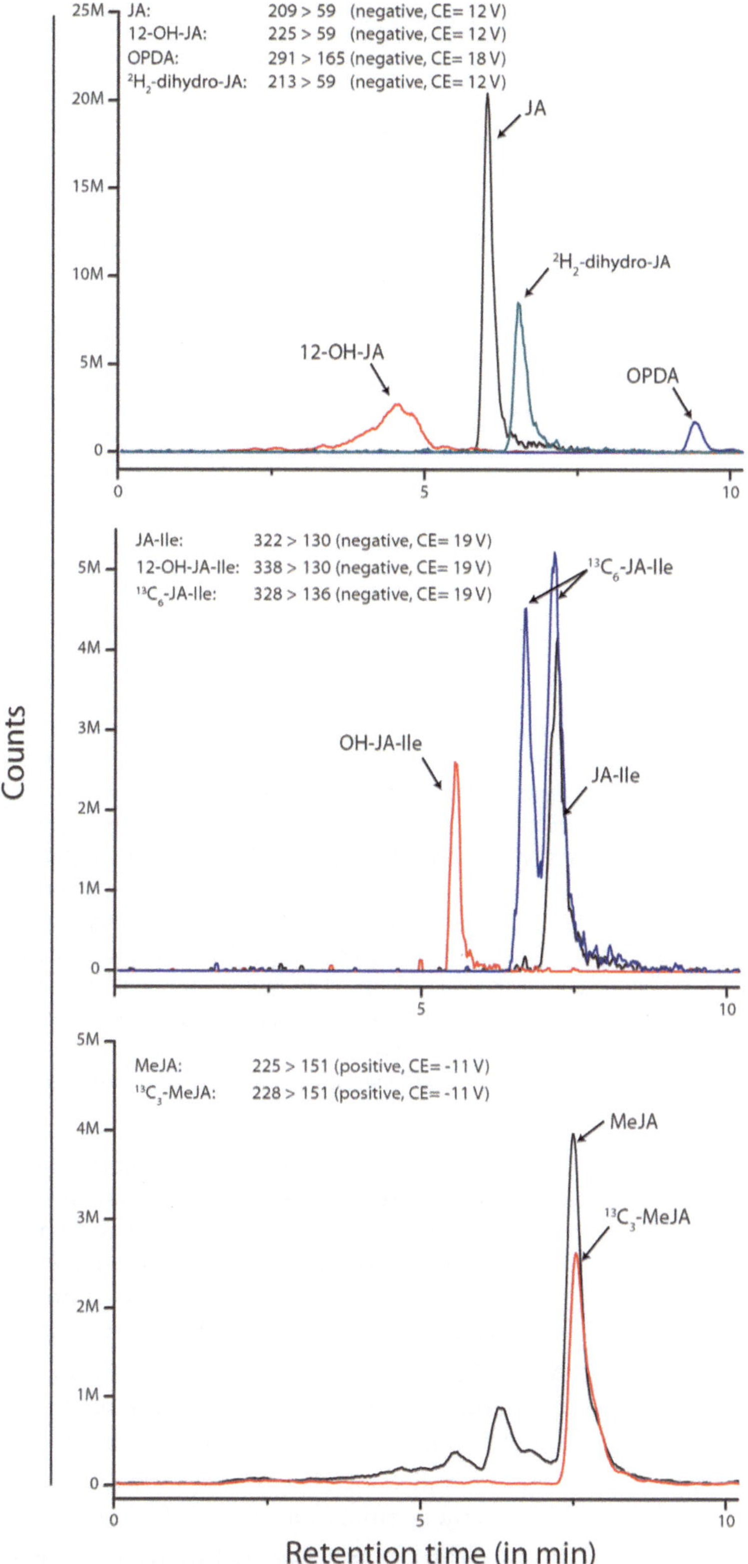

Fig. 2 Precursor ion/fragment ion transitions, collision energies, and ionization mode used for multiple reaction monitoring of natural and isotopically labeled jasmonates. Extracted ion chromatograms are presented for natural jasmonate and their respective internal standards. Extraction procedure and instrumental parameters are described in Subheadings 3.3 and 3.4. MS/MS measurement of the $^{13}C_6$-JA-Ile internal standard returns two peaks corresponding to the two isomers produced during synthesis

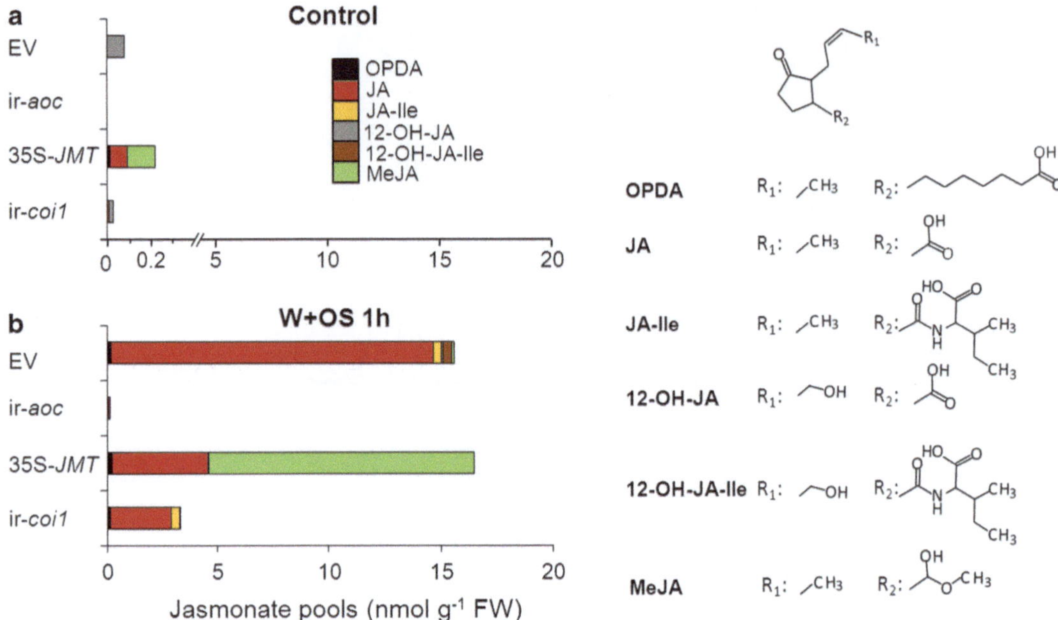

Fig. 3 Differential effects on the size and relative composition of jasmonate pools of field-grown plants by manipulation of specific nodes of the jasmonate metabolic and signaling network. *M. sexta* OS application onto mechanically wounded *N. attenuata* leaves represents a convenient means of eliciting plant's responses to *M. sexta* herbivory and inducing endogenous jasmonate production, thereby testing the jasmonate biosynthetic capacity of a given *N. attenuata* transformant. OS-induced jasmonate levels were measured 1 h after treatment. As expected, field-grown ir-*aoc* have almost undetectable constitutive and *M. sexta* OS-induced jasmonate levels due to the disruption of the cyclization step necessary for the penta-ring formation of jasmonates. Silencing *COI1*, the JA-Ile receptor, reduced by half OS-induced jasmonate levels and slightly increased JA-Ile due to attenuated JA-Ile catabolism—hydroxylated forms of JA-Ile were not detected in ir-*coi1* plants. Ectopical expression of 35S-*jmt* redirects the JA metabolic flux toward MeJA at the expense of other JA metabolites. Importantly, this redirection of the jasmonate flux does not alter total jasmonate pools, indicating probably that this approach does not influence feedback/forward mechanisms that normally regulate jasmonate flux

to be used and is calculated by spiking increasing amounts and analyzing the responses of the two analytes. Response factors are specific to the analyte and the biological matrix from which it is quantified. In the case of *N. attenuata* leaves, the response factor used for correcting the quantification of OPDA is 1.28, according to the $^{2}H_{2}$-dihydro-JA internal standard.

3. At our field station, *M. sexta* larvae used for OS collection are frequently encountered feeding on native *N. attenuata* or *Datura wrightii* (sacred datura) populations.
4. Liquid wood smoke is required to break the dormancy and to activate the germination of *N. attenuata* seeds [26–28].
5. Borax fertilization is essential to avoid "brown tip disease" in which rapidly growing plants suffer loss of tip meristematic tissue, probably due to boron deficiencies. This is a particular problem for plants growing in high N soils that for various

reasons may have insufficient boron to meet the metabolic needs of these fast-growing plants.

6. Stably transformed plants with an EV provide ideal controls of the fitness effect of silencing a gene of interest. EV plants control for possible changes that result from transformation, although despite extensive study [29], no fitness differences have been identified between EV and WT plants.
7. A wooden stick with the APHIS identification code can be placed by each plant and the field plot must be additionally posted with signs with the APHIS notification number code provided after authorization of the transgenic plant release.
8. In addition to the above-ground labeling system involving plastic colored flags and wooden sticks, we bury a plastic label carrying the identification number of each genotype under the roots of each plant to ensure unambiguous genotype identifications.
9. OS can be gathered from the larvae twice a day and should be pooled to reduce variability due to potential differences among collections.
10. The active components of *M. sexta* OS have been shown to elicit plant responses even in dilutions up to 1:1,000 [30]. To obtain larger volumes of OS solution, OS can be diluted 1/5 (v/v) with deionized water and stored in 0.5–1-mL aliquots under argon at −20 °C.
11. For instance, in *N. attenuata*, the first fully elongated leaf at node +1 of rosette-stage plants is commonly used because the elicitation of this leaf activates strong systemic responses in unelicited leaves and roots.
12. Optionally, a ball mill instrument can be used to grind samples. Homogenize 0.3 g of frozen material in 2-mL microcentrifuge tubes containing two steel balls by shaking the tubes in a Geno/Grinder for 30 s at 200–300 strokes/min. It is important to ensure that tissue is thoroughly ground to a powder prior to extraction. Multiple rounds of Geno/Grinder shaking (with samples placed back to liquid nitrogen in between) or additional grinding by hand might be required. Higher strike rates can break the lids of the microcentrifuge tubes and cause sample losses.

References

1. Kessler A, Halitschke R, Baldwin IT (2004) Silencing the jasmonate cascade: induced plant defenses and insect populations. Science 305:665–668
2. Allmann S, Baldwin IT (2010) Insects betray themselves in nature to predators by rapid isomerization of green leaf volatiles. Science 329:1075–1078
3. Pandey SP, Baldwin IT (2007) RNA-directed RNA polymerase 1 (RdR1) mediates the resistance of *Nicotiana attenuata* to herbivore attack in nature. Plant J 50:40–53

4. Pandey SP, Baldwin IT (2008) Silencing RNA-directed RNA polymerase 2 increases the susceptibility of *Nicotiana attenuata* to UV in the field and in the glasshouse. Plant J 54:845–862
5. Pandey SP, Gaquerel E, Gase K, Baldwin IT (2008) *RNA-directed RNA polymerase3* from *Nicotiana attenuata* is required for competitive growth in natural environments. Plant Physiol 147:1212–1224
6. Wünsche H, Baldwin IT, Wu J (2011) Silencing *NOA1* elevates herbivory-induced jasmonic acid accumulation and compromises most of the carbon-based defense metabolites in *Nicotiana attenuata*. J Integr Plant Biol 53: 619–631
7. von Dahl CC, Winz RA, Halitschke R, Kühnemann F, Gase K, Baldwin IT (2007) Tuning the herbivore-induced ethylene burst: the role of transcript accumulation and ethylene perception in *Nicotiana attenuata*. Plant J 51:293–307
8. Halitschke R, Baldwin IT (2003) Antisense LOX expression increases herbivore performance by decreasing defense responses and inhibiting growth-related transcriptional reorganization in *Nicotiana attenuata*. Plant J 36:794–807
9. Jander G, Howe G (2008) Plant interactions with arthropod herbivores: state of the field. Plant Physiol 146:801–803
10. Kallenbach M, Alagna F, Baldwin IT, Bonaventure G (2010) *Nicotiana attenuata* SIPK, WIPK, NPR1, and fatty acid–amino acid conjugates participate in the induction of jasmonic acid biosynthesis by affecting early enzymatic steps in the pathway. Plant Physiol 152: 96–106
11. Xie D-X, Feys BF, James S, Nieto-Rostro M, Turner JG (1998) *COI1*: an *Arabidopsis* gene required for jasmonate-regulated defense and fertility. Science 280:1091–1094
12. Yan J, Zhang C, Gu M, Bai Z, Zhang W, Qi T, Cheng Z, Peng W, Luo H, Nan F, Wang Z, Xie D (2009) The *Arabidopsis* coronatine insensitive1 protein is a jasmonate receptor. Plant Cell 21:2220–2236
13. Bonaventure G, VanDoorn A, Baldwin IT (2011) Herbivore-associated elicitors: FAC signaling and metabolism. Trends Plant Sci 16: 294–299
14. Halitschke R, Schittko U, Pohnert G, Boland W, Baldwin IT (2001) Molecular interactions between the specialist herbivore *Manduca sexta* (Lepidoptera, Sphingidae) and its natural host *Nicotiana attenuata*. III. Fatty acid–amino acid conjugates in herbivore oral secretions are necessary and sufficient for herbivore-specific plant responses. Plant Physiol 125:711–717
15. Yao K, De Luca V, Brisson N (1995) Creation of a metabolic sink of tryptophan alters the phenylpropanoid pathway and the susceptibility of potato to *Phytophthora infestans*. Plant Cell 7:1787–1799
16. Tieman D, Zeigler M, Schmelz E, Taylor MG, Rushing S, Jones JB, Klee HJ (2010) Functional analysis of a tomato salicylic acid methyl transferase and its role in synthesis of the flavor volatile methyl salicylate. Plant J 62:113–123
17. Qin G, Gu H, Zhao Y, Ma Z, Shi G, Yang Y, Pichersky E, Chen H, Liu M, Chen Z, Qu L-J (2005) An indole-3-acetic acid carboxyl methyltransferase regulates *Arabidopsis* leaf development. Plant Cell 17:2693–2704
18. Stitz M, Gase K, Baldwin IT, Gaquerel E (2011) Ectopic expression of *AtJMT* in *Nicotiana attenuata*: creating a metabolic sink has tissue-specific consequences for the jasmonate metabolic network and silences downstream gene expression. Plant Physiol 157:341–354
19. Stitz M, Baldwin IT, Gaquerel E (2011) Diverting the flux of the JA pathway in *Nicotiana attenuata* compromises the plant's defense metabolism and fitness in nature and glasshouse. PLoS One 6:e25925
20. Gase K, Weinhold A, Bozorov T, Schuck S, Baldwin IT (2011) Efficient screening of transgenic plant lines for ecological research. Mol Ecol Resour 11:890–902
21. McCloud ES, Baldwin IT (1997) Herbivory and caterpillar regurgitants amplify the wound–induced increases in jasmonic acid but not nicotine in *Nicotiana sylvestris*. Planta 203: 430–435
22. Hui D, Iqbal J, Lehmann K, Gase K, Saluz HP, Baldwin IT (2003) Molecular interactions between the specialist herbivore *Manduca sexta* (Lepidoptera, Sphingidae) and its natural host *Nicotiana attenuata*: V. Microarray analysis and further characterization of large-scale changes in herbivore-induced mRNAs. Plant Physiol 131:1877–1893
23. Giri AP, Wünsche H, Mitra S, Zavala JA, Muck A, Svatoš A, Baldwin IT (2006) Molecular interactions between the specialist herbivore *Manduca sexta* (Lepidoptera, Sphingidae) and its natural host *Nicotiana attenuata*. VII. Changes in the plant's proteome. Plant Physiol 142:1621–1641
24. Gaquerel E, Heiling S, Schoettner M, Zurek G, Baldwin IT (2010) Development and validation of a liquid chromatography-electrospray ionization-time-of-flight mass spectrometry method for induced changes in *Nicotiana attenuata* leaves during simulated herbivory. J Agric Food Chem 58:9418–9427

25. Gaquerel E, Weinhold A, Baldwin IT (2009) Molecular interactions between the specialist herbivore *Manduca sexta* (Lepidoptera, Sphigidae) and its natural host *Nicotiana attenuata*. VIII. An unbiased GC×GC-ToFMS analysis of the plant's elicited volatile emissions. Plant Physiol 149:1408–1423
26. Schwachtje J, Baldwin IT (2004) Smoke exposure alters endogenous gibberellin and abscisic acid pools and gibberellin sensitivity while eliciting germination in the post-fire annual, *Nicotiana attenuata*. Seed Sci Res 14:51–60
27. Baldwin IT, Staszak-Kozinski L, Davidson R (1994) Up in smoke: I. Smoke-derived germination cues for postfire annual, *Nicotiana attenuata* Torr. Ex. Watson. J Chem Ecol 20:2345–2371
28. Baldwin IT, Morse L (1994) Up in smoke: II. Germination of Nicotiana attenuata in response to smoke-derived cues and nutrients in burned and unburned soils. J Chem Ecol 20: 2373–2391
29. Schwachtje J, Kutschbach S, Baldwin IT (2008) Reverse genetics in ecological research. PLoS One 3(2):e1543
30. Schittko U, Preston CA, Baldwin IT (2000) Eating the evidence? *Manduca sexta* larvae cannot disrupt jasmonate induction in *Nicotiana attenuata* through rapid consumption. Planta 210:343–346

Chapter 8

Jasmonate Signaling in the Field, Part II: Insect-Guided Characterization of Genetic Variations in Jasmonate-Dependent Defenses of Transgenic and Natural *Nicotiana attenuata* Populations

Emmanuel Gaquerel, Michael Stitz, Mario Kallenbach, and Ian T. Baldwin

Abstract

The introduction of genetically modified plants into natural habitats represents a valuable means to determine organismic level functions of a gene and its effects on a plant's interaction with other organisms. *Nicotiana attenuata*, a wild tobacco species native of the southwestern USA that grows in the immediate postfire environment, is one of the important host plants for herbivore populations recolonizing recently burned habitats in the Great Basin Desert. Here, we provide detailed guidelines for the analysis, under field conditions, of jasmonate-dependent defense and its impact on the plant's native herbivore community. The procedures are based on the field release of transgenic lines silenced for jasmonate biogenesis, metabolism, or perception to conduct association studies between defense trait expression (secondary metabolite and trypsin proteinase inhibitor accumulation) and insect infestations. Additionally, because some insects have evolved mechanisms to "eavesdrop" on jasmonate signaling when selecting their host plants, we describe how leafhoppers of the species *Empoasca*, which selectively colonize jasmonate-deficient plants, can be used as "bloodhounds" for identifying natural variations in jasmonate signaling among natural *N. attenuata* populations.

Key words Jasmonate, Plant–insect interaction, Induced defenses, Field work, Proteinase inhibitors, Secondary metabolites

1 Introduction

Whether a gene is pseudogenized and lost from a plant's genome or maintained by natural selection depends critically on the contribution of the gene to the plant's Darwinian fitness. Since Darwinian fitness can only be adequately quantified in the habitats that an organism evolved in, field work is therefore an essential component for the characterization of a gene's function. The functions of most genes have been assigned based on their transcriptional behavior after elicitor studies or their sequence similarity to other better

Alain Goossens and Laurens Pauwels (eds.), *Jasmonate Signaling: Methods and Protocols*, Methods in Molecular Biology, vol. 1011, DOI 10.1007/978-1-62703-414-2_8, © Springer Science+Business Media, LLC 2013

characterized genes. In fewer cases, function has been evaluated by repressing or knocking down the expression of a gene of interest and challenging the resulting transformants with defined laboratory-based stresses. However, relying on only one of these approaches may result in misinterpretation of the gene function and of its fitness value for an organism. This particularly holds true for regulatory and structural genes that contribute to tolerance and resistance to herbivores, because function and regulation of these genes, shaped over long periods of (co)evolution between plants and insects [1], can only be fully revealed when these interactions take place in the natural settings in which they evolved.

Following this approach, we have developed tools to genetically dissect ecologically important traits activated by the wild tobacco *Nicotiana attenuata* in response to herbivore attack. The field station where all these ecological experiments are conducted is located at the Lytle Ranch Preserve in the Southwest of Utah in the USA. This preserve is owned and operated by the Brigham Young University and located in the Great Basin Desert. In its native habitat, *N. attenuata* plants occur ephemerally in large populations after fires and germinate from long-lived seed banks in response to germination cues in wood smoke [2–4]. Established rosette-stage plants represent important hosts for herbivores from different feeding guilds that colonize the environment after fires. The large year-to-year variability of this insect community but also the strong selection for rapid growth as plants compete with conspecifics after their synchronized germination from long-lived seed banks after fires have shaped strong growth–defense trade-offs controlling metabolic adjustments required for the production of defensive compounds [5].

To avoid wasting energy on superfluous metabolite production, the biosynthesis of most defenses deployed by rosette-stage plants against herbivores is not constitutively expressed, but induced in a timely manner after herbivore attack. During the last two decades of research on this plant, the regulation, impact on the plant's resistance to different herbivores, and the fitness value of many herbivory-inducible metabolic classes have been investigated in laboratory and field conditions. The best characterized and most active of these induced direct defense compounds include, among others, the neurotoxic alkaloid nicotine [6] that acts synergistically with anti-digestive proteinase inhibitors [7, 8], and large families of acyclic diterpene glycosides (DTGs) [9] and phenolic derivatives [10]. The defense efficiency of these secondary metabolites is to a large extent conditioned by the timing and magnitude of their production, elements controlled by the jasmonate signaling pathway [11]. Field experiments performed with transgenic *N. attenuata* plants deficient in jasmonate production or signaling demonstrated that jasmonate signaling is a critical determinant of the plant's resistance to folivores, such as grasshoppers, lepidopteran larvae, beetles, and leafhoppers [11, 12] as well as stem-boring weevils [13].

N. attenuata populations can be highly genetically diverse, namely, plants within a population are likely to be as genetically distinct as plants from different populations [14], a structural characteristic that is, in part, a consequence of the plant's long-lived seed bank and its "fire-chasing" germination behavior. Most individuals in native *N. attenuata* populations reliably produce a wound-induced jasmonate burst that is amplified by fatty acid amino acid conjugates in the OS from *M. sexta* larvae [15, 16] to reach maximum levels approximately 60 min after induction [17]. However, recent work has shown that there exist quantitative variations in the size of these transient herbivory-elicited jasmonate pools produced by different plants in native *N. attenuata* populations [18]. We hypothesize that low jasmonate-producing genotypes are maintained within native populations, despite the clear disadvantages of being defense impaired, by the growth–defense trade-offs incurred by JA-mediated defenses.

In this chapter, we provide guidelines for the qualitative and quantitative evaluation of damages resulting from the attack of native herbivores on untransformed and jasmonate signaling-deficient transgenic plants under field conditions and for the measurement of underlying alterations in herbivory-induced and jasmonate-regulated defense compound production. The described procedures for the field-based assessment of jasmonate-dependent antiherbivore defenses can be applied to any plant for which sufficient historical records are available of its interaction with the plant's natural herbivore communities as well as genetic resources. Additionally, we describe how to use leafhoppers (*Empoasca* spp.) that colonize jasmonate signaling-deficient transgenic *N. attenuata* plants [18] to identify natural variation in jasmonate levels within natural *N. attenuata* populations.

2 Materials

2.1 Growing *N. attenuata* Plants in Field Conditions

1. Field plot in the plant's native habitat.
2. Seeds from untransformed and jasmonate-deficient transgenic *N. attenuata* plants which can be obtained upon request from Ian T. Baldwin (baldwin@ice.mpg.de), Department of Molecular Ecology, Max Planck Institute for Chemical Ecology in Jena (http://www.ice.mpg.de/ext/).
3. Sterilization solution: 2 % (w/v) of dichloroisocyanuric acid sodium salt and 0.05 % (v/v) polysorbate 20 (Tween-20) in deionized water.
4. 1 mM gibberellic acid A3 (GA3).
5. Liquid smoke (House of Herbs, Passiac, NJ, USA).
6. Gamborg's B5 medium (Duchefa, Haarlem, The Netherlands).

7. 7.5 mg/L $Na_2[B_4O_5(OH)_4]\cdot 8H_2O$ (borax) in water. Approval from legal authorities to work with genetically modified plants in native habitats (described in Chapter 7).
8. Automatic watering system to grow plants used in common-garden experiments.
9. Shade house.
10. Jiffy 703 pots (Jiffy International AS, Stange, Norway).
11. Hand tools (shovel, rake, hoe, etc.).

2.2 Plant Elicitations

1. Preparation of *M. sexta* OS (described in Chapter 7).
2. *M. sexta* larvae for collecting herbivore OS.
3. Deionized water.
4. Fabric pattern tracing wheel with serrated edge.
5. Support for leaf to be treated (such as plastic discs with a diameter of 5–10 cm).
6. Dry ice blocks for sample collection.

2.3 Secondary Metabolite Analysis

1. Liquid nitrogen for freezing and grinding of samples.
2. Mortars and pestles or mechanical grinder (ball mill).
3. Extraction solvent: 40/60 (v/v) methanol/deionized water, acidified with 0.1 % acetic acid.
4. Nicotine, chlorogenic acid, and rutin calibration standards (range 10–250 ng/μL) dissolved in 40/60 (v/v) methanol/deionized water.
5. HPLC instrument with an analytical reverse-phase (RP) column and UV detector.
6. HPLC solvents: Methanol HPLC grade (solvent A), 0.5 % (v/v) acetic acid; 0.5 % (v/v) ammonia (concentrated) in deionized water (solvent B).
7. Refrigerated microcentrifuge.

2.4 Trypsin Proteinase Inhibitor Activity Determination

1. Liquid nitrogen for freezing and grinding of samples.
2. Mortars and pestles or mechanical grinder (ball mill).
3. Protein extraction buffer (1 L): 0.1 M tris(hydroxymethyl) aminomethane Tris–HCl (pH 7.6), 2 g of phenylthiourea (Sigma-Aldrich, St. Louis, MO, USA), 5 g of diethyldithiocarbamate, 18.6 g of Na_2-ethylenediaminetetraacetic acid (EDTA), and 50 g of polyvinylpyrrolidone (Sigma-Aldrich).
4. Refrigerated microcentrifuge.
5. Reagent for micro-Bradford assay: 25/75 (v/v) dye reagent concentrate (Bio-Rad, Hercules, CA, USA)/deionized water filtered through a Whatman #1 filter (GE-Healthcare, Little Chalfont, UK) to remove particles.

6. Bovine serum albumin (BSA, Sigma-Aldrich) (immunoglobulin) calibration standard (range 0.01–0.50 mg/mL) dissolved in the protein extraction buffer.
7. Flat-bottom 96-well plate.
8. Microplate reader GENios (TECAN, Männedorf, Switzerland).
9. Freshly prepared trypsin enzyme stock solution: 1 mg/mL in 0.1 M Tris–HCl, pH 7.6.
10. Square plastic Petri dish (small 12 × 12 cm or big ca. 24 × 24 cm) with 2 % plant agar-supported 0.1 M Tris–HCl, pH 7.6.
11. 4-mm diameter cork borer.
12. Soybean trypsin inhibitor calibration standard (range 0.0187–0.3 mg/mL) dissolved in 0.1 M Tris–HCl, pH 7.6.
13. Freshly prepared staining solution (25 mL for a small plate): 12 mg of Fast Blue, 6 mg of acetyl phenylalanine naphthyl ester (APNE), 5 mL of dimethylformamide, 20 mL 0.1 M Tris–HCl, pH 7.6.
14. Desktop illuminator.
15. Manual or digital stainless steel caliper.

2.5 Evaluating Damage from Native Herbivores in the Field

1. If the herbivores are infrequent at the field plot, specific herbivores can be collected on other plant species from native populations to perform choice assays. Additionally, *M. sexta* eggs can be ordered from laboratory cultures (Carolina Biological Supply Co., Burlington, NC, USA) (*see* **Note 1**).
2. Manual or digital stainless steel caliper to measure leaf and damage sizes.
3. Digital camera.
4. Global positioning system.
5. Plastic boxes for collecting insects.

3 Methods

3.1 Field Release of Genetically Modified Plants into the Field

Procedures for germinating and gradually adapting plants to high-light field conditions as well as legal requirements for the release of genetically modified *N. attenuata* plants—here empty vector (EV), *allene oxide cyclase* (*AOC*)-silenced, *coronatine insensitive1* (*COI1*)-silenced, and *Arabidopsis jasmonic acid O-methyltransferase* (*JMT*)-overexpressing transformants (Fig. 1)—are described in Chapter 7. Changes in defensive secondary metabolites elicited by *N. attenuata* during *M. sexta* feeding can be simulated by mechanically wounding the leaf lamina and immediately applying *M. sexta* OS to freshly produced puncture wounds [17]. This procedure is described in details in Chapter 7.

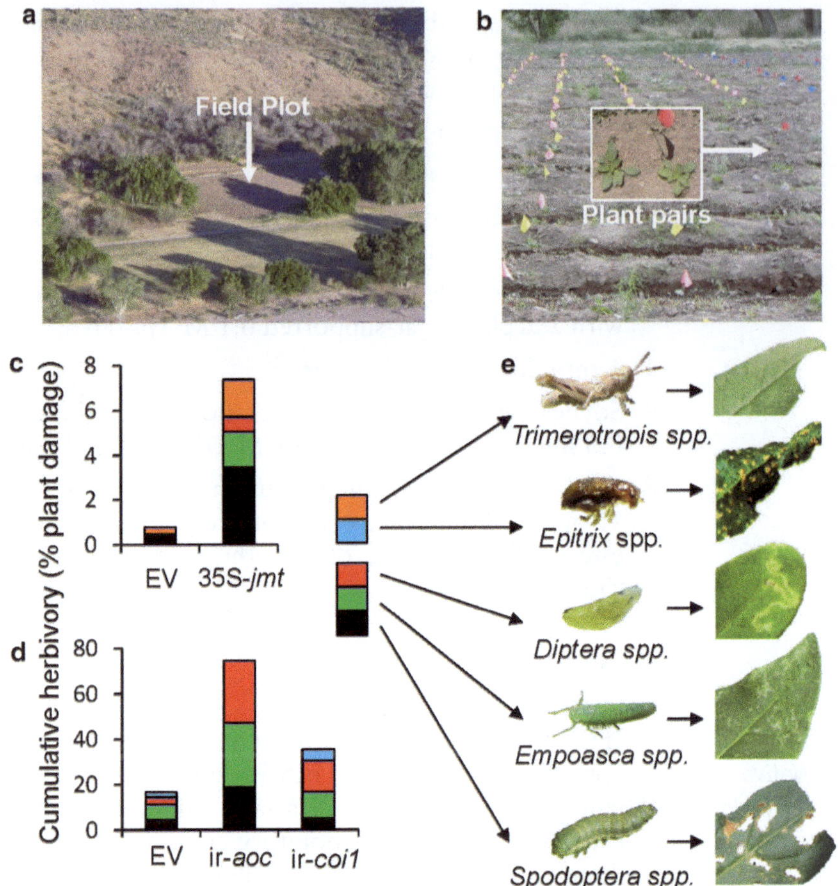

Fig. 1 Increased susceptibility to herbivores in the field of *N. attenuata* transgenic plants with altered jasmonate signaling. (**a**, **b**) Wild-type *N. attenuata* plants transformed with an EV or with a compromised jasmonate signaling (35S-*jmt*, ir-*aoc* and ir-*coi1*) are transplanted in a paired design with size-matched plants (EV/jasmonate signaling-deficient line) into a field plot at the Lytle Ranch Preserve, Great Basin Desert (Utah, USA). (**c**, **d**) Cumulative damages consistently more important inflicted by native herbivores on *N. attenuata* 35S-*jmt* and ir-*aoc* and ir-*coi1* plants 33 and 45 days post transplantation to the field, respectively, than those on EV plants. (**e**) Feeding activity of herbivores monitored by quantifying the damage signatures that are characteristic of each herbivore species on *N. attenuata* leaves

3.2 UV-Based Quantification of Defensive Secondary Metabolites

Nicotine, phenolic amines, and TPIs (Table 1) have been shown—using genetic manipulations of their levels and complementation experimentations—to act as direct defense traits against *M. sexta* larvae under optimized greenhouse conditions [5, 7, 10] and to actively reduce the herbivore load on *N. attenuata* plants in nature [6, 19]. Synergistic interactions among the defense traits that are strongly activated as a result of elevated jasmonate levels control the plant's resistance to most insect herbivores encountered in *N. attenuata* habitat [6].

1. Harvest leaves elicited with *M. sexta* OS after 72 h.
2. Collect at least three to five biologically replicated samples.

Table 1
Constitutive and herbivory-induced levels of direct defense traits in field-grown empty vector (EV) and 35S-*jmt N. attenuata* plants

Defense trait	Treatment[a]	EV	35S-*jmt*
TPI (nmol/mg protein)	Ctrl	20.4 ± 6.3	2.3 ± 0.3
	W + OS	41.2 ± 9.4	1.1 ± 0.8
Nicotine (mg/g FW)	Ctrl	0.5 ± 0.0	0.2 ± 0.0
	W + OS	0.5 ± 0.1	0.2 ± 0.0
Rutin (mg/g FW)	Ctrl	0.4 ± 0.1	0.5 ± 0.0
	W + OS	0.4 ± 0.0	0.4 ± 0.1
Chlorogenic acid (mg/g FW)	Ctrl	1.8 ± 0.4	2.5 ± 0.2
	W + OS	1.3 ± 0.1	1.5 ± 0.1
Dicaffeoylspermidine isomers (mg/g FW chlorogenic acid equivalents)	Ctrl	0.6 ± 0.1	ND
	W + OS	0.9 ± 0.1	0.1 ± 0.0

TPI activity was quantified according to the radial diffusion assay procedure described in Subheading 3.3. Secondary metabolites were quantified based on HPLC-UV profiling as described in Subheading 3.4. Levels of TPI, nicotine, and dicaffeoylspermidine isomers are severely compromised in 35S-*jmt*, whereas levels of rutin and chlorogenic acid, which have minor defensive activities, were not statistically changed. *ND* not detected. Data reported from ref. 19

[a]Ctrl, uninduced leaves of the same age as those used for water (Ctrl) and OS (W + OS) induction used to simulate feeding of chewing insect feeding

3. Crush plant tissue under liquid nitrogen and aliquot 100–150 mg of frozen leaf tissue into 2-mL microcentrifuge tubes containing two metal balls.
4. Record fresh mass of the samples. Prevent thawing of the samples at all times.
5. Grind the material by shaking frozen samples with 2 steel balls for 40 s at 1,000 strokes/min with Geno/Grinder (*see* **Note 2**).
6. Add 10 μL of extraction solvent per each mg of sample.
7. Extract samples by vortexing for 10 min at 4 °C.
8. Centrifuge for 20 min at 13,200 × *g* (4 °C).
9. Transfer supernatant to a clean 1.5-mL tube and spin again to remove all the remaining particles.
10. Transfer approximately 0.5 mL of the supernatant to a fresh 1.5-mL tube or appropriate glass vial, if HPLC is equipped with an autoinjector.
11. Inject 1–10 μL into an HPLC column to separate leaf analytes (*see* **Note 3**).
12. Calculate concentrations of secondary metabolites (μg/g fresh mass) from the external calibration curves detected at maximum absorbance of each of the target analyte (nicotine, 254 nm; chlorogenic acid, 320 nm; rutin, 360 nm) [20] (*see* **Note 4**) (Table 1).

3.3 Determination of TPI Activity Levels

1. Harvest leaves elicited with *M. sexta* OS after 72 h.
2. Collect at least five biologically replicated samples.
3. For each sample, aliquot 100 mg of frozen leaf tissue into 2-mL microcentrifuge tubes containing two metal balls and grind the material for 40 s at 1,000 strokes/min with a Geno/Grinder (*see* **Note 2**).
4. Add 0.3 mL of ice-cold extraction buffer (0.1 M of Tris–HCl (pH 7.6), 5 % (w/v) polyvinylpolypyrrolidone, 2 mg/mL phenylthiourea, 5 mg/mL diethyldithiocarbamate, and 0.05 M Na_2EDTA).
5. Thoroughly vortex the samples and centrifuge at 4 °C for 20 min at 12,000 × *g*.
6. Transfer supernatants into fresh 2-mL microcentrifuge tubes and keep them on ice.
7. Take an aliquot to determine the protein concentration by means of a protein assay kit with BSA as a standard.
8. For the preparation of the agar plates, melt 1.8 % (w/v) agar in 0.1 M Tris–HCl buffer (pH 7.6).
9. Cool this solution down to 50 °C with the proteinase stock solution to a final concentration of 42 nM.
10. Mix quickly and pour 100 mL of the solution in a 24 × 24 cm square Petri dish.
11. Allow solidification for 2 h at 4 °C.
12. Use a cork borer (4 mm diameter) connected to an aspiration apparatus to punch wells at intervals of 3 cm into the agar.
13. Fill each well completely with sample solution and allow diffusion for 16–18 h at 4 °C.
14. Develop each plate with 25 mL freshly prepared staining solution.
15. Incubate the plate for 55 min at 37 °C.
16. Place plates on the desktop illuminator to enhance contrast and measure the diameters of TPI inhibition zones with a caliper.
17. Extrapolate the TPI activity in the samples from a reference curve.

3.4 Determination of Damage from Native Herbivores

Feeding marks caused by native insects attacking *N. attenuata* have been extensively studied in our group. Even though herbivores may attack plants only at night or other times when they are not readily observed, the evidence of their feeding damage is unequivocal (*see* **Notes 5** and **6**).

1. Examine rosette leaves at intervals of 3–5 days (*see* **Note 5**) for the characteristic damages of the various herbivores that

commonly attack *N. attenuata* in Utah (Fig. 1), such as mirids, grasshoppers, and caterpillars (*see* **Note 6**).

2. After identifying damages from specific herbivores by their characteristic feeding patterns (Fig. 1e), determine the "normalized leaf number" for the plant as follows: the largest leaves on a plant count as one leaf, and smaller leaves count as 1/5 to 1/2 of a leaf based on leaf area (*see* **Notes** 7 and **8**).
3. For each herbivore damage type, estimate the percentage leaf area damage for each normalized leaf, sum up the total, and divide by the normalized leaf number.
4. Round leaf area damage estimates in categories of 1 %, 5 %, 10 %, 15 %, and so on, in steps of 5 % (*see* **Note** 7).

3.5 Insects as "Bloodhounds" for the Identification of Natural Variations in Jasmonate Signaling in N. attenuata Populations

During multiple field seasons, we have observed that leafhoppers of the species *Empoasca* spp. specifically colonize genetically modified *N. attenuata* with disrupted jasmonate biosynthesis or signaling. We have demonstrated that *Empoasca* leafhoppers "eavesdrop" on JA levels rather than on the known JA-dependent defense traits to select their host plant and can, therefore, be reliably used to identify *N. attenuata* plants with reduced jasmonate flux within native populations [18] (Fig. 2). We suspect that these natural jasmonate mutants, with presumably attenuated defenses, are maintained in native populations by the growth–defense trade-offs incurred by JA-mediated defenses and the intense selection for rapid growth that results from the synchronized seed germination that occurs after fires.

1. Prior to the insect release, carefully inspect all plants to screen for existing leaf damages.
2. Find suitable plants for insect collection (*see* **Note 9**).
3. Use an exhauster to collect a sufficient number of living *Empoasca* spp. (approximately equal to the plant number) on the day of the experiment (*see* **Note 10**).
4. Release leafhoppers into the area to be screened.
5. After 48 h, screen all plants for new *Empoasca* spp. damage as described in Subheading 3.4 (*see* **Note 11**).
6. From all candidate plants and one or more undamaged nearest neighbors in the same growth state (as control), choose the least damaged non-senescing leaves and randomly select them for analysis of constitutive or induced jasmonate levels.
7. On *Empoasca*-damaged plants and undamaged neighbors, elicit the three leaves chosen for analysis of inducible jasmonates at 60 min after OS elicitation.
8. Harvest leaf material from elicited and control leaves as described in Chapter 7.

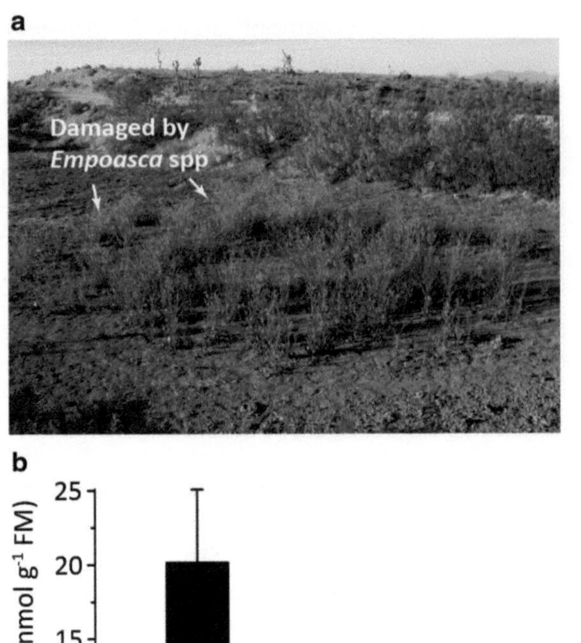

Fig. 2 *Empoasca* spp. used as "bloodhounds" to identify natural variations in JA accumulation in genetically diverse native *N. attenuata* populations. (**a**) Approximately 50 *Empoasca* spp. individuals were released into a native *N. attenuata* population. Feeding damage caused by *Empoasca* spp. monitored after 48 h was observed on only two plants. (**b**) To evaluate the importance of induced JA levels in *Empoasca* spp. host choice, leaves from both the two attacked plants and from one non-attacked plant (control) were elicited by treating mechanical wounds with *M. sexta* OS and harvesting for jasmonate analysis after 60 min. Both plants showing *Empoasca* spp. damage had lower OS-elicited amounts of JA than those of the noninfested plant (mean ± SE, $n = 3$; bars sharing the *same letters* are not significantly different, Student *t*-test, $P < 0.05$) (Figures redrawn from ref. 18)

9. Extract and analyze jasmonate levels as described in Chapter 7.
10. To collect self-pollinated seeds from *Empoasca*-damaged plants, first choose a branch for collection.
11. Remove all open and wilted flowers and existing seed capsules.
12. Cover the branch with a fine meshed net to prevent pollinator visits, but allow airflow. This is to ensure that the seeds produced

will be self-pollinated and increase the chance of preserving the phenotype for later inbreeding in the glasshouse.

13. Collect seed capsule when they are matured (usually 14–19 days).

4 Notes

1. Put the eggs on a moistened piece of cloth inside a sandwich box. Place the closed sandwich box inside a growth chamber with 28 °C, 50–70 % relative humidity, and 16-h light/8-h dark cycle as settings. The next day, place a wild-type leaf of *N. attenuata* next to the eggs. The eggs hatch 1–2 days after arrival. To delay emergence, place the sandwich box in a cooler area. Larvae should be reared on wild-type *N. attenuata* plants until reaching third- to fifth-instar stages for OS collection. Alternatively, larvae can be reared on the same transgenic plant that is to be treated with *M. sexta* OS. This approach allows the chemistry of OS to match that produced and deposited onto wounds during direct feeding.
2. Optionally, a ball mill instrument can be used to grind samples. Homogenize 0.3 g of frozen material in 2-mL microcentrifuge tubes containing two steel balls by shaking the tubes in a Geno/Grinder for 30 s at 200–300 strokes/min. It is important to ensure that tissue is thoroughly ground to a powder prior to extraction. Multiple rounds of Geno/Grinder shaking (with samples placed back to liquid nitrogen in between) or additional grinding by hand might be required. Higher strike rates can break the lids of the microcentrifuge tubes and cause sample losses.
3. Example for an Agilent 1100 series HPLC (Agilent Technologies, Santa Clara, CA, USA): 1 μL sample is injected in a Chromolith FastGradient RP 18-e column (50 × 2 mm; monolithic silica with bimodal pore structure, macropores with 1.6 μm diameter, EMD Millipore, Darmstadt, Germany) attached to a precolumn (Gemini NX RP18, 2 × 4.6 mm, 3 μm) (Phenomenex). The mobile phases (0.1 % formic acid + 0.1 % ammonium water, pH 3.5) as solvent A and methanol as solvent B are used in a gradient mode with the following conditions: time/concentration (min/%) for B: 0.0/0; 0.5/0; 6.5/80; 9.5/80; reconditioning 5 min to 0 % B. The flow rate is 0.8 mL/min and column oven temperature 40 °C.
4. Two abundant phenylpropanoid-polyamine conjugates in tobacco—caffeoylputrescine and dicaffeoylspermidine—[20, 21] can be detected at 320 nm and concentrations calculated using the chlorogenic acid calibration curve (expressed as chlorogenic acid equivalents) (Table 1).

5. It is important to keep in mind that too frequent screens on the field plot may disturb or even chase insects off the plants!
6. Larvae of *Spodoptera* spp. feed only on the mesophyll and the cuticle of one leaf side, leaving behind a shiny transparent cuticle on the other side of the leaf. Larvae or silky web traces can be found frequently on the ground under damaged plants. Grasshoppers exclusively feed at the leaf margins. Leafminers produce species-typical mines in which the growing larvae can often be observed. Piercing-sucking herbivores, such as *Tupiocoris* spp. (a single-cell feeder) and *Empoasca* spp. (a phloem feeder), can be encountered on *N. attenuata* plants. Infestations by these two insects can be distinguished by the typical black frass left by *Tupiocoris* and the characteristic hopper-burn caused by *Empoasca* spp. Additionally, imagos as well as larval stages of these two herbivores can frequently be found on the damaged plants.
7. Estimates of herbivore damage should be conducted only by one researcher and ideally the genotype identity should not be revealed before the end of the experiment to avoid biased estimations. Unambiguous genotype identification is determined by plastic labels carrying the identification number of each genotype buried in close proximity to the plant roots.
8. If the herbivore population is low at the field plot, specific herbivores can be collected on other plant species from native populations to perform choice assays. Additionally, eggs of *M. sexta* can be ordered from laboratory cultures.
9. *Empoasca* leafhoppers can be found in sufficient amounts on *Cucurbita* spp. plants, but also *Helianthus annuus* and *Medicago sativa* might be a host for these leafhoppers.
10. Leafhoppers should be collected before sunrise when they are relatively immobile.
11. This screening should be done before sunrise because immobile leafhoppers on plants can help to identify the damaged ones.

References

1. Pichersky E, Lewinsohn E (2011) Convergent evolution in plant specialized metabolism. Annu Rev Plant Biol 62:549–566
2. Schwachtje J, Baldwin IT (2004) Smoke exposure alters endogenous gibberellin and abscisic acid pools and gibberellin sensitivity while eliciting germination in the post-fire annual, *Nicotiana attenuata*. Seed Sci Res 14:51–60
3. Baldwin IT, Staszak-Kozinski L, Davidson R (1994) Up in smoke: I. Smoke-derived germination cues for postfire annual, *Nicotiana attenuata* Torr. Ex. Watson. J Chem Ecol 20: 2345–2371
4. Baldwin IT, Morse L (1994) Up in smoke: II. Germination of *Nicotiana attenuata* in response to smoke-derived cues and nutrients in burned and unburned soils. J Chem Ecol 20:2373–2391
5. Steppuhn A, Baldwin IT (2008) Induced defenses and the cost-benefit paradigm. In: Schaller A (ed) Induced plant resistance to herbivory. Springer, Berlin, pp 61–83

6. Steppuhn A, Gase K, Krock B, Halitschke R, Baldwin IT (2004) Nicotine's defensive function in nature. PLoS Biol 2:e217
7. Steppuhn A, Baldwin IT (2007) Resistance management in a native plant: nicotine prevents herbivores from compensating for plant protease inhibitors. Ecol Lett 10:499–511
8. Zavala JA, Patankar AG, Gase K, Hui D, Baldwin IT (2004) Manipulation of endogenous trypsin proteinase inhibitor production in *Nicotiana attenuata* demonstrates their function as antiherbivore defenses. Plant Physiol 134:1181–1190
9. Heiling S, Schuman M, Schöttner M, Mukerjee P, Berger B, Schneider B, Jassbi A, Baldwin IT (2010) Jasmonate and ppHsystemin regulate key malonylation steps in the biosynthesis of 17-hydroxygeranyllinalool diterpene glycosides, an abundant and effective direct defense against herbivores in *Nicotiana attenuata*. Plant Cell 22:273–292
10. Kaur H, Heinzel N, Schöttner M, Baldwin IT, Gális I (2010) R2R3-NaMYB8 regulates the accumulation of phenylpropanoid-polyamine conjugates, which are essential for local and systemic defense against insect herbivores in *Nicotiana attenuata*. Plant Physiol 152: 1731–1747
11. Paschold A, Halitschke R, Baldwin IT (2007) Co(i)-ordinating defenses: NaCOI1 mediates herbivore-induced resistance in *Nicotiana attenuata* and reveals the role of herbivore movement in avoiding defenses. Plant J 51: 79–91
12. Kessler A, Halitschke R, Baldwin IT (2004) Silencing the jasmonate cascade: induced plant defenses and insect populations. Science 305: 665–668
13. Diezel C, Kessler D, Baldwin IT (2011) Pithy protection: *Nicotiana attenuata*'s jasmonic acid-mediated defenses are required to resist stem-boring weevil larvae. Plant Physiol 155: 1936–1946
14. Bahulikar RA, Stanculescu D, Preston CA, Baldwin IT (2004) ISSR and AFLP analysis of the temporal and spatial population structure of the post-fire annual, *Nicotiana attenuata*, in SW Utah. BMC Ecol 4:12
15. Halitschke R, Schittko U, Pohnert G, Boland W, Baldwin IT (2001) Molecular interactions between the specialist herbivore *Manduca sexta* (Lepidoptera, Sphingidae) and its natural host *Nicotiana attenuata*. III. Fatty acid-amino acid conjugates in herbivore oral secretions are necessary and sufficient for herbivore-specific plant responses. Plant Physiol 125:711–717
16. Bonaventure G, VanDoorn A, Baldwin IT (2011) Herbivore-associated elicitors: FAC signaling and metabolism. Trends Plant Sci 16:294–299
17. McCloud ES, Baldwin IT (1997) Herbivory and caterpillar regurgitants amplify the wound-induced increases in jasmonic acid but not nicotine in *Nicotiana sylvestris*. Planta 203: 430–435
18. Kallenbach M, Bonaventure G, Gilardoni PA, Wissgott A, Baldwin IT (2012) *Empoasca* leafhoppers attack wild tobacco plants in a jasmonate-dependent manner and identify jasmonate mutants in natural populations. Proc Natl Acad Sci USA 109:1548–1557
19. Stitz M, Gase K, Baldwin IT, Gaquerel E (2011) Ectopic expression of *AtJMT* in *Nicotiana attenuata*: creating a metabolic sink has tissue-specific consequences for the jasmonate metabolic network and silences downstream gene expression. Plant Physiol 157: 341–354
20. Keinänen M, Oldham NJ, Baldwin IT (2001) Rapid HPLC screening of jasmonate-induced increases in tobacco alkaloids, phenolics, and diterpene glycosides in *Nicotiana attenuata*. J Agric Food Chem 49:3553–3558
21. Gaquerel E, Heiling S, Schoettner M, Zurek G, Baldwin IT (2010) Development and validation of a liquid chromatography-electrospray ionization-time-of-flight mass spectrometry method for induced changes in *Nicotiana attenuata* leaves during simulated herbivory. J Agric Food Chem 58:9418–9427

Part III

Molecules

Chapter 9

Profiling of Jasmonic Acid-Related Metabolites and Hormones in Wounded Leaves

Yusuke Jikumaru, Mitsunori Seo, Hideyuki Matsuura, and Yuji Kamiya

Abstract

The endogenous concentration of *N*-jasmonoyl-L-isoleucine (JA-Ile) is regulated by the balance between biosynthesis and deactivation and controls plant developmental processes and stress responses. Therefore, profiling of its precursors and metabolites is required to understand the mechanism by which the JA-Ile concentration is regulated. Also, other hormones, such as indole-3-acetic acid, abscisic acid, salicylic acid, and ethylene, have been suggested to interact with JA-Ile signaling. Profiling of these hormones and their metabolites should give us insights into their interaction mode. Liquid chromatography-electrospray ionization-tandem mass spectrometry has enabled us to develop a highly sensitive and high-throughput comprehensive quantification analysis of phytohormones.

Key words Plant hormone, Quantification, Liquid chromatography-electrospray ionization-tandem mass spectrometry, Hormone interaction, Stable isotope-labeled internal standard

1 Introduction

N-jasmonoyl-L-isoleucine (JA-Ile) is involved in the responses of plants to environmental stimuli, such as wounding and pathogen infections, through the regulation of gene expression. Early steps of the JA-Ile biosynthesis occur in the plastid membrane where linolenic acid is converted to 12-oxo-phytodienoic acid (OPDA) by the consecutive actions of lipoxygenase, allene oxide synthase, and allene oxide cyclase. OPDA is then transported to the peroxisome and converted into jasmonic acid (JA) by successive reactions catalyzed by OPDA reductase and β-oxidation enzymes. In turn, JA is converted into JA-Ile by the enzyme jasmonate-resistant1 to function as a signaling molecule. Among the jasmonate-related compounds tested (JA-amino acid conjugates, JA, OPDA, and methyl-JA), only JA-Ile has been shown to promote the interaction between CORONATINE INSENSITIVE1 and the JA ZIM-domain (JAZ) repressor proteins to regulate the downstream gene

Alain Goossens and Laurens Pauwels (eds.), *Jasmonate Signaling: Methods and Protocols*, Methods in Molecular Biology, vol. 1011, DOI 10.1007/978-1-62703-414-2_9, © Springer Science+Business Media, LLC 2013

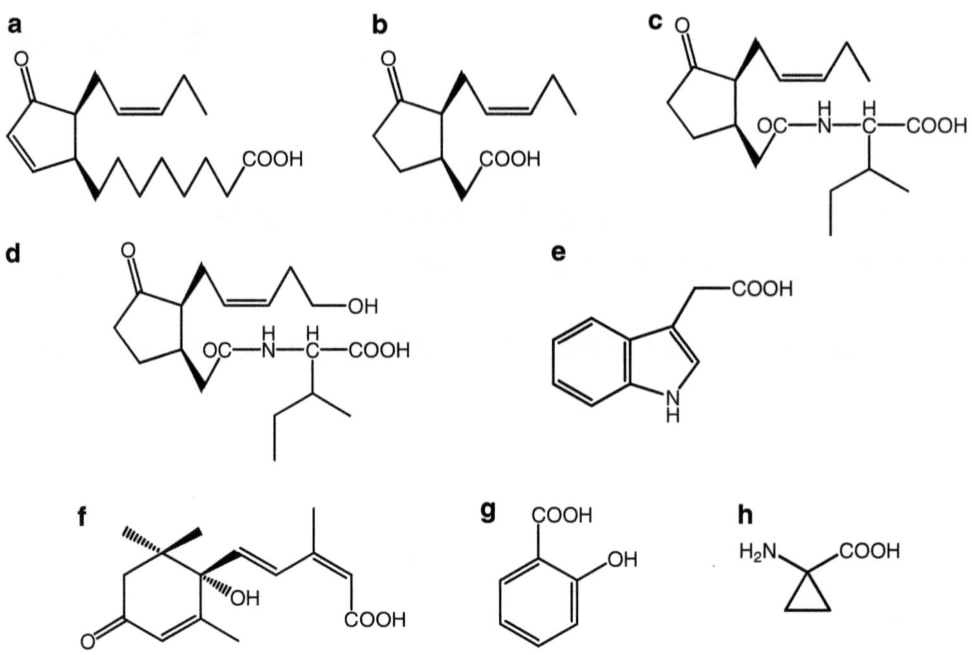

Fig. 1 Structures of target compounds. (**a**) OPDA. (**b**) JA. (**c**) JA-Ile. (**d**) 12-OH-JA-Ile. (**e**) IAA. (**f**) ABA. (**g**) SA. (**h**) ACC

expression [1–3]. Recently, a cytochrome P450 monooxygenase CYP94B3 that converts JA-Ile into 12-hydroxy-JA-Ile (12-OH-JA-Ile) had been identified [4, 5]. The loss-of-function mutant *cyp94b3* had enhanced sensitivity to exogenous JA-Ile, whereas overexpression of *CYP94B3* resulted in insensitivity to applied JA, indicating that this enzyme plays a crucial role in regulating endogenous JA-Ile levels and, hence, the JA-Ile-mediated responses. As physiological responses triggered by JA-Ile depend on its endogenous concentration, its precise determination is important. The endogenous levels of JA-Ile are regulated by the balance between biosynthesis and deactivation. Thus, to understand the detailed regulatory mechanisms that control JA-Ile concentration, it is necessary to analyze its metabolites (JA-Ile and its related compounds, OPDA, JA, and 12-OH-JA-Ile) as well (hereafter, referred to as jasmonates) (*see* Fig. 1).

Responses to environmental stimuli are not merely regulated by JA-Ile alone, but rather by the combination or the interaction with other hormones, such as indole-3-acetic acid (IAA), abscisic acid (ABA), salicylic acid (SA), and ethylene (ET). IAA suppresses jasmonate signals through the induction of JAZ1/TIFY10 [6]; the ABA receptor PYR/PYL/RCAR proteins mediate jasmonate signaling [7]; SA suppresses jasmonate-responsive gene expression downstream of jasmonate biosynthesis [8]; and the synergistic effects by ET and jasmonates are regulated by the ET-stabilized transcription factors ETHYLENE INSENSITIVE3/ETHYLENE INSENSITIVE-LIKE1 [9]. Therefore, to obtain more insight into these interactions, analysis of the concentrations of multiple hormones is required. In comparison to gas chromatography-electron impact ionization-mass spectrometry, liquid chromatography-electrospray ionization-tandem mass spectrometry (LC-ESI-MS/MS) is more suitable to analyze simultaneously plant hormones and their related compounds. ESI can produce protonated/deprotonated molecules to obtain further specific product ions in collision-induced dissociation, allowing a highly specific determination of target compounds in impurity-rich plant extracts [10].

2 Materials

2.1 Sample Extraction and Purification

1. Methanol (MeOH).
2. Acetonitrile (MeCN).
3. Distilled water.
4. 80 % (v/v) MeCN.
5. 80 % (v/v) MeCN containing 1 % (v/v) acetic acid.
6. Water containing 1 % (v/v) acetic acid.
7. Water containing 1 % (w/v) ammonia.
8. Oasis HLB (30 mg, 1 cc) (Waters, Milford, MA, USA).
9. Oasis MCX (30 mg, 1 cc) (Waters).
10. Oasis WAX (30 mg, 1 cc) (Waters).
11. Internal standards: D_6-ABA (Icon Isotopes, Summit, NJ, USA); d_5-OPDA, d_4-ACC (Olchemim Ltd, Olomouc, The Czech Republic); d_2-IAA, d_6-SA (Sigma-Aldrich, St. Louis, MO, USA); d_2-JA (Tokyo Kasei, Tokyo, Japan). $^{13}C_6$-JAIle and $^{13}C_6$-12-OH-JA-Ile were prepared as described [5, 11].

2.2 LC-ESI-MS/MS Analysis

1. Liquid chromatographer 1200 (Agilent, Santa Clara, CA, USA).
2. Triple Quad LC/MS 6410 (Agilent).
3. Column ZORBAX Eclipse XDB-C18 (2.1 mm × 50 mm, 1.8 μm) (Agilent).
4. Column ZORBAX AQUA (2.1 mm × 50 mm, 1.8 μm) (Agilent).
5. Spectrometry software MassHunter™ version B.01.03 (Agilent).
6. MeCN containing 0.05 % acetic acid.
7. Water containing 0.01 % acetic acid.
8. MeCN containing 0.1 % formic acid.
9. Water containing 0.1 % formic acid.

3 Methods

Here we describe a fundamental procedure to quantify jasmonates, IAA, ABA, SA, and the ET precursor, 1-aminocyclopropane-1-carboxylic acid (ACC) (*see* Fig. 1). Because of its volatility and small molecular weight, LC-ESI-MS/MS is not suitable for the analysis of ET itself, but, instead, the precursor ACC can be analyzed to determine the ET biosynthesis rate (*see* **Note 1**). Extraction and partial purification are based on the procedure as described [12] (*see* Fig. 2). A set of results obtained from *Arabidopsis thaliana* leaves collected at 0 or 1 h after wounding is shown in Table 1. The endogenous concentrations of jasmonates increased as reported previously [13], whereas those of other hormones did not change after 1 h of wounding. For the analysis of cytokinins (*trans*-zeatin, isopentenyladenine, and dihydrozeatin), gibberellins (GA_{12}, GA_{15}, GA_{24}, GA_{9}, GA_{4}, GA_{51}, GA_{34}, GA_{53}, GA_{44}, GA_{19}, GA_{20}, GA_{1}, GA_{29}, and GA_{8}), and brassinosteroids (castasterone and brassinolide) from vegetative tissues, large amounts of samples and additional purification steps are required (*see* **Note 2**). It is important to avoid cross-contaminations (*see* **Note 3**).

3.1 Sample Extraction and Purification

1. Prepare plant material as required (approximately 5 mg of freeze-dried samples were used for the analysis presented in Table 1) (*see* **Note 4**).

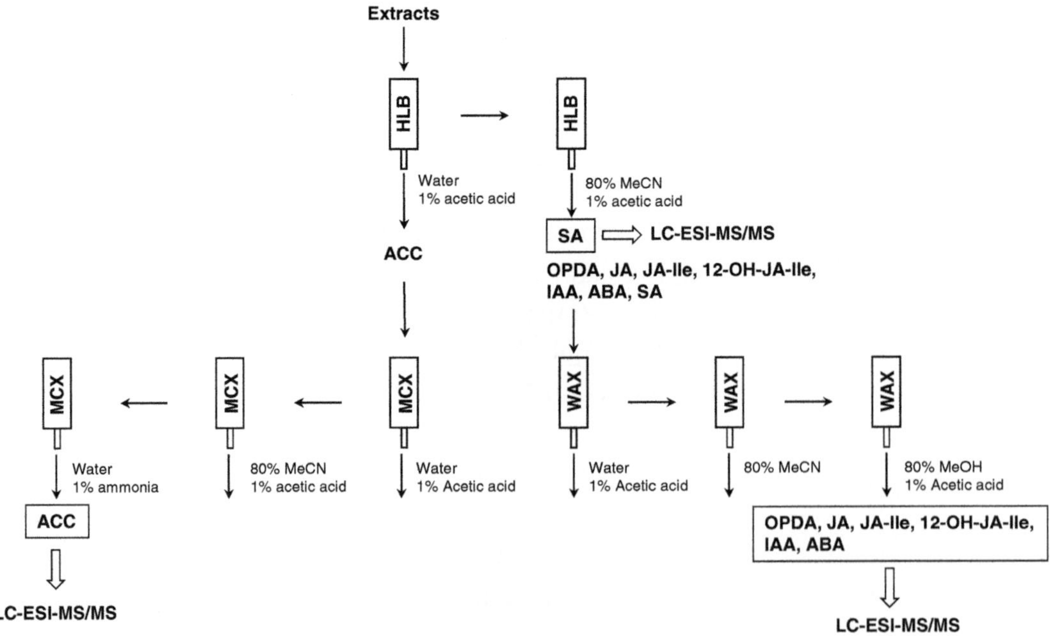

Fig. 2 Overview of the purification by cartridge columns

Table 1
Endogenous hormone concentrations in wounded leaves of *Arabidopsis* and concentrations of internal standards (in ng/g dry weight)

	Control		Wounded (1 h)	
Hormones	Endogenous	Internal standard	Endogenous	Internal standard
OPDA	12.3×10^3	10.0×10^3	11.4×10^6	10.0×10^6
JA	8.13	10.0	9.23×10^3	10.0×10^3
JA-Ile	1.09	1.00	1.00×10^3	1.00×10^3
12-OH-JA-Ile	4.34	5.00	15.5×10^3	15.0×10^3
IAA	58.9	50.0	56.1	50.0
ABA	33.7	50.0	50.9	50.0
SA	1.66×10^3	2.00×10^3	1.58×10^3	2.00×10^3
ACC	2.39×10^3	2.00×10^3	2.43×10^3	2.00×10^3

2. Freeze the materials immediately in liquid nitrogen and store at −80 °C until use. If required, material dry weights can be measured after lyophilization (*see* **Note 5**).
3. Grind the plant material into powder and add 1 mL of 80 % MeCN containing 1 % acetic acid as an extraction solvent, to, at least, 100 times the volume of the dry weight material.
4. Add internal standards so that their ratio to the endogenous compounds will be approximately 1 (Table 1) (*see* **Note 6**) and extract for 1 h at room temperature.
5. Centrifuge the extracts at 14,000 × *g* for 10 min at room temperature and collect the supernatant.
6. Extract the pellets with the same volume of extraction solvent again for 10 min and collect the supernatants after centrifugation.
7. Evaporate MeCN in the combined supernatants to obtain extracts in acetic acid-containing water (*see* **Note 7**).
8. To obtain ACC-containing fractions, apply the extracts after **step 3** to an Oasis HLB column cartridge (*see* **Note 8** and Subheading 3.2 for equilibration) and collect together the flow-through and 1 mL of water containing 1 % acetic acid.
9. Elute the other target compounds (OPDA, JA, JA-Ile, 12-OH-JA-Ile, IAA, ABA, and SA) with 2 mL of 80 % MeCN containing 1 % acetic acid. Keep 100 μL of this eluate for SA analysis.
10. Apply the fraction containing ACC to an Oasis MCX column cartridge (*see* Subheading 3.2, for equilibration and regeneration).

11. Wash the column with 1 mL of water containing 1 % acetic acid and then with 80 % MeCN containing 1 % acetic acid.
12. Elute ACC with 2 mL of water containing 1 % ammonia.
13. Evaporate MeCN in 1.9 mL of eluate after **step 4** to obtain extracts in water containing acetic acid.
14. Apply the extracts after **step 7** to the Oasis WAX column cartridge (*see* Subheading 3.2, for equilibration and regeneration).
15. Wash the column with 1 mL of water containing 1 % acetic acid and then with 2 mL of 80 % MeCN.
16. Elute the remainder of the target compounds (OPDA, JA, JA-Ile, 12-OH-JA-Ile, IAA, and ABA) with 2 mL of 80 % MeCN containing 1 % acetic acid.
17. Dry the fractions containing the target compounds after **step 4** (SA), **step 6** (ACC), and **step 9** (OPDA, JA, JA-Ile, 12-OH-JA-Ile, IAA, and ABA).
18. Dissolve the residue in 100 μL of MeOH (*see* **Note 9**) and transfer to the vials for LC-ESI-MS/MS.
19. Evaporate MeOH in the vials and dissolve the extracts in 100 μL (for ACC) or 30 μL (for SA, and for OPDA, JA, JA-Ile, 12-OH-JA-Ile, IAA, and ABA) of water containing 1 % acetic acid.
20. Inject 1 μL (for ACC) or up to 15 μL (for SA, and for OPDA, JA, JA-Ile, 12-OH-JA-Ile, IAA, and ABA) to LC-ESI-MS/MS for the analysis (*see* **Note 10**).

3.2 Column Cartridge Equilibration

1. For the Oasis HLB column, wash with 1 mL of MeCN and then with MeOH.
2. Equilibrate with 1 mL of initial solvent (water containing 1 % acetic acid).
3. For the Oasis MCX column, wash with 1 mL of MeCN and then with MeOH. After regeneration with 0.5 mL of 0.1 M HCl, equilibrate with 1 mL of initial solvent (water containing 1 % acetic acid).
4. For the Oasis WAX column, wash with 1 mL of MeCN and then with MeOH. After regeneration with 0.5 mL of 0.1 M NaOH, equilibrate with 1 mL of initial solvent (water containing 1 % acetic acid).

3.3 LC-ESI-MS/MS Analysis

1. Set the LC conditions as follows: flow rate, 200 μL/min. Composition and gradients of solvents are listed in Table 2.
2. Set the MS/MS conditions: Capillary, 4,000 V; desolvation temperature, 300 °C; gas flow, 9 L/min; and nebulizer, 30 psi. Other parameters for analysis are listed in Table 3.
3. Determine the areas of peaks from each compound with the spectrometer software (MassHunter™ version B.01.03) and

Table 2
LC conditions

Method no.	Column	Solvent A	Solvent B	Composition of solvent B
1	XDB-C18	Water containing 0.01 % acetic acid	MeCN containing 0.05 % acetic acid	3–70 % B over 30 min
2	AQUA	Water containing 0.1 % formic acid	MeCN containing 0.1 % formic acid	3–98 % B over 10 min
3	XDB-C18	Water containing 0.1 % formic acid	MeCN containing 0.1 % formic acid	0–3 % B over 5 min

Table 3
Parameters for LC-ESI-MS/MS analysis

Hormone	LC method no.	Retention time (min)[a]	Charge	MS/MS (*m/z*)	Collision energy	Fragmentor
OPDA	1	24.7	–	291/165	12	160
D_5-OPDA	1	24.7	–	296/170	12	160
JA	1	15.6	–	209/59	10	150
D_2-JA	1	15.6	–	211/59	10	150
JA-Ile	1	18.6	–	322/130	14	160
$^{13}C_6$-JA-Ile	1	18.6	–	328/136	14	160
12-OH-JA-Ile	1	12.2	–	338/130	20	150
$^{13}C_6$-12-OH-JA-Ile	1	12.2	–	344/136	20	150
IAA	1	11.8	–	174/130	18	110
D_2-IAA	1	11.8	–	176/132	18	110
ABA	1	13.4	–	263/153	8	140
D_6-ABA	1	13.3	–	269/159	8	140
SA	3	6.3	–	137/93	16	100
D_6-SA	3	6.3	–	141/97	16	100
ACC	2	0.8	+	102/56	12	50
D_4-ACC	2	0.7	+	106/60	12	50

[a]Retention time of deuterium-labeled internal standards is slightly shorter than that of nonlabeled endogenous compounds (*see* **Note 11**)

calculate the endogenous concentrations of the target compounds.

Typical MS/MS chromatograms obtained in *Arabidopsis* leaves 1 h after wounding are presented in Fig. 3.

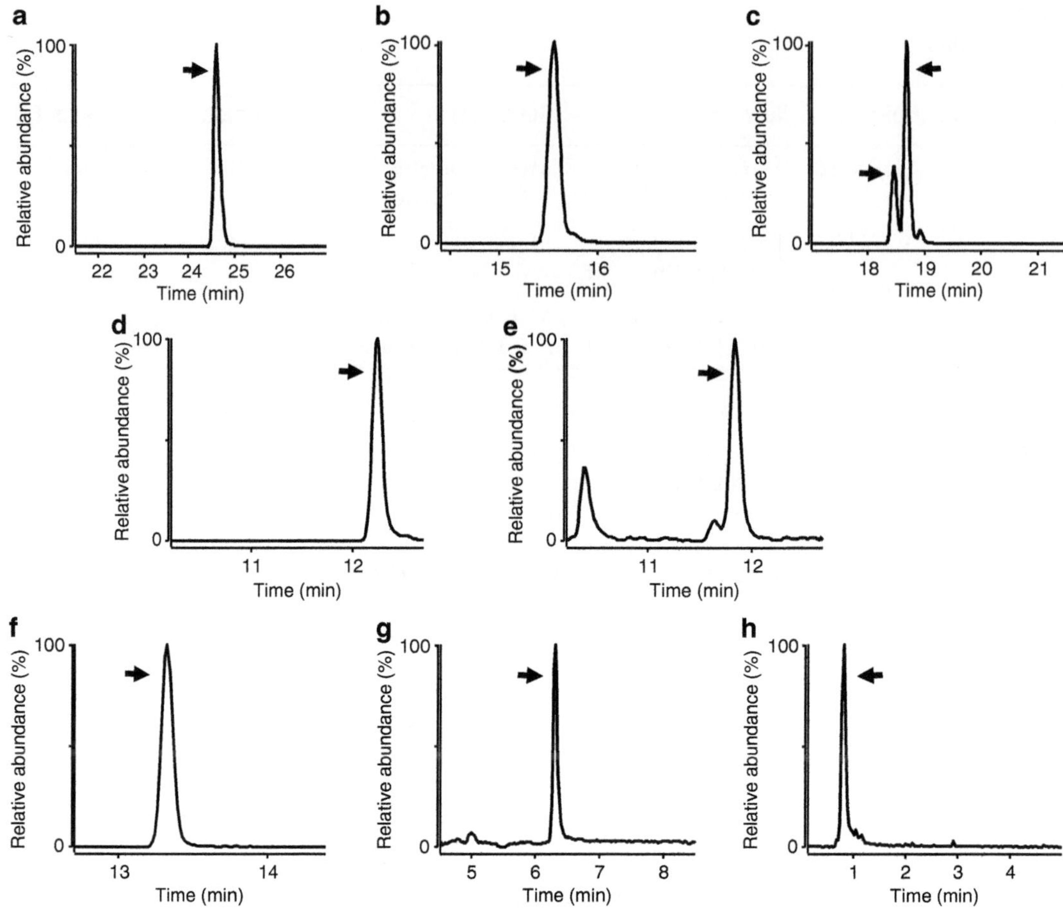

Fig. 3 MS chromatograms of target compounds. (**a**) OPDA. (**b**) JA. (**c**) JA-Ile. (**d**) 12-OH-JA-Ile. (**e**) IAA. (**f**) ABA. (**g**) SA. (**h**) ACC. JA-Ile gives two peaks (*see* **Note 12**)

4 Notes

1. ACC is generated by the reaction of ACC synthase and then converted to ET by ACC oxidase. ACC synthase has been shown to be a key regulatory enzyme in the ET biosynthesis [14].
2. Cation exchange cartridge purification is required to analyze basic compounds, such as cytokinins as described [10]. Because of their low concentrations, large amounts of plant material and additional purifications are required for analysis of gibberellins [15] and brassinosteroids [16].
3. To avoid cross-contamination, all cartridges, glassware, and plasticwares must not be reused.
4. The amount of plant material needed for the analysis depends on the endogenous concentrations of the target compounds and on the type of material (species, tissue, physiological state,

and treatment). A preliminary analysis is carried out to estimate the amount of samples needed.

5. Endogenous concentrations of jasmonates have been reported to increase rapidly within a few minutes after wounding [17]. For accurate determination of the concentration of jasmonates, it is recommended to flash-freeze plant materials just after sampling without measuring their fresh weight. Lyophilized samples are stable at room temperature under dry conditions and hormone concentrations per dry weight can be calculated.
6. The amount of internal standards added here depended on the dry weight of the samples. The following amounts of internal standards were added to approximately 5 mg of dry weight. For control leaves, 50 ng of d_5-OPDA, 50 pg of d_2-JA, 5 pg of $^{13}C_6$-JA-Ile, 25 pg of $^{1}3C_6$-12-OH-JA-Ile, 250 pg of d_2-IAA, 250 pg of d_6-ABA, 10 ng of d_6-SA, and 10 ng of d_4-ACC were added to the extracts, whereas for wounded leaves, 50 μg of d_5-OPDA, 50 ng of d_2-JA, 5 ng of $^{13}C_6$-JA-Ile, 75 ng of $^{13}C_6$-12-OH-JA-Ile, 250 pg of d_2-IAA, 250 pg of d_6-ABA, 10 ng of d_6-SA, and 10 ng of d_4-ACC were supplemented. The accumulation of wound-induced impurities must not be ignored. When endogenous concentrations of JA increase, accumulation of unknown impurities that have a 211/59 of MS/MS transition can also be observed. This peak is overlapping and cannot be discriminated from d_2-JA that is added as the internal standard. To minimize the effect of this overlap, d_2-JA has to be added a 1,000 times higher than the amounts of the control sample. Other internal standards for jasmonates have to be added in the same manner.
7. Because of the volatility of JA, the eluate should not be kept for a long time under a negative pressure after it is completely dry.
8. The amount of HLB resin must be adapted to the dry weight of the starting material. Generally, the dry weight of the sample is less than five times that of the resin.
9. Sonicate the extracts for a short time in MeOH to help dissolution. Do not transfer debris to the vial for LC-ESI-MS/MS analysis. If necessary, the MeOH solution is filtrated.
10. Because of its high polar nature, the amount of ACC retained on the column used here is limited. The injection volume must not be increased, because nonretained ACC will give a broad peak that is not identical. If necessary, dilute the extract.
11. Due to the isotope effect, the retention time of the deuterium-labeled internal standard is slightly shorter than that of the corresponding nonlabeled compounds (e.g., d_6-ABA is 4 s shorter than ABA and d_2-JA is 2 s shorter than JA). This information is important to be sure of the identity of the compounds in impurity-rich mass chromatograms.

12. JA-Ile gives two peaks of diastereomers: (−)-JA-Ile at a short retention time and (+)-7-iso-JA-Ile at a long retention time. (+)-7-iso-JA-Ile is converted to (−)-JA-Ile during extraction and purification in protic solvent. Both peaks are used for the quantification of endogenous JA-Ile.

References

1. Staswick PE (2008) JAZing up jasmonate signaling. Trends Plant Sci 13:66–71
2. Fonseca S, Chico JM, Solano R (2009) The jasmonate pathway: the ligand, the receptor and the core signalling module. Curr Opin Plant Biol 12:539–547
3. Thines B, Katsir L, Melotto M, Niu Y, Mandaokar A, Liu G, Nomura K, He SY, Howe GA, Browse J (2007) JAZ repressor proteins are targets of the SCF^{COI1} complex during jasmonate signalling. Nature 448:661–665
4. Koo AJK, Cooke TF, Howe GA (2011) Cytochrome P450 CYP94B3 mediates catabolism and inactivation of the plant hormone jasmonoyl-L-isoleucine. Proc Natl Acad Sci USA 108:9298–9303
5. Kitaoka N, Matsubara T, Sato M, Takahashi K, Wakuta S, Kawaide H, Matsui H, Nabeta K, Matsuura H (2011) *Arabidopsis* CYP94B3 encodes jasmonoyl-L-isoleucine 12-hydroxylase, a key enzyme in the oxidative catabolism of jasmonate. Plant Cell Physiol 52:1757–1765
6. Grunewald W, Vanholme B, Pauwels L, Plovie E, Inzé D, Gheysen G, Goossens A (2009) Expression of the *Arabidopis* jasmonate signalling repressor *JAZ1/TIFY10A* is stimulated by auxin. EMBO Rep 10:923–928
7. Lackman P, González-Gusmán M, Tilleman S, Carqueijeiro I, Cuéllar Pérez A, Moses T, Seo M, Kanno Y, Häkkinen ST, Van Montagu MCE, Thevelein JM, Maaheimo H, Oksman-Caldentey K-M, Rodriguez PL, Rischer H, Goossens A (2011) Jasmonate signaling involves the abscisic acid receptor PYL4 to regulate metabolic reprogramming in *Arabidopsis* and tobacco. Proc Natl Acad Sci USA 108: 5891–5896
8. Leon-Reyes A, Van der Does D, De Lange ES, Delker C, Wasternack C, Van Wees SCM, Ritsema T, Pieterse CMJ (2010) Salicylate-mediated suppression of jasmonate-responsive gene expression in Arabidopsis is targeted downstream of the jasmonate biosynthesis pathway. Planta 232:1423–1432
9. Zhu Z, An F, Feng Y, Li P, Xue L, Mu A, Jiang Z, Kim J-M, To TK, Li W, Zhang X, Yu Q, Dong Z, Chen W-Q, Seki M, Zhou J-M, Guo H (2011) Derepression of ethylene-stabilized transcription factors (EIN3/EIL1) mediates jasmonate and ethylene signaling synergy in Arabidopsis. Proc Natl Acad Sci USA 108: 12539–12544
10. Seo M, Jikumaru Y, Kamiya Y (2011) Profiling of hormones and related metabolites in seed dormancy and germination studies. Methods Mol Biol 773:99–111
11. Jikumaru Y, Asami T, Seto H, Yoshida S, Yokoyama T, Obara N, Hasegawa M, Kodama O, Nishiyama M, Okada K, Nojiri H, Yamane H (2004) Preparation and biological activity of molecular probes to identify and analyze jasmonic acid-binding proteins. Biosci Biotechnol Biochem 68:1461–1466
12. Ohkama-Ohtsu N, Sasaki-Sekimoto Y, Oikawa A, Jikumaru Y, Shinoda S, Inoue E, Kamide Y, Yokoyama T, Hirai MY, Shirasu K, Kamiya Y, Oliver DJ, Saito K (2011) 12-Oxo-phytodienoic acid-glutathione conjugate is transported into the vacuole in Arabidopsis. Plant Cell Physiol 52:205–209
13. Koo AJK, Gao X, Jones AD, Howe GA (2009) A rapid wound signal activates the systemic synthesis of bioactive jasmonates in Arabidopsis. Plant J 59:974–986
14. Chae HS, Kieber JJ (2005) *Eto Brute*? Role of ACS turnover in regulating ethylene biosynthesis. Trends Plant Sci 10:291–296
15. Varbanova M, Yamaguchi S, Yang Y, McKelvey K, Hanada A, Borochov R, Yu F, Jikumaru Y, Ross J, Cortes D, Ma CJ, Noel JP, Mander L, Shulaev V, Kamiya Y, Rodermel S, Weiss D, Pichersky E (2007) Methylation of gibberellins by *Arabidopsis* GAMT1 and GAMT2. Plant Cell 19:32–45
16. Yoshimitsu Y, Tanaka K, Fukuda W, Asami T, Yoshida S, K-i H, Kamiya Y, Jikumaru Y, Shigeta T, Nakamura Y, Matsuo T, Okamoto S (2011) Transcription of *DWARF4* plays a crucial role in auxin-regulated root elongation in addition to brassinosteroid homeostasis in *Arabidopsis thaliana*. PLoS One 6:e23851
17. Glauser G, Grata E, Dubugnon L, Rudaz S, Farmer EE, Wolfender J-L (2008) Spatial and temporal dynamics of jasmonate synthesis and accumulation in *Arabidopsis* in response to wounding. J Biol Chem 283:16400–16407

Chapter 10

A Non-targeted Approach for Extended Liquid Chromatography-Mass Spectrometry Profiling of Free and Esterified Jasmonates After Wounding

Gaëtan Glauser and Jean-Luc Wolfender

Abstract

Upon wounding or herbivory, plants quickly react by activating various defense mechanisms. A major part of these defenses is thought to be regulated by the jasmonate pathway through the induction of jasmonic acid and its biologically active jasmonoyl-isoleucine conjugate. Yet, these well-known phytohormones are only two among the numerous compounds that compose the jasmonate family. Here, we describe a method based on ultrahigh-pressure liquid chromatography coupled to quadrupole-time-of-flight mass spectrometry that can potentially profile the full range of known free and esterified jasmonates in a non-targeted manner. The developed approach is illustrated by the analysis of *Arabidopsis thaliana* leaves after mechanical wounding.

Key words Jasmonates, Arabidopsides, UHPLC-Q-TOFMS, Metabolite profiling, *Arabidopsis*, Wounding

1 Introduction

Jasmonates are cyclic oxygenated fatty acid derivatives formed from the enzymatic oxygenation of 18- and 16-carbon triunsaturated fatty acids [1]. The best known member of this family is jasmonic acid (JA) that was originally described as a plant growth regulator [2, 3]. Later on, in *Arabidopsis thaliana*, JA was found to trigger wound responses [4]. The jasmonate pathway regulates an estimated 67–84 % of transcripts induced by wounding and herbivores [5]. Although JA had long been thought to be the major player involved in the jasmonate-dependent response to stress, in *Arabidopsis*, its isoleucine conjugate, (+)-7-iso-jasmonoyl-L-isoleucine (JA-Ile), has been shown to be much more active than JA itself [6, 7]. JA-Ile, unlike JA, binds its receptor CORONATINE INSENSITIVE1 (COI1), which is part of an SCF ubiquitin E3 ligase and the formed complex targets JA ZIM-domain (JAZ)

Alain Goossens and Laurens Pauwels (eds.), *Jasmonate Signaling: Methods and Protocols*, Methods in Molecular Biology, vol. 1011, DOI 10.1007/978-1-62703-414-2_10,

repressors for protein degradation, allowing gene expression. Other biologically active JA derivatives have been identified in *Arabidopsis* and other plants. For example, JA conjugated to tryptophan inhibits auxin in *Arabidopsis* roots [8]. Leucine and valine conjugates seem as active as JA-Ile in *Solanum lycopersicum* (tomato) [9], whereas 12-oxo-phytodienoic acid (OPDA) and its 16-carbon homolog dinor-OPDA (dn-OPDA), both precursors of JA, have signaling properties either independent [10, 11] or partially dependent [12] on the jasmonate signaling pathway. Moreover, some JA derivatives have been suggested to contribute to the switch-off of JA and JA-Ile signaling, in addition to inactivation by epimerization, including hydroxylated (11-HO-JA and 12-HO-JA) [13], sulfonated [14], and glycosylated [15] forms of JA as well as hydroxylated [13] and carboxylated [16] isoleucine conjugates. Besides "free" jasmonates, galactolipid derivatives containing OPDA and/or dn-OPDA esterified to glycerol, referred to as arabidopsides, have been found in *Arabidopsis* and a few other closely related species [17–20]. Arabidopsides are induced in large amounts upon wounding [21, 22] and may play a role either as direct defense metabolites or as supplies for the rapid release of jasmonates after wounding [19, 23].

Because of the large diversity of jasmonates and their various biological roles, methods that provide comprehensive profiling of this class of compounds are needed for a more complete understanding of the plant's physiological responses to stress [24]. In most of the studies involving jasmonate analysis, only JA and, occasionally, OPDA and JA-Ile are monitored. For this purpose, gas chromatography mass spectrometry (GC-MS) after derivatization has been extensively used and protocols are well established [25]. With the advent of highly selective, robust, and sensitive liquid chromatography-tandem mass spectrometry (LC-MS/MS) methods, protocols requiring less complex sample preparation have been recently developed for the quantification of selected jasmonates or other phytohormones [26, 27]. The latest advances in chromatography involve the use of ultrahigh-pressure liquid chromatography (UHPLC), a technique that employs columns packed with very small particles (<2 μm) for increased resolution and speed [28]. For detection, high-resolution mass spectrometers able to work at high acquisition rate, i.e., compatible with rapid chromatography, represent the state of the art [29].

In this chapter, we propose a non-targeted method for jasmonate and arabidopside profiling, based on UHPLC coupled to quadrupole-time-of-flight mass spectrometry (Q-TOFMS). Using the developed generic approach, 17 free jasmonates and at least 13 arabidopsides could be monitored in *Arabidopsis* leaves after wounding.

2 Materials

2.1 Solvents and Reagents

For extraction, solvents of analytical or high-performance liquid chromatography (HPLC) grade should be used, as well as water of milli-Q quality (18.2 MΩ). For UHPLC-Q-TOFMS analysis, LC-MS-grade solvents and additives are recommended (*see* **Note 1**).

1. Extraction solvent: Isopropanol–formic acid, 99.5:0.5 (v/v; *see* **Note 2**).
2. Reconstitution solvent: Methanol–water, 85:15 (v/v).
3. Mobile phases: A, formic acid 0.05 % in water and B, formic acid 0.05 % in acetonitrile.
4. Internal standard solution: d_5-JA (C/D/N Isotopes, Pointe-Claire, Québec, Canada) at a concentration of 10 μg/mL in isopropanol (*see* **Note 3**).

2.2 Equipment

2.2.1 Extraction

1. Liquid nitrogen dewar and cryo-gloves.
2. Mortars, pestles, and spatulas.
3. Analytical balance.
4. Microcentrifuge tubes of 2 mL.
5. Pipettes and pipette tips.
6. Glass beads (2–3 mm diameter).
7. Vortex mixer.
8. Mixer mill (such as Retsch MM300, Haan, Germany) with holders for 1.5- or 2-mL microcentrifuge tubes.
9. Microcentrifuge.
10. Centrifugal or nitrogen evaporator.
11. Ultrasonic bath (optional).
12. Solid-phase extraction (SPE) cartridges Sep-Pak C18 (1 cc, 50 mg) (Waters, Milford, MA, USA) and vacuum manifold.
13. Glass vials (5 mL).
14. HPLC vials and caps.

2.2.2 UHPLC-Q-TOFMS Analysis

1. UHPLC-Q-TOFMS system with electrospray source: The system must consist of a binary or a quaternary pump able to withstand a maximal pressure of at least 600 bars (*see* **Note 4**). The mass spectrometer should be preferably a last-generation Q-TOFMS with high dynamic range (>10^4; *see* **Note 5**).
2. UHPLC column: Acquity BEH C18, 50 × 2.1 mm i.d., 1.7 μm particle size (Waters) coupled to an Acquity BEH C18 guard column, 5 × 2.1 mm i.d., 1.7 μm particle size (Waters).

3 Methods

All steps are carried out at room temperature unless otherwise specified. In the case presented here, all analyses were done on control and wounded *Arabidopsis* plants harvested 3 h after wounding (*see* **Note 6**). The method should be applicable to any experiment in which jasmonate induction has to be monitored.

3.1 Jasmonate Extraction

1. Harvest control and wounded leaves.
2. Immediately put them in liquid nitrogen (*see* **Note 7**).
3. Store the samples at −80 °C until extraction.
4. Grind the tissue to a fine powder in a mortar with a pestle under liquid nitrogen (*see* **Note 8**).
5. If necessary, gently pour liquid nitrogen into the mortar after grinding to maintain a very low temperature during the subsequent weighing phase (*see* **Note 9**).
6. Weigh 200 mg of powder and transfer it to a 2.0-mL microcentrifuge tube.
7. Make sure that the plant powder does not thaw (*see* **Note 10**) and put the samples back into liquid nitrogen immediately after weighing.
8. Add 1.5 mL of extraction solvent and 10 μL of internal standard solution to the tube.
9. Mix the solution with a vortex for a few seconds.
10. Add approximately 5–10 glass beads to the microcentrifuge tube.
11. Extract in a mixer mill at a frequency of 30 Hz for 3 min.
12. Centrifuge the mixture at 14,000 × *g* for 3 min.
13. Transfer the supernatant to another 2.0-mL microcentrifuge tube (*see* **Note 11**).
14. Evaporate the solvent to dryness, e.g., with a centrifugal evaporator or under a gentle nitrogen flow (*see* **Note 12**).
15. Resuspend the residue in 1 mL of reconstitution solvent (*see* **Note 13**).
16. Centrifuge the suspension at 14,000 × *g* for 90 s and take the whole supernatant for further SPE.
17. Condition and equilibrate the SPE cartridge with 1 mL of 100 % MeOH and 1 mL of reconstitution solvent, respectively (*see* **Note 14** and Fig. 1).
18. Collect the liquid flow in a waste glass vial.
19. Open the vacuum manifold and replace the waste vial by a clean 5-mL glass vial.

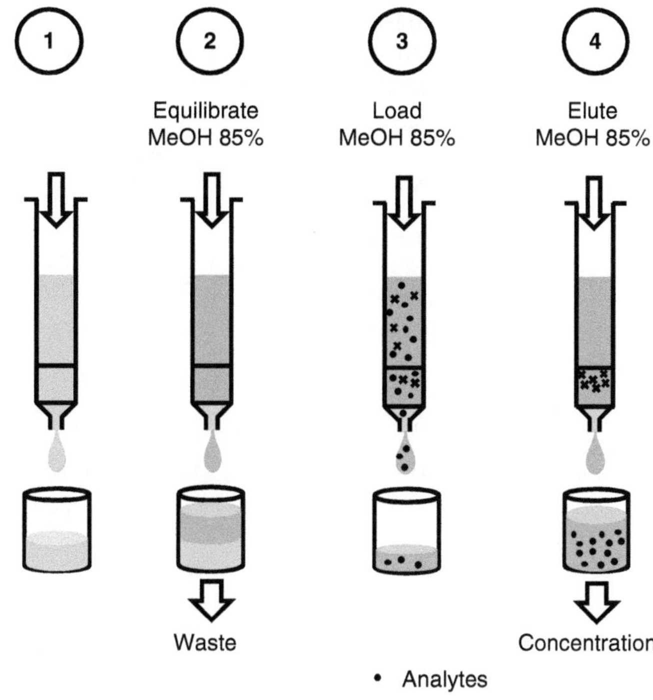

Fig. 1 Scheme of the solid-phase extraction (SPE) procedure used. Contaminants, such as pigments, triglycerides, and lipophilic vitamins, are retained on the hydrophobic C18 material, whereas jasmonates and arabidopsides are eluted with the appropriate mobile phase

20. Load the sample (1 mL) and pass 1 mL of reconstitution solvent through the cartridge.
21. Save the eluate (*see* **Note 15** and Fig. 1).
22. Evaporate the solvent to dryness, either with a centrifugal evaporator or under a gentle nitrogen flow.
23. Resolubilize the residue in 400 μL of reconstitution solvent (*see* **Note 16**) and transfer the solution into an HPLC vial.
24. Store the samples at −80 °C until analysis.

3.2 UHPLC-Q-TOFMS Analysis

3.2.1 UHPLC Gradient Conditions

The parameters presented are appropriate for the Acquity UPLC™ system (Waters) with an Acquity BEH C18 column (50 × 2.1 mm i.d., 1.7 μm): 5–100 % B in 10 min, holding at 100 % B for 2 min, and reequilibration at 5 % B for 2 min. The flow rate was set to 400 μL/min. The column and autosampler temperatures were maintained at 30 and 15 °C, respectively. The injection volume was 3 μL. Retention times for all jasmonates and arabidopsides are given in Tables 1 and 2, respectively. Typical chromatograms obtained for control and wounded *Arabidopsis* leaves are presented in Fig. 2.

Table 1
Chromatographic and mass spectrometric data of free jasmonates identified in wounded *Arabidopsis* (accession Columbia 0)

No	Name	Molecular formula	$(M-H)^-$	Fragments	RT (min)
1	HO-JA	$C_{12}H_{18}O_4$	225.1129	–	1.87
2	HO-JA-Val	$C_{17}H_{27}NO_5$	324.1813	116.0711	2.43
3	JA-Gln	$C_{17}H_{26}N_2O_5$	337.1758	145.0612	2.53
4	JA-Glc	$C_{18}H_{28}O_8$	417.1765[a]	209.1179	2.74
5	HO-JA-Ile	$C_{18}H_{29}NO_5$	338.1967	130.0870	2.87
6	HOOC-JA-Ile	$C_{18}H_{27}NO_6$	352.1763	130.0871	2.87
7	OPC-4-Gln	$C_{19}H_{30}N_2O_5$	365.2071	145.0610	3.18
8	OPC-4-Glc	$C_{20}H_{32}O_8$	445.2071[a]	237.1492	3.41
9	JA	$C_{12}H_{18}O_3$	209.1180	59.0134	3.66
IS	d_5-JA	$C_{12}H_{13}D_5O_3$	214.1494	62.0322	3.66
10	dn-OPDA-Glc	$C_{22}H_{34}O_8$	471.2227[a]	263.1646	3.86
11	JA-Val	$C_{17}H_{27}NO_4$	308.1860	116.0712	4.01
12	Unknown OPC-4 derivative	$C_{18}H_{26}O_7$	353.1599	237.1491	4.14
13	OPC-4	$C_{14}H_{22}O_3$	237.1490	165.1282	4.44
14	JA-Ile	$C_{18}H_{29}NO_4$	322.2020	130.0870	4.51
15	OPDA-Glc	$C_{24}H_{38}O_8$	499.2547[a]	291.1962	4.61
16	dn-OPDA	$C_{16}H_{24}O_3$	263.1647	165.1281	4.94
17	OPDA	$C_{18}H_{28}O_3$	291.1962	165.1280	5.85

RT retention time, *JA* jasmonic acid, HO- hydroxy-, HOOC- carboxy-, *Val* valine, *Gln* glutamine, *Glc* glucose, *Ile* isoleucine, *OPC*-4, 3-oxo-2-(2Z-pentenyl) cyclopentane-1-butyric acid; *OPDA* 12-oxo-phytodienoic acid, *dn-OPDA* dinor-oxophytodienoic acid
[a]Formate adduct $(M+HCOO)^-$

3.2.2 Q-TOFMS Conditions

The negative electrospray ionization (ESI) mode must be used (*see* **Note 17**). An acquisition mode using alternating scans at low and high collision energies should be employed (such as MS^E) (*see* **Note 18**). For the Synapt G2 Q-TOFMS (Waters), optimal source parameters were as follows: capillary voltage −2,500 V, cone voltage −25 V, extraction cone voltage −4.5 V, source temperature 120 °C, desolvation temperature 350 °C, desolvation gas flow 800 L/h, and cone gas flow 20 L/h. Data were acquired over the range 85–1,200 Da with scans of 0.2 s at a collision energy of 4 eV and 0.2 s at a collision energy ramp of 10–30 eV, both applied on the transfer region of the collision cell. Argon was the collision gas at a flow

Table 2
Chromatographic and mass spectrometric data of predominant arabidopsides identified in wounded *Arabidopsis* (accession Columbia 0)

No	Name	Molecular formula	$(M+HCOO)^-$	Fragments	RT (min)
18	dn-OPDA-DGMG	$C_{31}H_{50}O_{15}$	707.3124	263.1650	3.35
19	dn-OPDA-MGMG	$C_{25}H_{40}O_{10}$	545.2594	263.1648	3.68
20	OPDA-DGMG	$C_{33}H_{54}O_{15}$	735.3434	291.1963	4.02
21	OPDA-MGMG	$C_{27}H_{44}O_{10}$	573.2905	291.1959	4.40
22	OPDA/dn-OPDA-DGDG	$C_{49}H_{76}O_{17}$	981.5052	291.1960/263.1649	6.16
23	OPDA/OPDA-DGDG	$C_{51}H_{80}O_{17}$	1,009.5368	291.1962	6.71
24	OPDA/dn-OPDA-MGDG	$C_{43}H_{66}O_{12}$	819.4527	291.1959/263.1649	6.90
25	OPDA/OPDA-MGDG	$C_{45}H_{70}O_{12}$	847.4841	291.1964	7.52
26	OPDA/16:3-MGDG	$C_{43}H_{68}O_{11}$	805.4733	291.1963/249.1858	8.80
27	OPDA/dn-OPDA-MGDG-dn-OPDA[a]	$C_{59}H_{88}O_{14}$	1,065.6147	291.1958/263.1649	8.88
28	OPDA/dn-OPDA-MGDG-OPDA	$C_{61}H_{92}O_{14}$	1,093.6452	291.1963/263.1651	9.35
29	OPDA/18:3-MGDG	$C_{45}H_{72}O_{11}$	833.5046	291.1962/277.2168	9.47
30	OPDA/OPDA-MGDG-OPDA	$C_{63}H_{96}O_{14}$	1,121.6769	291.1961	9.79

RT retention time, *OPDA* 12-oxo-phytodienoic acid, *dn-OPDA* dinor-oxophytodienoic acid, *MGMG* monogalactosyl-monoacylglyceride, *DGMG* digalactosyl-monoacylglyceride, *MGDG* monogalactosyl-diacylglyceride, *DGDG* digalactosyl-diacylglyceride

[a]Putative identification

rate of 2.1 mL/min (pressure inside the collision cell 7.0e^{-3} mbars). Internal calibration was done by means of the Lockspray interface (Waters) by infusing a 500 ng/mL solution of leucine-enkephalin in the mass spectrometer at a flow rate of 7.5 μL/min. The Lockspray scan time and frequency were set at 0.5 and 15 s, respectively. Data were averaged over five scans for mass correction (*see* **Note 19**). Pseudomolecular ions and fragments obtained for all jasmonates and arabidopsides are given in Tables 1 and 2, respectively.

3.3 Data Processing

1. To identify the various jasmonates, generate extracted ion chromatograms (EIC) with a mass window of ±0.005 Da (*see* **Note 20**) from the data obtained at low collision energy (Fig. 3).
2. Extract mass spectra produced at low and high collision energies for all chromatographic peaks corresponding to putative jasmonates.
3. Verify the mass accuracy of the pseudomolecular and fragment ions of interest (*see* Tables 1 and 2).

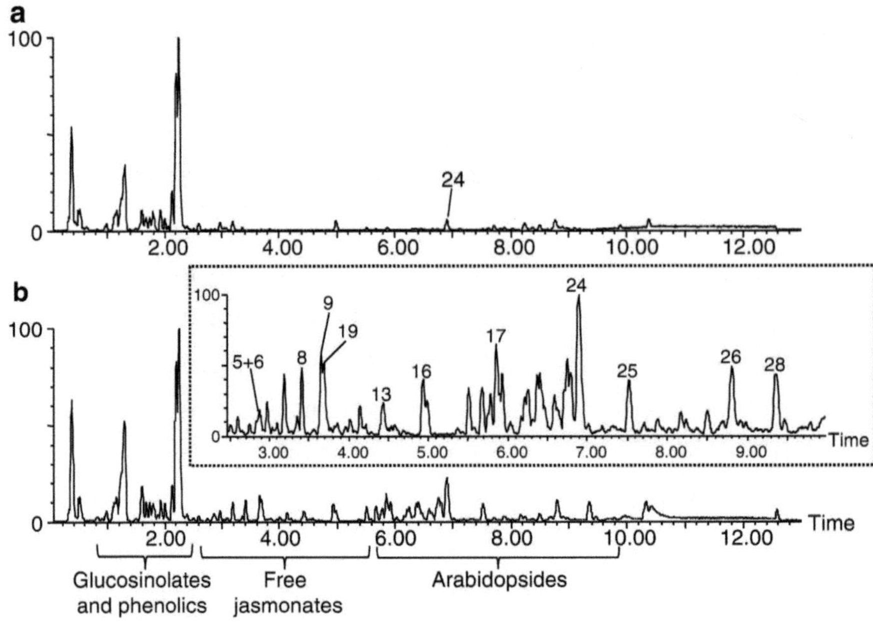

Fig. 2 Base peak intensity (BPI) UHPLC-Q-TOFMS chromatograms obtained at low collision energy. (**a**) Unwounded *Arabidopsis* leaf extract. (**b**) Wounded *Arabidopsis* leaf extract (leaves harvested 3 h after wounding). *Inset*, zoom of the 2.5–10.0 min part of wounded *Arabidopsis* leaf extract chromatogram. The predominant peaks corresponding to jasmonates and arabidopsides are labeled according to their numbers in Tables 1 and 2, respectively

4. In the low-collision-energy chromatogram, integrate the peaks of identified jasmonates and arabidopsides and divide the obtained areas by those of the internal standard d_5-JA (*see* **Note 21**).
5. Report results as mean ± standard deviation (SD) or standard error (SE).

4 Notes

1. Ultrahigh-purity solvents are required to ensure low background noise in the LC-MS chromatograms and to maintain good instrument performances.
2. Formic acid is essential to prevent the artifactual production of OPDA, dn-OPDA, and related arabidopsides during extraction. An alternative, but less practical, solution is to use boiling isopropanol without acid.
3. Alternatively, other isotopically labeled jasmonate or dihydrojasmonic acid may be used as internal standards.
4. The UHPLC conditions have been intentionally adapted not to exceed a pressure of 600 bars to be compatible with all UHPLC systems available on the market. In principle, HPLC

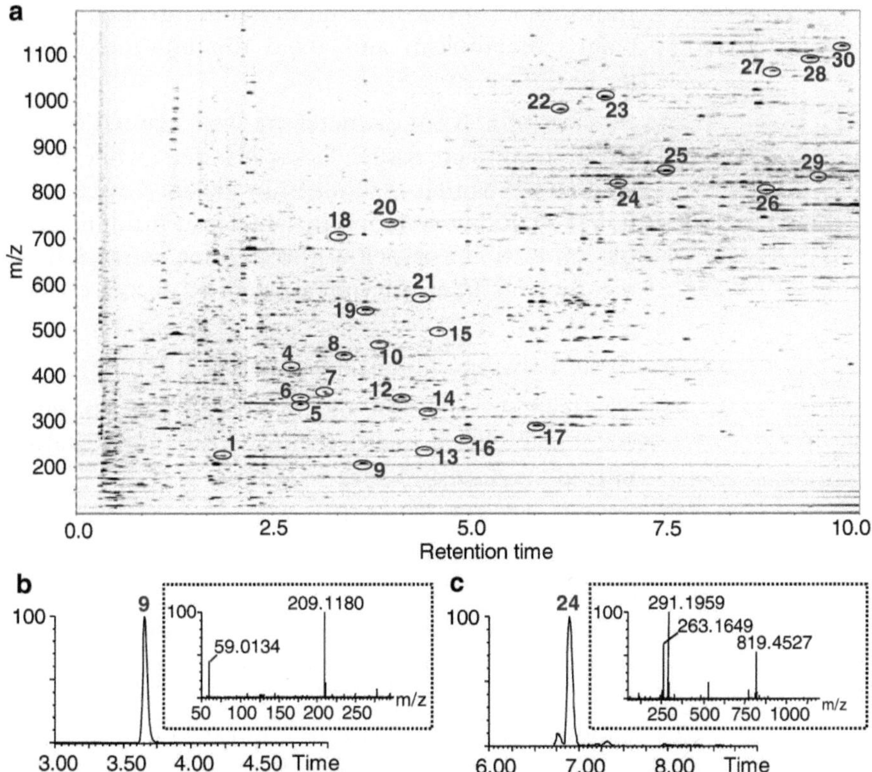

Fig. 3 Two-dimensional map and extracted ion chromatograms (EICs). (**a**) UHPLC-Q-TOFMS map of the wounded *Arabidopsis* leaf extract presented in Fig. 2b. *Visible dots* representing jasmonates and arabidopsides are labeled according to their number in Tables 1 and 2, respectively. (**b**) EIC at *m/z* 209.118 ± 0.005 Da corresponding to jasmonic acid. *Inset*, mass spectrum of jasmonic acid obtained at high collision energy. (**c**) EIC at *m/z* 819.453 ± 0.005 Da corresponding to arabidopside A (OPDA/dn-OPDA-MGDG). *Inset*, mass spectrum of arabidopside A obtained at high collision energy

systems may be used in which pressures lower than 400 bars are reached and the columns are packed with larger particles, provided that an appropriate gradient transfer is applied [30]. Chromatographic conditions from HPLC to UHPLC and vice versa can be transferred with the HPLC Calculator v. 3 that can be downloaded at http://www.unige.ch/sciences/pharm/fanal/lcap/telechargement-en.htm. Depending on the column used, a tee may be needed to split the HPLC flow prior to MS detection.

5. The Q-TOFMS detector may be replaced by an Orbitrap mass spectrometer that should acquire data at a compatible frequency with acceptable resolution. Single TOFMS might be considered as well.
6. In the presented experiment, approximately 60 % of the leaves were wounded with forceps.

7. In principle, shock-freezing in liquid nitrogen will quench the plant's metabolism and avoid the unwanted reactions that occur during storage.
8. Mortars of 5–8 cm diameter are well adapted for the required amounts of plant tissue. Tissue grinders, while appropriate for lyophilized samples, should not be used with fresh material unless a cooling system is installed that maintains the temperature at −20 °C or below throughout extraction. We do not recommend freeze-drying because it decreases JA extraction yields [27].
9. Avoid thawing of plant tissue during grinding.
10. To facilitate weighing and make sure that thawing does not occur, dip closed microcentrifuge tubes and spatula in liquid nitrogen before and after weighing.
11. During this step, it is important to pipet as much supernatant as possible, but without pipetting any solid particles from the pellet.
12. Samples take about 60 min to evaporate at 40 °C with both centrifugal or nitrogen concentrators. The maximal temperature should not be above 40 °C.
13. Ultrasounds may be used to improve solubilization.
14. A flow rate of approximately 1 drop per second is appropriate for cartridge conditioning, sample loading, and elution.
15. This SPE step is an efficient prepurification procedure. Hydrophobic components, such as chlorophyll, long-chain fatty acid-derived galactolipids, triglycerides, liposoluble vitamins, and carotenoids, will stick to the stationary phase of the cartridge, whereas more polar molecules, such as jasmonates and arabidopsides, will be eluted.
16. If the solution is not perfectly clear, transfer it first into a microcentrifuge tube and centrifuge for 90 s. A perfectly clear solution is a prerequisite for injection in the UHPLC system. If a less sensitive mass spectrometer is used, it may be appropriate to resolubilize the residue in a volume of 200 or even 100 μL of reconstitution solvent.
17. Most jasmonates monitored in this study are either not detected or poorly ionized in positive ionization mode.
18. This is a valuable option for metabolite identification because it provides both pseudomolecular ions and corresponding fragments at high mass accuracy within a single run.
19. Using the Synapt G2 Q-TOFMS, accuracy higher than 2 ppm should be obtained over the whole mass range.
20. If a slightly less accurate mass spectrometer is used, the mass window should be set larger, e.g., ±0.02 Da.

21. Normalization to the internal standard is important to compensate for extraction and analysis variability. The presented method is only semiquantitative because the large majority of jasmonates are not available as standards. Normalization of all peak areas to the internal standard provides a good means to assess relative changes resulting, for example, from wounding. In most cases, such an approach can be informative enough to investigate a given biological phenomenon that involves jasmonate induction. Finally, the presence of the internal standard can be used to assess the quality of both retention time and mass accuracy over the course of the analyses.

Acknowledgments

This work was supported by Swiss National Science Foundation grants no. 205320_135190 and CRSII3_127187. G.G. and J.L.W. are also grateful to the National Center of Competence in Research (NCCR) Plant Survival and to the Swiss Plant Science Web (SPSW).

References

1. Wasternack C, Kombrink E (2010) Jasmonates: structural requirements for lipid-derived signals active in plant stress responses and development. ACS Chem Biol 5:63–77
2. Dathe W, Rönsch H, Preiss A, Schade W, Sembdner G, Schreiber K (1981) Endogenous plant hormones of the broad bean, *Vicia faba* L. (−)-jasmonic acid, a plant growth inhibitor in pericarp. Planta 153:530–535
3. Yamane H, Takagi H, Abe H, Yokota T, Takahashi N (1981) Identification of jasmonic acid in three species of higher plants and its biological activities. Plant Cell Physiol 22:689–697
4. Farmer EE, Johnson RR, Ryan CA (1992) Regulation of expression of proteinase inhibitor genes by methyl jasmonate and jasmonic acid. Plant Physiol 98:995–1002
5. Reymond P, Bodenhausen N, Van Poecke RMP, Krishnamurthy V, Dicke M, Farmer EE (2004) A conserved transcript pattern in response to a specialist and a generalist herbivore. Plant Cell 16:3132–3147
6. Staswick PE, Tiryaki I (2004) The oxylipin signal jasmonic acid is activated by an enzyme that conjugates it to isoleucine in Arabidopsis. Plant Cell 16:2117–2127
7. Fonseca S, Chini A, Hamberg M, Adie B, Porzel A, Kramell R, Miersch O, Wasternack C, Solano R (2009) (+)-7-*iso*-Jasmonoyl-L-isoleucine is the endogenous bioactive jasmonate. Nat Chem Biol 5:344–350
8. Staswick PE (2009) The tryptophan conjugates of jasmonic and indole-3-acetic acids are endogenous auxin inhibitors. Plant Physiol 150:1310–1321
9. Katsir L, Schilmiller AL, Staswick PE, He SY, Howe GA (2008) COI1 is a critical component of a receptor for jasmonate and the bacterial virulence factor coronatine. Proc Natl Acad Sci USA 105:7100–7105
10. Stintzi A, Weber H, Reymond P, Browse J, Farmer EE (2001) Plant defense in the absence of jasmonic acid: the role of cyclopentenones. Proc Natl Acad Sci USA 98:12837–12842
11. Taki N, Sasaki-Sekimoto Y, Obayashi T, Kikuta A, Kobayashi K, Ainai T, Yagi K, Sakurai N, Suzuki H, Masuda T, K-i T, Shibata D, Kobayashi Y, Ohta H (2005) 12-Oxo-phytodienoic acid triggers expression of a distinct set of genes and plays a role in wound-induced gene expression in Arabidopsis. Plant Physiol 139:1268–1283
12. Ribot C, Zimmerli C, Farmer EE, Reymond P, Poirier Y (2008) Induction of the Arabidopsis *PHO1;H10* gene by 12-oxo-phytodienoic acid but not jasmonic acid via a CORONATINE INSENSITIVE1-dependent pathway. Plant Physiol 147:696–706

13. Miersch O, Neumerkel J, Dippe M, Stenzel I, Wasternack C (2008) Hydroxylated jasmonates are commonly occurring metabolites of jasmonic acid and contribute to a partial switch-off in jasmonate signaling. New Phytol 177:114–127
14. Gidda SK, Miersch O, Levitin A, Schmidt J, Wasternack C, Varin C (2003) Biochemical and molecular characterization of a hydroxyjasmonate sulfotransferase from *Arabidopsis thaliana*. J Biol Chem 278:17895–17900
15. wi tek A, Van Dongen W, Esmans EL, Van Onckelen H (2004) Metabolic fate of jasmonates in tobacco bright yellow-2 cells. Plant Physiol 135:161–172
16. Glauser G, Grata E, Dubugnon L, Rudaz S, Farmer EE, Wolfender J-L (2008) Spatial and temporal dynamics of jasmonate synthesis and accumulation in *Arabidopsis* in response to wounding. J Biol Chem 283:16400–16407
17. Stelmach BA, Müller A, Hennig P, Gebhardt S, Schubert-Zsilavecz M, Weiler EW (2001) A novel class of oxylipins, *sn*1-*O*-(12-oxophytodienoyl)-*sn*2-*O*-(hexadecatrienoyl)-monogalactosyl diglyceride, from *Arabidopsis thaliana*. J Biol Chem 276:12832–12838
18. Andersson MX, Hamberg M, Kourtchenko O, Brunnström Å, McPhail KL, Gerwick WH, Göbel C, Feussner I, Ellerström M (2006) Oxylipin profiling of the hypersensitive response in *Arabidopsis thaliana*. Formation of a novel oxo-phytodienoic acid-containing galactolipid, arabidopside E. J Biol Chem 281:31528–31537
19. Kourtchenko O, Andersson MX, Hamberg M, Brunnström Å, Göbel C, McPhail KL, Gerwick WH, Feussner I, Ellerström M (2007) Oxo-phytodienoic acid-containing galactolipids in Arabidopsis: jasmonate signaling dependence. Plant Physiol 145:1658–1669
20. Böttcher C, Weiler EW (2007) *cyclo*-Oxylipin-galactolipids in plants: occurrence and dynamics. Planta 226:629–637
21. Buseman CM, Tamura P, Sparks AA, Baughman EJ, Maatta S, Zhao J, Roth MR, Esch SW, Shah J, Williams TD, Welti R (2006) Wounding stimulates the accumulation of glycerolipids containing oxophytodienoic acid and dinor-oxophytodienoic acid in Arabidopsis leaves. Plant Physiol 142:28–39
22. Glauser G, Grata E, Rudaz S, Wolfender J-L (2008) High-resolution profiling of oxylipin-containing galactolipids in Arabidopsis extracts by ultra-performance liquid chromatography/time-of-flight mass spectrometry. Rapid Commun Mass Spectrom 22:3154–3160
23. Glauser G, Dubugnon L, Mousavi SAR, Rudaz S, Wolfender J-L, Farmer EE (2009) Velocity estimates for signal propagation leading to systemic jasmonic acid accumulation in wounded *Arabidopsis*. J Biol Chem 284:34506–34513
24. Erb M, Glauser G (2010) Family business: multiple members of major phytohormone classes orchestrate plant stress responses. Chemistry 16:10280–10289
25. Mueller MJ, Mène-Saffrané L, Grun C, Karg K, Farmer EE (2006) Oxylipin analysis methods. Plant J 45:472–489
26. Pan X, Welti R, Wang X (2008) Simultaneous quantification of major phytohormones and related compounds in crude plant extracts by liquid chromatography-electrospray tandem mass spectrometry. Phytochemistry 69: 1773–1781
27. Forcat S, Bennett MH, Mansfield JW, Grant MR (2008) A rapid and robust method for simultaneously measuring changes in the phytohormones ABA, JA and SA in plants following biotic and abiotic stress. Plant Methods 4:16
28. Nguyen DT-T, Guillarme D, Rudaz S, Veuthey J-L (2006) Fast analysis in liquid chromatography using small particle size and high pressure. J Sep Sci 29:1836–1848
29. Hopfgartner G (2011) Can MS fully exploit the benefits of fast chromatography? Bioanalysis 3:121–123
30. Guillarme D, Nguyen DTT, Rudaz S, Veuthey J-L (2008) Method transfer for fast liquid chromatography in pharmaceutical analysis: application to short columns packed with small particle. Part II: gradient experiments. Eur J Pharm Biopharm 68:430–440

Chapter 11

Cell-Specific Detection of Jasmonates by Means of an Immunocytological Approach

Bettina Hause, Kati Mielke, and Susanne Forner

Abstract

To determine the location of specific molecules within tissues or cells, immunological techniques are frequently used. However, immunolocalization of small molecules, such as jasmonic acid (JA) and its bioactive amino acid conjugate, JA-isoleucine, requires proper fixation and embedding methods as well as specific antibodies. In this chapter, we present a method to prepare plant tissues for the detection of jasmonates, including the chemical fixation to immobilize JA within the tissue, the subsequent embedding in a suitable medium, and the immunolabeling procedure itself.

Key words Immunocytology, Chemical fixation, Embedding into polyethylene glycol, Jasmonate-specific antibodies, Slide coating, Jasmonic acid-to-protein coupling

1 Introduction

Immunocytochemistry is based on a combination of immunochemistry and morphology and, therefore, identifies the cellular location of biochemically distinct antigens. This technique can basically be used for all types of cells. For most plant tissues, the so-called postembedding labeling is mostly favored, in which the specimen is fixed, embedded, and sectioned before the immunolabeling procedure is applied.

For in situ detection of phytohormones, only a few attempts turned out to be successful in respect to the discovery of their immunocytochemistry and cell specificity [1–3]. The two main challenges for the use of immunocytochemical techniques are the following: on the one hand, it is essential that such small molecules are immobilized within a cell or a tissue while the structural cell and tissue morphologies are preserved and, on the other hand, it is a prerequisite that antibodies with a high specificity against a particular antigen are available.

Alain Goossens and Laurens Pauwels (eds.), *Jasmonate Signaling: Methods and Protocols*, Methods in Molecular Biology, vol. 1011, DOI 10.1007/978-1-62703-414-2_11, © Springer Science+Business Media, LLC 2013

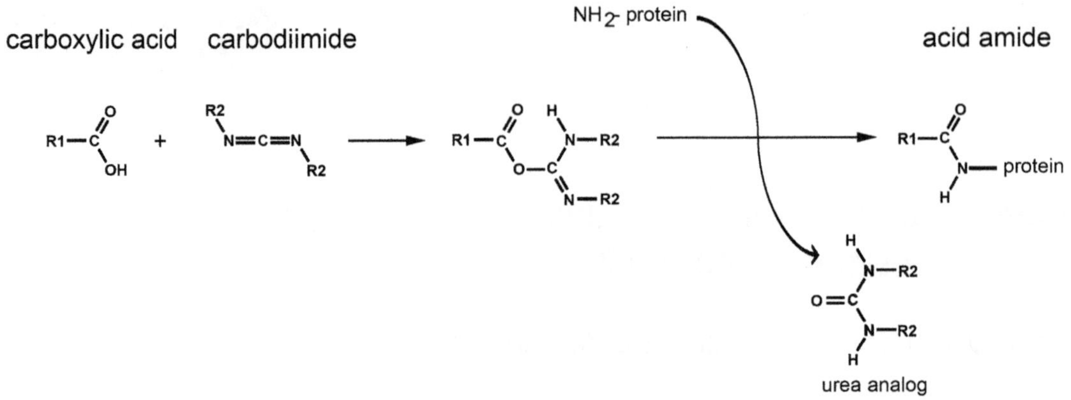

Fig. 1 Schematic representation of the reaction mechanism for coupling of carboxylic acids to proteins. The water-soluble carbodiimide EDC reacts with the acid group and condenses it with amino groups in the protein matrix to form stable conjugates

The ultimate aim of a proper fixation is to “freeze” the cellular and tissue organization in a short time frame so that every molecule in that cell or tissue remains in its original location during subsequent preparation, staining, and visualization. In immunochemistry, formaldehyde and paraformaldehyde are the most frequently used compounds for chemical fixation [4], but they are unsuitable for jasmonates, because aldehydes are only able to form addition products via connecting amino groups. In contrast, carbodiimides, such as 1-ethyl-3-(3-dimethylaminopropyl)-carbodiimide (EDC), are capable of binding small molecules harboring an acid group [2, 5] (*see* Fig. 1). This holds true also for jasmonic acid (JA) and its amino acid conjugate JA-isoleucine (JA-Ile). Both compounds are efficiently bound via their acid moiety to cellular proteins, resulting in a coupling product that can be recognized by an antibody. The fixed material is then embedded in polyethylene glycol 1500 (PEG 1500) [6] that provides not only fast infiltration at moderate temperatures but also easy removal from sections by washing with aqueous solutions/buffer with well-preserved structures as a result.

The antibody used for detection should not show any irrelevant cross-reactivity with an unrelated compound that shares a common epitope with the antigen of interest. To date, monoclonal antibodies raised against various phytohormones have been used with success [1, 3]. Regarding jasmonates, the production of polyclonal antibodies from rabbits against JA coupled to bovine serum albumin (BSA) was successful and delivered antibodies that bind specifically to JA, JA-Ile, and the JA methyl ester, but not to oxophytodienoic acid (OPDA), 12-hydroxy-JA, and coronatine [2].

Similarly to all types of immunolabeling, it is very important to use controls, both positive and negative ones (*see* Fig. 2). Positive controls are generated by infiltrating JA into the tissue and should

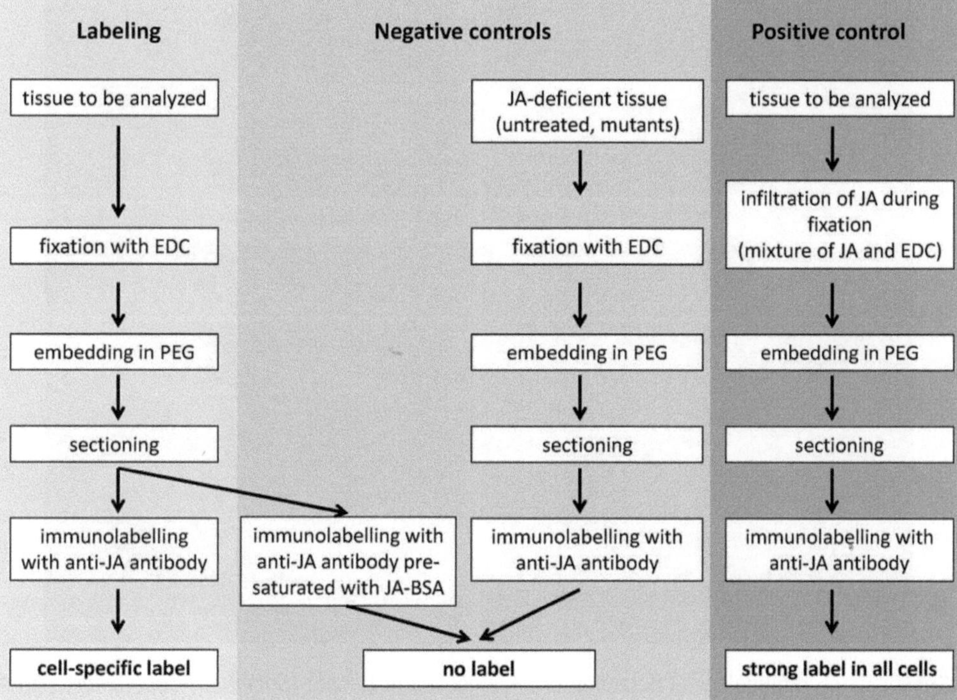

Fig. 2 Flowchart showing the steps for immunolabeling of JA in plant tissues, including the recommended negative and positive controls

result in an evenly distributed label, whereas for negative controls tissues from mutants lacking jasmonates are commonly used [7]. When no mutants are available, presaturation of the antibody with the antigen (JA) is recommended. Both approaches should lead to a label that is absent or, at least, clearly reduced (*see* Fig. 3).

2 Materials

Prepare all solutions with ultrapure water and analytical grade reagents. All reagents are made and stored at room temperature (unless indicated otherwise). Diligently follow all waste disposal regulations. No sodium azide is added to the reagents.

2.1 Special Equipment

1. Dialysis tubing, cellulose, molecular weight cutoff of 12–14 kDa.
2. Exsiccator equipped with vacuum pump.
3. Rotator.
4. Microwave oven.
5. Glass vials (5-mL volume, 2 cm diameter, and snap-on lid).
6. Incubator at constant temperature of 50 °C.

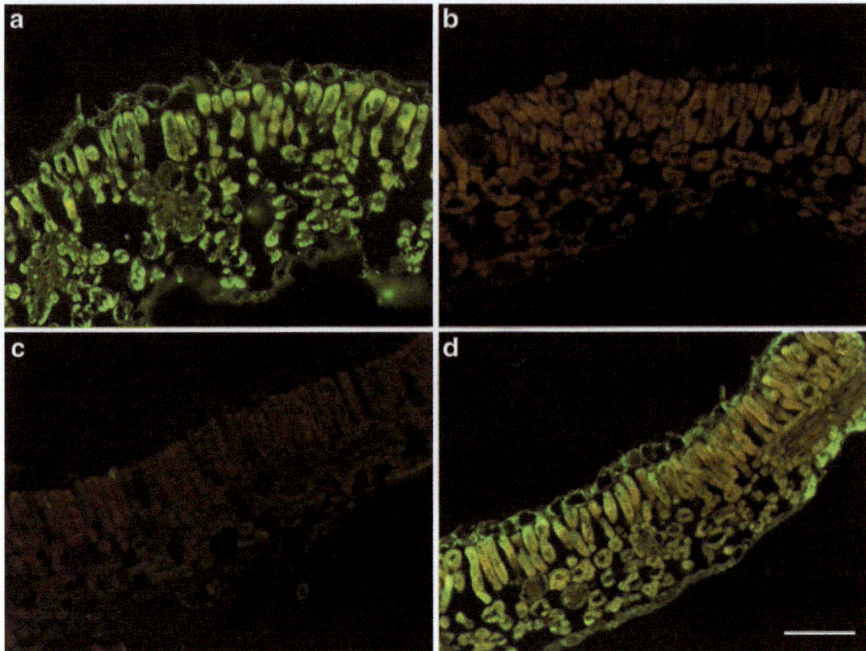

Fig. 3 Specificity of immunological detection of JA in leaves of tomato (*Solanum lycopersicum*). (**a**) and (**b**) Leaves of the wild type (cv. Castlemart) and (**c**) and (**d**) the JA-deficient mutant *suppressor of prosystemin-mediated response2* (*spr2*) [8] were wounded across the mid vein for 30 min. Standard processing of the wounded wild-type leaf (**a**) led to a strong label in all cells of the leaf, whereas incubation of the antibody with 8 μg/mL JA–BSA abolished the label (negative control). In the wounded leaf of the JA-deficient mutant *spr2* (**c**), the label is also absent after standard processing (negative control), but infiltration of JA during fixation with EDC resulted in a homogenously distributed label (positive control). *Bar* = 100 μm for all micrographs

7. Embedding molds.
8. Forceps, razor blades, ethanol burner, paintbrush, and metal loop.
9. Rotary microtome, including disposable microtome blades and specimen holder.
10. Paper boxes.
11. Slides labeled with numbers.
12. Hellendahl staining cuvettes.
13. Epifluorescence microscope equipped with proper filter modules and camera.
14. Eppendorf tubes.

2.2 Buffers and Solutions

1. 0.1 M sodium borate (pH 8.5).
2. 0.2 M Tris–HCl (pH 5.5).
3. 0.1 M Tris–HCl (pH 5.5).
4. JA solution for coupling with BSA: 0.05 mmol (±)-Jasmonic acid (JA) (Sigma-Aldrich, St. Louis, MO, USA) in 0.175 mL tetrahydrofuran and 0.05 mL dimethylformamide.

5. EDC solution for coupling JA and BSA: 0.0782 mM 1-Ethyl-3-(3-dimethylaminopropyl)-carbodiimide hydrochloride (EDC) (Merck KGaA, Darmstadt, Germany) in 0.2 M Tris–HCl (pH 5.5).
6. BSA solution for coupling with JA: 26.1 mg BSA in 1.7 mL 0.2 M Tris–HCl (pH 5.5).
7. Phosphate-buffered saline (PBS) (pH 7.2): 135 mM NaCl, 3 mM KCl, 1.5 mM KH_2PO_4, and 8 mM Na_2HPO_4.
8. Fixative: 4 % (v/v) EDC in PBS, freshly prepared (*see* **Note 1**).
9. Activated JA for infiltration of plant tissue: 4 % (v/v) EDC in PBS complemented with various amounts (e.g., 25, 50, 100, and 500 μM) of JA, incubated for 60 min before use.
10. Graded ethanol series: 10, 30, 50, 70, and 90 % (v/v) ethanol in H_2O.
11. Graded PEG 1500 series: 25, 50, and 75 % (v/v) melted polyethylene glycol (PEG) 1500 (*see* **Note 2**) in ethanol (kept all the time at 50 °C).
12. 45 % (w/v) PEG 6000 in PBS.
13. 0.1 M NH_4Cl in PBS.
14. Blocking solution: 5 % (w/v) BSA in PBS (*see* **Note 3**). Store at −20 °C.
15. Diluent solution: 1 % (v/v) acetylated BSA (BSA_{acet}) (Biotrend Chemikalien GmbH, Köln, Germany) and 5 % (w/v) BSA in PBS. Store at −20 °C.

2.3 Antibodies and Dyes

1. Anti-JA antibody, polyclonal from rabbit (*see* **Note 4**).
2. Anti-rabbit-IgG from goat coupled with AlexaFluor488 (Invitrogen, Carlsbad, CA, USA).
3. Counterstaining solution: 0.1 μg/mL 4,6 diamidino-2-phenylindole (DAPI) (Sigma-Aldrich) in PBS.
4. Nail polish.

3 Methods

Carry out all procedures at room temperature, unless otherwise specified.

3.1 Coupling of JA to BSA

1. Incubate the JA solution (*see* Subheading 2.2) together with the EDC solution (*see* Subheading 2.2) for 30 min.
2. Add the JA–EDC solution in 0.1 mL aliquots while stirring to the BSA solution.
3. Incubate under continuous stirring at room temperature for 2 h and subsequently at 4 °C overnight.

4. To purify the JA–BSA conjugate (all steps at 4 °C), put the mixture into dialysis tubing and dialyze it against 100 mL of 0.1 M Tris–HCl (pH 5.5) for 3 h and against 100 mL H_2O for 1 h, followed by four cycles alternating 250 mL of 0.1 M sodium borate (pH 8.5) and 250 mL H_2O for 12 h each.
5. Transfer the conjugate solution into Eppendorf tubes and remove possibly occurring precipitates by centrifugation at 9,600 × *g* for 1 min.
6. Add one volume of glycerol and store in 50 μL aliquots at −20 °C until use.

3.2 Coating of Slides

1. Clean slides by incubation with ethanol for 30 min.
2. Dry and clean slides by wiping.
3. Put 60 μL of 0.1 % poly-L-lysine (Sigma-Aldrich) on one slide and place a second slide on top (arranged "face by face"), making sure that the solution is distributed evenly between both slides and without any air bubbles.
4. Incubate in a humid chamber for 30 min (*see* **Note 5**).
5. Separate the slides by dunking them in distilled water.
6. Wash slides with water three times for 5 min each.
7. Remove excess water by shaking and let dry at room temperature (dust-free area).
8. Store slides at −20 °C.
9. Before use, warm slides up and let them dry again.

3.3 Fixation and Embedding of Plant Material

1. Cut plant material in small pieces (approximately 3 × 3 mm for leaf material and of 3–4 mm long for roots) (*see* **Note 6**).
2. Transfer pieces immediately into vials containing the fixative (4 % [v/v] EDC in PBS).
3. To generate positive controls (*see* Fig. 2 and **Note 7**), incubate JA with the fixative for 30 min and use this mixture to fix the specimen.
4. Infiltrate the fixative by vacuum application for 5–10 min.
5. Repeat two to three times until the specimens are completely infiltrated, but use only specimen sunken to the bottom after first vacuum application (*see* **Note 8**).
6. Put the vials on a rotator and allow the specimen fix for 2 h (*see* **Note 9**).
7. Wash the specimen twice with PBS for 15 min each (*see* **Note 10**).
8. Dehydrate the specimen in a graded ethanol series: 10 % EtOH (30 min), 30 % (60 min), 50 % (60 min), 70 % (overnight at 4 °C), 90 % (30 min), and 100 % (30 min, twice).

9. Transfer the vials to 50 °C (all the following have to be done at 50 °C).
10. Infiltrate PEG 1500 as follows: 25 % PEG in ethanol (60 min), 50 % PEG in ethanol (60 min), 75 % PEG in ethanol (60 min), and 100 % PEG (60 min, twice).
11. For the embedding, put some melted PEG in each well of the mold and place each specimen separately into the middle of one well.
12. Let the material harden at room temperature overnight.
13. Store embedded material at 4 °C within closed tubes until use.

3.4 Sectioning and Transfer of Sections to Slides

1. Fix specimen-containing blocks on a holder with melted PEG.
2. Cut semi-thin sections of 3–5 μm thickness with a microtome.
3. Collect sections in paper boxes and store for a few hours only.
4. Perform immunolabeling always on the same day.
5. Transfer the sections to poly-L-lysine-coated slides via the "hanging-drop method." Usually, we use a wire loop and load it with 45 % (w/v) PEG 6000 in PBS (*see* **Note 11**).
6. Transfer the slides into a PBS-containing staining cuvette.
7. Incubate the slides in PBS for 10 min, leading to the removal of PEG from the sections (*see* **Note 12**).

3.5 Immunolabeling and Visualization

1. Incubate the slides in 0.1 M NH_4Cl (in PBS) for 5 min to block free aldehydes (*see* **Note 13**).
2. Wash the slides with PBS for 5 min.
3. Block unspecific binding sites by incubation with 5 % BSA in PBS for 60 min.
4. For incubation with the primary antibody, dilute the anti-JA antibody according to the supplier's instructions in 5 % BSA/1 % BSA_{acet}/PBS and add approximately 200 μL solution per slide.
5. Incubate in a humid chamber (*see* **Note 5**) at 4 °C overnight (*see* **Note 14**).
6. For the negative controls, add 8 μg/mL of JA–BSA conjugate to the diluted antibody.
7. Incubate this mixture for 30 min and use it instead of the anti-JA antibody alone (*see* **Note 15**).
8. Wash the slides with 0.1 % BSA in PBS three times for 10 min each.
9. Wash the slides with 1 % BSA in PBS for 10 min.
10. For incubation with the secondary antibody, dilute anti-rabbit-IgG conjugated with AlexaFluor488 1:500 in 5 % BSA/PBS and add approximately 200 μL solution per slide.

11. Incubate in a humid chamber at 37 °C for 120 min (*see* **Note 16**).
12. Wash slides with PBS four times for 10 min each.
13. As an optional step (*see* **Note 17**), counterstain with DAPI (1 μg/mL in PBS) for 15 min followed by two washing steps with PBS for 10 min each.
14. Put the slides in anti-fading reagent and use nail polish to arrest the coverslip.
15. For examination of the sections after labeling, use a regularly available epifluorescence microscope, for which usually, all the standard filter combinations for the most common dyes are available, among which are AlexaFluor488 (blue excitation, green fluorescence) and DAPI (UV excitation, blue fluorescence).

4 Notes

1. To maximize the buffering capacity, it is recommended to use a volume of fixative larger than that of the tissue to be fixed. This avoids uncontrolled fixative dilution with the cell material.
2. PEG 1500 can be melted easily in a microwave oven. Take care that it does not get too hot, because sometimes residual sugars become brownish. Stir the melted PEG continuously on a heating plate set at 60 °C.
3. Dissolve the appropriate amount of BSA directly in PBS by gentle stirring. The blocking solution can be used several times. We always use BSA for blocking unspecific binding sites. Milk powder commonly used for immunoblots is not recommended due to the high amount of solid particles.
4. The crucial step for the immunolabeling of jasmonates is the availability of a specific anti-JA antibody. To date, there is no anti-JA antibody commercially available. We generated the antibody described [2] by immunization of rabbits with a JA–BSA conjugate. Its specificity was checked by competitive ELISA [2]. However, the company Agrisera (www.agrisera.com) announced an anti-JA antibody to become available in 2012.
5. The humid chamber consists of a closed plastic container with a moist paper towel at the bottom. The container must be large enough to contain the slides in a horizontal position without being in contact with the towel.
6. A very important point is the fast transfer of the specimen into the fixative. To prevent degeneration, drying-out of the material during sampling, and undesired movements of jasmonates within the tissues, the specimens should be prepared directly in

the fixative. This preparation requires special safety conditions, such as working in a fume hood, due to the poisoning activity of the fixative itself. As the samples should be small, the materials have to be dissected with tools (e.g., razor blades, biopsy punches, and scalpels) that are very sharp to avoid mechanical damage of the biological material. In general, to be fast and careful, an intensive training of the sampling procedure is a great advantage.

7. Activation of JA with EDC leads to an efficient binding of JA to proteins as described [2]. Fixation with the mixture of EDC and JA will result in an even distribution and immobilization of JA in all cells, leading to a uniform label after immunostaining.
8. The vacuum infiltration of the fixative into the tissue is a very important and crucial step. Make sure that the fixative can reach all cells in the shortest possible time. If the samples did not sink down after the first vacuum treatment, changes in membrane integrity might lead to unspecific binding of the anti-JA antibody. The same effect is observed after floatation of leaves on aqueous solutions to incubate them with other stressors or hormones. Therefore, it is highly recommended to use only freshly harvested material and to fix it as quickly as possible.
9. The fixation should be done within a few hours to avoid "overfixation."
10. All changes of solutions are done by pipetting the "old" solution/buffer out of the vial and by rapidly adding the next solution/buffer, excluding drying of the specimen in all embedding steps and, most importantly, for solutions containing a high percentage of ethanol (drying) or melted PEG (solidification by cooling).
11. This "hanging-drop method" is crucial for an optimal transfer and sticking of sections to the coated slides. Sections floating on the bottom side of the drop are allowed to stretch and will—after contact of the loop with the slide—immediately connect with the poly-L-lysine coat. Here too, training of this technique is recommended.
12. Due to the solubility of PEG in water, no dewaxing and rehydratation steps are necessary.
13. For all washing steps, we use Hellendahl-type staining cuvettes. One cuvette can take up eight slides in a vertical position. Washing solutions are changed by careful decantation and filling. For one filling, approximately 80 mL of solution is necessary.
14. The reaction with the first antibody can also be carried out at room temperature or at 37 °C for 1 h. However, in our hands,

incubation at 4 °C delivers the best labelings and the lowest background.

15. For a negative staining control, the presaturation of the antibody with its antigen is strongly advised. Therefore, we included the preparation of the JA–BSA conjugate that turned out to be the best agent for this approach. Treatment of the antibody with the JA–BSA conjugate should saturate all JA-binding sites of the antibody that should not bind to the section anymore. When the label is still visible, the antibody binds nonspecifically to some compounds of the tissue.
16. Protect the slides from strong light to prevent fading of the dye.
17. The DAPI staining is optional. DAPI binds to DNA and, therefore, stains mainly the nuclei. Their visualization is sometimes helpful to recognize specific tissues.

Acknowledgments

This work was supported by a grant of the Deutsche Forschungsgemeinschaft (Project HA2655/7-2 within SPP1212) to B.H.

References

1. Dewitte W, Van Onckelen H (2001) Probing the distribution of plant hormones by immunocytochemistry. Plant Growth Regul 33:67–74
2. Mielke K, Forner S, Kramell R, Conrad U, Hause B (2011) Cell-specific visualization of jasmonates in wounded tomato and Arabidopsis leaves using jasmonate-specific antibodies. New Phytol 190:1069–1080
3. Aloni R, Schwalm K, Langhans M, Ullrich CI (2003) Gradual shifts in sites of free-auxin production during leaf-primordium development and their role in vasculat differentiation and leaf morphogenesis in *Arabidopsis*. Planta 216:841–853
4. Hause B, Frugier F, Crespi M (2006) Immunolocalization. In: Mathesius U, Journet E-P, Sumner LW (eds) The Medicago truncatula handbook. The Samuel Noble Foundation, Ardmore, pp 1–11. ISBN 0-9754303-1-9 (http://www.noble.org/MedicagoHandbook/)
5. Pena JTG, Sohn-Lee C, Rouhanifard SH, Ludwig J, Hafner M, Mihailovic A, Lim C, Holoch D, Berninger P, Zavolan M, Tuschl T (2009) miRNA in situ hybridization in formaldehyde and EDC-fixed tissues. Nat Methods 6:139–141
6. Tretner C, Huth U, Hause B (2008) Mechanostimulation of *Medicago truncatula* leads to enhanced levels of jasmonic acid. J Exp Bot 59:2847–2856
7. Browse J (2009) The power of mutants for investigating jasmonate biosynthesis and signaling. Phytochemistry 70:1539–1546
8. Li C, Liu G, Xu C, Lee GI, Bauer P, Ling H-Q, Ganal MW, Howe GA (2003) The tomato *suppressor of prosystemin-mediated response2* gene encodes a fatty acid desaturase required for the biosynthesis of jasmonic acid and the production of a systemic wound signal for defense gene expression. Plant Cell 15:1646–1661

Chapter 12

Jasmonic Acid–Amino Acid Conjugation Enzyme Assays

Martha L. Rowe and Paul E. Staswick

Abstract

Jasmonic acid (JA) is activated for signaling by its conjugation to isoleucine (Ile) through an amide linkage. The *Arabidopsis thaliana* JASMONIC ACID RESISTANT1 (JAR1) enzyme carries out this Mg-ATP-dependent reaction in two steps, adenylation of the free carboxyl of JA, followed by condensation of the activated group to Ile. This chapter details the protocols used to detect and quantify the enzymatic activity obtained from a glutathione-*S*-transferase:JAR1 fusion protein produced in *Escherichia coli*, including an isotope exchange assay for the adenylation step and assays for the complete reaction that involve the high-performance liquid chromatography quantitation of adenosine monophosphate, a stoichiometric by-product of the reaction, and detection of the conjugation product by thin-layer chromatography or gas chromatography/mass spectrometry.

Key words Adenylation, Conjugation, Jasmonic acid, Hormone, GH3 protein, Enzyme

1 Introduction

Jasmonic acid (JA) is an important signaling compound involved in both plant development and defense responses. The active form of JA in signaling reactions is an amino acid conjugate, primarily JA-isoleucine (JA-Ile). The major enzyme responsible for the JA-Ile formation in *Arabidopsis thaliana* is JASMONATE RESISTANT1 (JAR1), also designated GH3.11 [1, 2]. JAR1 is a member of the firefly luciferase superfamily that forms an adenylated JA intermediate prior to condensation to the amino acid (Fig. 1).

The GH3 family is widely distributed in plants and other enzymes of the GH3 family conjugate amino acids to other plant hormones, especially indole-3-acetic acid [3], or small carboxylic acids [4]. The methods presented here for the JAR1 assay would be applicable to the related enzymes with minor modifications. In addition to a rapid qualitative assessment of conjugate synthesis by thin-layer chromatography (TLC) [2], we have used two procedures to quantify the conjugation activity: one measures the production of the

Alain Goossens and Laurens Pauwels (eds.), *Jasmonate Signaling: Methods and Protocols*, Methods in Molecular Biology, vol. 1011, DOI 10.1007/978-1-62703-414-2_12, © Springer Science+Business Media, LLC 2013

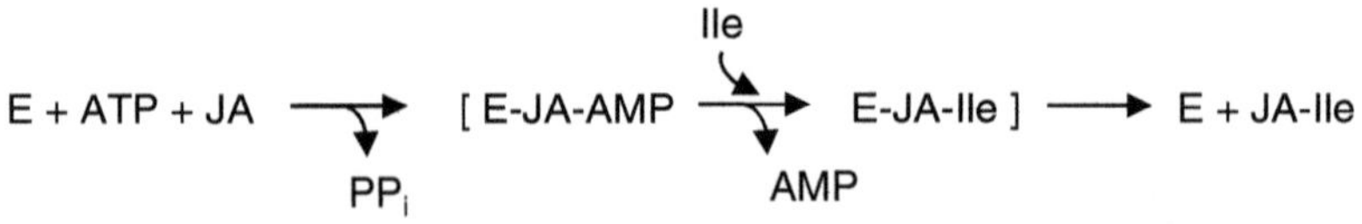

$$E + ATP + JA \xrightarrow[PP_i]{} [\,E\text{-}JA\text{-}AMP \xrightarrow[AMP]{Ile} E\text{-}JA\text{-}Ile\,] \longrightarrow E + JA\text{-}Ile$$

Fig. 1 Mechanism of JAR1 enzyme activity. The enzyme performs two distinct steps, adenylation of the carboxyl group of JA followed by condensation of the amino acid to the activated carboxyl of JA. *JA* jasmonic acid, *E* enzyme, *Ile* isoleucine

conjugation product by gas chromatography/mass spectrometry (GC/MS) [5] and the other the production of adenosine monophosphate (AMP) by high-performance liquid chromatography (HPLC) [6]. We also included an adenosine triphosphate (ATP)-pyrophosphate (PPi) isotope exchange protocol that detects the enzyme-adenylating activity by carboxylic acid-dependent exchange of ^{32}P-PPi into ATP (substrate + Mg-ATP ↔ substrate-AMP + PPi) [1]. This assay can be used to test the adenylating activity of an enzyme on carboxylic acids without requiring the knowledge of which amino acid the enzyme prefers [1]. An enzyme-coupled assay for adenylation has been described elsewhere [4].

The glutathione-*S*-transferase (GST):*JAR1* construct made from JAR1 cDNA makes production and purification of proteins expressed in *Escherichia coli* simple and fast. Alternate protein expression systems could be used as well. The fusion protein is isolated by binding with glutathione-agarose beads. The target enzyme activity can be assessed qualitatively with the fusion protein still bound to the glutathione-agarose beads or eluted proteins of known concentrations can be used for kinetic studies.

It should be noted that the JAR1 enzyme of both *Arabidopsis* and tomato (*Solanum lycopersicum*) have a strong preference for (3R,7S)-JA, also called (+)-7-iso-jasmonic acid, which is the isomer synthesized in plants [5]. However, this *cis*-isomer is unstable and constitutes only a small fraction of commercially available preparations of (±)-JA. Therefore, the kinetic activity will probably be underestimated by using the isomeric mixture as a substrate. Although (3R,7S)-JA is not generally commercially available, it can be prepared through a laborious process [5].

2 Materials

In our laboratory, a pGEX-4T expression vector in *E. coli* BL21 cells is used according to the general procedures outlined by the supplier (GE Healthcare, Little Chalfont, UK). Prepare solutions in ultrapure water and use HPLC-grade reagents for HPLC and GC/MS. Dispose of all hazardous materials appropriately.

2.1 Glutathione-Agarose

1. 1× phosphate-buffered saline (PBS) (140 mM NaCl, 2.7 mM KCl, 10 mM Na_2HPO_4, 1.8 mM KH_2PO_4): 8.2 g NaCl, 0.2 g KCl, 2.68 g $Na_2HPO_4·7H_2O$, and 0.24 g KH_2PO_4 added to 1 L of water. The pH should approximately be 7.3 without adjustment.
2. Glutathione agarose (Sigma-Aldrich, St. Louis, MO, USA): For preparation, *see* Subheading 3.1.

2.2 E. coli Growth and Sonication

1. LB medium: 25 g/L LB Miller (EMD Millipore, Billerica, MA, USA). In 1-L flasks, place 250 mL LB medium and autoclave.
2. Ampicillin: 50 mg/mL stock, filter sterilized before use and aliquoted into sterile microfuge tubes for storage at −20 °C. Add 250 μL of 50 mg/mL ampicillin to 250 mL LB medium for a final concentration of 50 μg/mL.
3. 1 M isopropyl β-D-1-thiogalactopyranoside (IPTG): Add 2.38 g of IPTG to 10 mL of water and vortex until dissolved. Filter sterilize and store at −20 °C.
4. Sonicator: 130-W high-intensity ultrasonic processor with 6-mm diameter probe (Autotune Model GE130; Sonics and Materials, Newtown, CT, USA) (*see* Subheading 3.3 for use).

2.3 Protein Purification and Elution

1. 20 % (v/v) Triton X-100: 4 mL of Triton X-100 dissolved in 16 mL water (*see* **Note 1**).
2. 10 mM reduced glutathione in 50 mM Tris–HCl (pH 8): Prepare 50 mM Tris–HCl (pH 8) from 1 M stock solution or dissolve 0.12 g Tris base in 15 mL water, adjust pH to 8.0, and bring to final volume of 20 mL. Add 0.06 g reduced glutathione (Sigma-Aldrich) to 20 mL of 50 mM Tris–HCl (pH 8), dissolve, and filter sterilize before storing at 4 °C.

2.4 Isotope Exchange Assays

1. 1 M Tris–HCl (pH 8.0): Dissolve 6.06 g Tris base in 40 mL water, adjust pH to 8.0, and bring to final volume of 50 mL.
2. 25 mM ATP: Dissolve 0.552 g ATP in 10 mL water for a 100-mM stock solution and dilute with water to obtain 25 mM ATP. Divide into small aliquots in microfuge tubes and store at −20 °C (*see* **Note 2**).
3. 25 mM $MgCl_2$: Dissolve 0.092 g $MgCl_2$ in 10 mL water for a 100-mM stock solution and dilute with water to obtain 25 mM.
4. 5 mM JA: Dissolve 21 mg (±)-JA (Sigma-Aldrich) in 1 mL of 100 % ethanol to make a 100-mM stock solution and dilute this stock solution with ethanol.
5. 100 mM dithiothreitol (DTT): Dissolve 0.55 g DTT in 10 mL water.
6. 10 mM pyrophosphate (PP_i): Dissolve 0.045 g $NaP_2O_7·10H_2O$ (Sigma-Aldrich) in 10 mL water.

7. ^{32}P-PP$_i$: Dilute ^{32}P-tetrasodium pyrophosphate (1 mCi at 10 μCi/μL) to 10× with water for a final activity of 1 μCi/μL.
8. PEI-F cellulose TLC Sheets: 20×20 cm sheets can be purchased (J.T. Baker Baker-flex Pre-Coated Flexible TLC Sheets, PEI-F Cellulose; Avantor Performance Materials, Center Valley, PA, USA) and cut into strips (10 cm long) for TLC. Spots can also be cut out of developed chromatograms for quantitative scintillation counting.
9. X-ray film: Kodak Bio-Max MS film or equivalent.
10. Chromatogram developer (1.9 M formic acid and 1 M LiCl): Dissolve 4.24 g LiCl in 80 mL water, add 8 mL of 88 % formic acid, and bring to a final volume of 100 mL.

2.5 JA–Amino Acid-Conjugating Enzyme Assays

1. 50 mM Tris–HCl (pH 8.5): Dissolve 0.30 g Tris in 40 mL water. Adjust pH to 8.5 and bring to a final volume of 50 mL.
2. 25 mM ATP: *See* Subheading 2.4.
3. 25 mM $MgCl_2$: *See* Subheading 2.4.
4. 25 mM JA: *See* Subheading 2.4.
5. 2 mM amino acids for most amino acid solutions: Dissolve 0.05 mmol amino acid in 1 mL water to obtain a 50-mM stock solution. For a few amino acids that are not readily soluble in water (aspartic acid, glutamic acid, and tyrosine), dissolve them in a minimum amount of 1 N NaOH, and dilute with water to a final volume of 1 mL for the stock solution. Cysteine should optimally always be made fresh because it converts to cystine in solution. Make 2 mM solutions by adding 40 μL of the 50 mM stock to 960 μL water.
6. 100 μM AMP: Dissolve 0.0035 g AMP in 1 mL of water to make a 10-mM stock and dilute with water to obtain 100 μM.
7. Stop buffer: Add 0.34 g KH_2PO_4 to 50 mL water to make a 50-mM solution.

2.6 Thin-Layer Chromatography

1. Silica gel plates: TLC Silica Gel 60F$_{254}$ glass plates, 20×20 cm (EMD Millipore).
2. Developing solution: Ethyl acetate (55 %), chloroform (35 %), formic acid (10 %) [v/v].
3. Vanillin reagent: Dissolve 3 g vanillin in 50 mL ethanol and add 500 μL H_2SO_4.
4. We use a tube-type chromatography sprayer (Sigma-Aldrich) to wet the TLC plate.

2.7 HPLC Detection of AMP

All solutions should be filtered prior to use (Whatman nylon membrane filters, 0.45 μm, 47 mm diameter) (GE-Healthcare).

1. Ultra IBD reverse-phase column, length, 150 mm; diameter 4.6 mm dia; particle size 5.0 μm (Restek, Bellefonte, PA, USA).
2. 100-μL limited-volume polypropylene HPLC vials (Restek) or regular vials with microvolume insert.
3. Solution A: 50 mM potassium phosphate (pH 5.5) with 0.01 % sodium azide to impede microbial contamination.
4. Solution B: 50 % methanol.

2.8 Assay and Sample Preparation for GC/MS

1. 0.5 M Tris–HCl (pH 8.8): Dissolve 3.02 g Tris base in 40 mL water. Adjust pH to 8.8 and bring final volume to 50 mL.
2. 10 mM JA (*see* Subheading 2.4).
3. 100 mM ATP, 100 mM $MgCl_2$, and 100 mM DTT (*see* Subheading 2.4).
4. 5 mM amino acid (*see* Subheading 2.4).
5. Internal standards: Will vary with the specific amino acid to be assayed. We use a stable isotope of Ile to synthesize $[^{13}C_6]$-JA-Ile, or amino acid conjugates can be synthesized with dihydrojasmonic acid with a mixed anhydride reaction [2].
6. 2 M (trimethylsilyl)diazomethane (Sigma-Aldrich).
7. Isooctane (2,2,4-trimethylpentane; EMD Millipore).

3 Methods

3.1 Preparation of Glutathione-Agarose

1. Add a small amount (approximately 0.5 mL) of dry glutathione-agarose to a clean 15-mL polypropylene conical centrifuge tube and add approximately 5 mL cold 1× PBS buffer.
2. Thoroughly suspend the glutathione-agarose beads in the buffer and allow to hydrate overnight at 4 °C to obtain a final volume of 1–2 mL. Centrifuge (250 × *g*) the hydrated beads briefly in a benchtop centrifuge and discard the PBS supernatant.
3. Add fresh PBS (5–10 mL), resuspend beads, and centrifuge again.
4. Repeat washing of the glutathione-agarose beads three to four times.
5. After final wash, leave enough PBS buffer so that the bead bed volume is approximately equal to the free buffer volume for a 50 % suspension that can be kept refrigerated for at least 2 weeks.

3.2 Growth and Harvest of Cells

1. Inoculate 5 mL of LB broth supplemented with 50 μg/mL ampicillin to the pGEX expression vector-containing *E. coli.*
2. Grow overnight at 37 °C with shaking.

3. In the morning, add the 5 mL of overnight culture to 250 mL of the 25-μg/mL ampicillin-containing LB medium in a 1-L flask.
4. Grow at 37 °C with shaking for approximately 2 h, until $OD_{600} = 0.6–0.8$.
5. Induce production of the fusion protein by adding IPTG to a final concentration of 0.1 mM and by shaking at room temperature for 5–6 h.
6. Pour culture into a 250-mL polypropylene centrifuge bottle and centrifuge for 10 min at 5,000 × *g*.
7. Discard the supernatant back into the culture flask for autoclaving.
8. Drain the bottle upside down briefly on paper towels.
9. Freeze bottles with bacterial pellets for later use or transfer them to an ice bucket for immediate use.

3.3 Cell Sonication

1. Add 10 mL of cold PBS to each pellet generated in Subheading 3.2.
2. Keep cells on ice and resuspend pellets gently by agitating the ice bucket on a rotary shaker at low speed or by using a pipettor to break up the pellet.
3. Transfer the cell suspensions to cold clean 50-mL open-mouth polypropylene centrifuge tubes and keep in the ice bucket.
4. Prepare a 250-mL beaker filled with ice and a little water.
5. To disrupt the bacterial cells and release the produced proteins without damage, sonicate with settings as follows: amplitude, 40 %; pulse, 3 s; and time, 3 min (*see* **Note 3**).
6. Put centrifuge tube with suspended cells in beaker of ice/water.
7. Rinse probe with water.
8. Lower probe into the tube so that it does not quite touch the bottom (*see* **Note 3**).
9. Push START button on the machine with pulse *ON* for 3 s, and *OFF* for 1 s continually for 3 min (*see* **Note 4**).
10. Repeat procedure with the remaining samples.
11. Rinse probe with 70 % ethanol.
12. Turn dials OFF, and then power OFF.
13. Add 500 μL of 20 % Triton X-100 to each tube (1 % final concentration).

3.4 Binding Fusion Protein to Glutathione-Agarose

1. Spin sonicated cells in a refrigerated high-speed centrifuge at 12,000 × *g* for 10 min at 4 °C.
2. With a 200-μL pipettor with trimmed tip, transfer 100–200 μL of washed glutathione-agarose (50 % suspension) into a 15-mL conical polypropylene centrifuge tube with cap.

3. Pour the supernatant carefully into the conical tube with the glutathione-agarose suspension.
4. Close tube tightly and place on rotator for 30 min (*see* **Note 5**).
5. Spin tube briefly in a benchtop centrifuge at approximately 250 × *g*, just long enough to pull down the glutathione-agarose beads.
6. Pour off supernatant into the used empty centrifuge bottle.
7. Add 1 mL of cold 1× PBS to the tube and transfer the entire volume into a 1.5-mL microfuge tube for further washings.
8. Centrifuge the microfuge tube briefly and discard the supernatant.
9. Add 1 mL of cold 1× PBS to the tube, resuspend the beads, and centrifuge briefly (250 × *g*).
10. Discard the supernatant and repeat washings with 1× PBS three times. Beads can be stored at 4 °C for at least 2 weeks, as long as microbial growth does not occur.
11. Check the binding of the protein to the beads on sodium dodecyl sulfate-polyacrylamide gel electrophoresis (SDS-PAGE) gels (*see* **Notes 6** and **7**).

3.5 Elution of Bound Proteins with Reduced Glutathione

1. After a final 1× PBS wash of the fusion protein-bound glutathione-agarose beads, remove as much buffer as possible from the beads (*see* **Note 8**).
2. Add a volume of reduced glutathione (10 mM in 50 mM Tris, pH 8 buffer) approximately equal to the bed volume of the beads.
3. Put on rotator for 10 min.
4. Open lid of microfuge tube and hold it over another clean microfuge tube (*see* **Note 9**).
5. Very carefully puncture a very small hole at the end of the bead-containing tube with the sharp end of a jeweler's forceps or syringe needle.
6. Set this tube into the clean tube and tape them together.
7. In the benchtop centrifuge with whatever adaptor that will hold them, spin at approximately 500 × *g* for approximately 30 s, until the agarose beads in the top tube are white and look dry (*see* **Note 10**).
8. Use the eluted fusion protein for enzyme assays. Purified enzymes can be stored at 4 °C for a few days or indefinitely in a freezer, although solubility of individual proteins after freezing should be tested first.

3.6 Isotope Exchange Assay

1. Prepare the reaction buffer minus the $^{32}P\text{-}PP_i$ as follows: 75.4 μL of water, 10.0 μL of 1 M Tris–HCl (pH 8.0), 5.6 μL of 25 mM $MgCl_2$, 5.6 μL of 25 mM ATP, 2.0 μL of 100 mM DTT, and 4 μL of 10 mM PP_i.

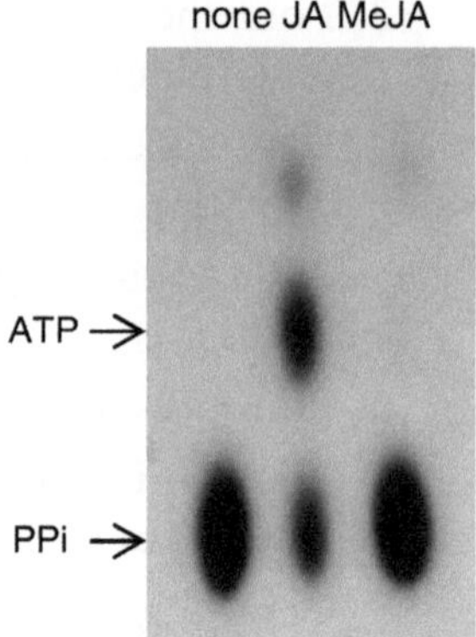

Fig. 2 Autoradiogram of a thin-layer chromatograph from the isotope exchange assay. Origin is at the bottom and the ATP and PPi positions are indicated by *arrows*. The reactions contained JA, the methyl ester of JA (MeJA), or no carboxylic acid as substrate, as indicated above. MeJA is unreactive because it lacks a free carboxyl group

2. Add 5 μL of 5 mM JA to the reaction tube.
3. In the isotope work area, add 1–2 μL of $^{32}P\text{-}PP_i$ (1 μCi/μL) for every 100 μL of reaction buffer.
4. Add 10 μL of reaction buffer with the isotope to each reaction tube.
5. Add 5 μL of glutathione-agarose-fusion protein or eluted fusion protein to each reaction tube.
6. Let the reaction proceed for 10 min. If the bead-bound enzyme is used, gently flick tubes occasionally during the reaction to resuspend the agarose beads that have settled to the bottom.
7. Mark PEI-F Cellulose strips with a pencil by placing dots where the reaction mixes will be spotted, approximately 1.5 cm from the bottom edge; place another dot on the right side of the strip for orienting the thin-layer strip on the X-ray film.
8. At the end of the reaction time, spot 1.5 μL of the reaction mixtures on the dots at the bottom of PEI-F Cellulose strip.
9. Use a small amount of the last reaction mixture to spot the orienting dot on the right side of the strip.
10. Let the thin-layer strip dry.
11. Develop the strip in 1.9 M formic acid and 1 M LiCl.
12. Let solvent rise to within approximately 2 cm of the strip top in approximately 20 min.
13. Remove the strip and let dry on paper towel.
14. Wrap the thin-layer strip in a plastic wrap and expose to X-ray film for approximately 3 h.
15. Develop film (Fig. 2) which is sufficient for qualitative activity assessment.

16. For quantitative activity measurements, use the developed X-ray film as a guide to mark the location of the radioactive spots that can be cut out and counted in a scintillation counter.

3.7 Enzyme Conjugation Reactions

This procedure is based on the use of eluted fusion proteins for the enzyme assay. Products can be detected by TLC (*see* Subheading 3.8) or, because AMP is a product of the conjugation reaction, the released AMP can be quantified by HPLC (*see* Subheading 3.9). For TLC product detection, reactions can also be done with enzymes bound to the agarose beads (*see* **Note 11**). For HPLC, AMP control reactions should be run with each set of conjugation assays to develop a standard curve that compensates for possible AMP metabolism during the reactions.

To assess the conjugating enzyme activity with different amino acids, the following reaction is set up, but the procedure may be modified to measure the activity at different amino acid or hormone concentrations.

1. Immediately before use, prepare a stock solution for four reactions by mixing together 30 μL of water, 25 μL of 100 mM Tris–HCl (pH 8.5), 4 μL of 5 mM $MgCl_2$, 4 μL of 5 mM ATP, 2 μL of 25 mM JA, and 10 μL of enzyme preparation. For each reaction, 18 μL of the stock solution plus 6 μL of 2 mM amino acid will be needed.
2. For control reactions, substitute water for the amino acid; for the AMP measurement by HPLC, substitute the amino acid with dilutions of the 100 μM AMP preparation to construct a standard curve.
3. Place 6 μL of the variable substrate (amino acid or AMP standard) in a microfuge tube.
4. Add 18 μL of the reaction mix at time 0 (*see* **Note 12**).
5. At the end of the desired reaction time, add 96 μL of stop buffer (50 mM KH_2PO_4) and vortex both to dilute the reaction and lower the pH to a point where the reaction will essentially stop (*see* **Note 13**).
6. For analysis by TLC, spot the reaction directly to the plate without addition of stop buffer to maintain the product concentration.

3.8 Thin-Layer Chromatography

1. Use a pencil to mark spot locations on a silica gel plate approximately 1.5 cm from the bottom edge.
2. Apply 2 μL of the reaction mixture to a spot and let dry.
3. Reapply 2 μL twice more for a total of 6 μL (*see* **Notes 14** and **15**).
4. When the plate is dry, develop in a mixture of ethyl acetate:chloroform:formic acid (55:35:10 %).

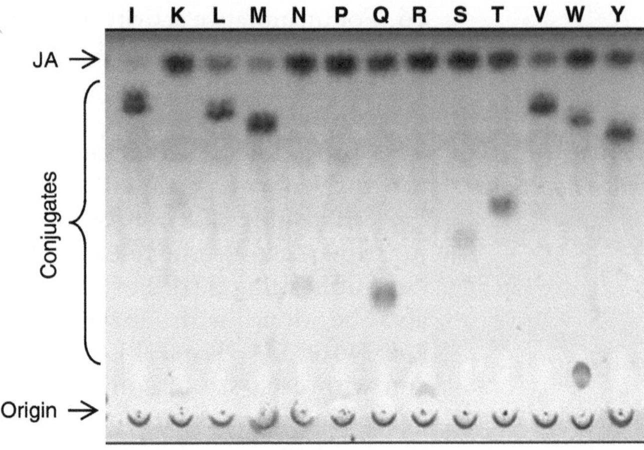

Fig. 3 Thin-layer chromatogram of JAR1 enzyme reactions with JA and indicated amino acids. The origin and migration point for JA are indicated with *arrows*, where JA conjugates migrate to positions below JA. Individual amino acids included in each enzyme reaction are indicated with standard single-letter abbreviations

5. Remove the plate from the developing tank and allow to dry approximately for 5 min.
6. Spray with vanillin reagent, blot lightly, and bake at 100 °C for 10–15 min (*see* **Note 16** and Fig. 3).

3.9 Reaction Quantification of AMP Production by HPLC Detection

1. Place the reaction samples (120 μL) in microvolume HPLC vials, making sure that no air bubbles are caught at the vial bottom.
2. Set injection volume of HPLC at 50 μL, detector wavelength at 260 nm, and pump at 1 mL per minute. Use the following HPLC cycle to separate AMP from ATP and other possible products: 90 % solution A and 10 % solution B isocratic for 5 min, 100 % solution B for 5 min, and 90 % solution A and 10 % solution B for 10 min (*see* **Note 17**).
3. Wash the column and equilibrate for the next run.
4. Use the area under the AMP peak with a standard AMP curve to calculate the enzyme activity or the fraction (AMP/AMP + ATP) (*see* **Note 18**).

3.10 Reaction Quantification of Conjugates by GC/MS Detection

1. For each reaction, prepare 130 μL of 2× reaction mix: 80 μL of water, 20 μL of 0.5 M Tris–HCl (pH 8.8), 20 μL of 10 mM (−)JA, 4 μL of 100 mM $MgCl_2$, 4 μL of 100 mM ATP, and 2 μL of 100 mM DTT.
2. Prepare the stop solution: 540 μL of water, 90 μL of glacial acetic acid, and 45 μL of the appropriate internal standard at the desired concentration.

3. Use a known quantity of internal standard to account for any sample loss during the subsequent processing.
4. Add 50 μL of 5 mM amino acid to 130 μL of 2× reaction mix.
5. Start reaction by adding 20 μL of the enzyme; mix well, and note the time.
6. At each desired time point, withdraw 45 μL of the assay mixture into the tube containing 150 μL of stop solution and vortex.
7. Add 200 μL of chloroform to each sample, vortex, centrifuge (250 × *g*) briefly, and pipette chloroform (lower phase) into a 6-mL disposable glass tube (100 × 13 mm).
8. Repeat the chloroform extraction of the assay mixture and combine with the first.
9. Dry sample under N_2 in fume hood (*see* **Note 19**).
10. Add 100 μL of 100 % methanol.
11. Vortex well and roll tube slightly to make sure that the sample is fully dissolved from the tube walls.
12. In a fume hood, add 5 μL of 2 M (trimethylsilyl)diazomethane (derivatizing agent); cover the sample with parafilm and leave for 30 min.
13. Add 1 μL of glacial acetic acid to stop the reaction and inactivate the remaining derivatizing agent.
14. Dry the sample again under N_2 and resuspend it in 100 μL isooctane.
15. Vortex and roll tubes slightly.
16. Transfer the sample to a glass vial with a vial insert appropriate for the GC instrument used.
17. For the quantification by GC/MS, follow the procedure depending on the available instruments. The Finnigan Trace GC with a DB-5ht column (15 m, 0.25 mm, 0.1 L) coupled to a Finnigan DSQ mass spectrometer was operated in electron ionization mode; the injector port was at 280 °C and the column gradient was 60–300 °C in 25 min. The amount of JA-Ile was obtained from the molecular ion-integrated peak areas with the isotope dilution method described previously [7].

4 Notes

1. First, a gel-like product will be formed, but heating and gentle agitation will eventually bring it into solution.
2. ATP can break down after multiple freeze–thaw cycles. It is best to prepare a large amount and to subdivide it into small portions to be frozen and used as needed.

3. Use ear protection when operating the sonicator.
4. At the end of the sonication, the suspension in tubes should be slightly paler because the cells have burst open.
5. Fusion proteins with GST moiety will bind to the glutathione bound to the agarose beads.
6. To check the protein availability and approximate amounts, use a trimmed pipette tip to transfer 10 μL of the fusion protein-bound beads to a microfuge tube. Add an equal amount of 2× SDS-PAGE loading buffer and boil for 5 min. Place on ice until ready to load on an SDS-PAGE gel. The fusion protein will now be free of the beads and only the liquid need to be loaded.
7. A rough measure of the enzyme activity can also be determined while the protein is still bound to the glutathione-agarose beads (*see* Subheading 2.7), but quantification of the protein is difficult. For specific activities, the protein must first be eluted from the beads.
8. The volume of the beads should be approximately half the volume of the suspension originally used.
9. Be sure to open the microfuge tube before puncturing. Positive pressure from warming may have built up in the tube and the enzyme-containing liquid will leak out as the hole is punctured.
10. If the beads have come into the catch tube, the hole was too large.
11. A more rapid qualitative endpoint assessment of the enzyme activity or substrate preferences can be done with the GST fusion protein still attached to the glutathione-agarose beads. After Subheading 3.4, **step 5**, and without the final centrifugation, allow beads to settle for a few minutes, and remove most of the PBS so that only beads loosely remain in the solution. Add 5 μL of the suspended beads to the 20-μL reactions in a 1.5-mL conical centrifuge tube. The end of the micropipette tips may need to be cut to produce a hole large enough for beads to enter the tips. Place the tube on a slowly rotating tube mixer so that the long axis of the microfuge tube is parallel with the rotation axis. For this purpose, pieces of foam with holes for tubes have been attached to the rotator. The small sample volume will stay at the "bottom" of the tube, while the beads will tumble into the solution, ensuring that the enzyme is mixed with the solution over an extended time, as desired. For endpoint assays, we often incubate the reaction for several hours or overnight.
12. Addition of the reaction mix to the subsequent tubes is spaced at fixed intervals of approximately 15–30 s.

13. The reaction time should be determined empirically when doing kinetics so that product formation is proportional to enzyme concentration.
14. If too much assay mixture is applied to the TLC plate at once, the spot becomes too large and diffuse; hence 2 μL is applied at a time after drying between applications.
15. JA and the desired conjugates can be used as standards on the plate.
16. JA will move near the solvent front and will appear as a dark brown or a purple spot. Conjugates will have lower retention values, depending on the amino acid (Fig. 3).
17. Exact parameters may need to be determined for other instruments and columns.
18. A typical chromatogram will show ATP coming off the column early, followed by AMP.
19. Placing the tubes into a 50 °C water bath or dry bath will speed the solvent evaporation.

Acknowledgments

This research has been supported in part by funds from the Hatch Act and additionally by the National Science Foundation (Award IOS-0744758).

References

1. Staswick PE, Tiryaki I, Rowe ML (2002) Jasmonate response locus *JAR1* and several related Arabidopsis genes encode enzymes of the firefly luciferase superfamily that show activity on jasmonic, salicylic, and indole-3-acetic acids in an assay for adenylation. Plant Cell 14:1405–1415
2. Staswick PE, Tiryaki I (2004) The oxylipin signal jasmonic acid is activated by an enzyme that conjugates it to isoleucine in Arabidopsis. Plant Cell 16:2117–2127
3. Staswick PE, Serban B, Rowe M, Tiryaki I, Maldonado MT, Maldonado MC, Suza W (2005) Characterization of an Arabidopsis enzyme family that conjugates amino acids to indole-3-acetic acid. Plant Cell 17:616–627
4. Okrent RA, Brooks MD, Wildermuth MC (2009) *Arabidopsis* GH3.12 (PBS3) conjugates amino acids to 4-substituted benzoates and is inhibited by salicylate. J Biol Chem 284: 9742–9754
5. Suza WP, Rowe ML, Hamberg M, Staswick PE (2010) A tomato enzyme synthesizes (+)-7-*iso*-jasmonoyl-L-isoleucine in wounded leaves. Planta 231:717–728
6. Guranowski A, Miersch O, Staswick PE, Suza W, Wasternack C (2007) Substrate specificity and products of side-reactions catalyzed by jasmonate:amino acid synthetase (JAR1). FEBS Lett 581:815–820
7. Cohen JD, Baldi BG, Slovin JP (1986) $^{13}C_6$-[benzene ring]-indole-3-acetic acid. A new internal standard for quantitative mass spectral analysis of indole-3-acetic acid in plants. Plant Physiol 80:14–19

Chapter 13

Pull-Down Analysis of Interactions Among Jasmonic Acid Core Signaling Proteins

Sandra Fonseca and Roberto Solano

Abstract

Pull-down assays are key tools to test specific protein–protein interactions and have been particularly fruitful in jasmonate signaling research. Here, we describe a standard protocol in which a matrix-bound "bait" protein, expressed in *Escherichia coli*, pulls down a "prey" protein that is soluble in a protein extract obtained from *Arabidopsis thaliana* plant tissues. The pulled-down protein can be detected by immunoblot with protein-specific or epitope-specific antibodies. Such a pull-down method was used to reveal interactions among components of the jasmonic acid signaling, including hormone-dependent coreceptor interaction, homodimerization and heterodimerization of JASMONATE ZIM DOMAIN repressors, and interactions among other corepressor components and with transcription factors. Pull-down assays contributed not only to shape this signaling pathway but also to identify the active jasmonate hormone.

Key words Pull-down, Protein–protein interaction, *E. coli* protein expression, MBP fusion, Hormone-dependent interaction, Jasmonate-isoleucine, Jasmonic acid, Coronatine, COI1, JAZ

1 Introduction

Pull-down assays are a versatile and straightforward in vitro technique that allows the detection or confirmation of protein–protein interactions. These assays are a kind of affinity purification that relies on the fusion of the "bait" protein to a matrix affinity tag that immobilizes the "bait" protein into the matrix. Once immobilized, the "bait" protein is incubated with a protein extract or another purified protein and captures the "prey" protein by matrix centrifugation. In this protocol, we also describe an *Escherichia coli* protein expression method of the maltose-binding protein (MBP)-fused "bait" protein that is purified by affinity to an amylose resin. The MBP-"bait" fusion protein is incubated with a transgenic plant extract overexpressing a "prey" protein of interest, also fused to a tag, in this case, FLAG. Amylose beads containing the MBP-"bait"–"prey"-FLAG complex (when the interaction is positive) will be recovered by centrifugation and washed from non-bound

Alain Goossens and Laurens Pauwels (eds.), *Jasmonate Signaling: Methods and Protocols*, Methods in Molecular Biology, vol. 1011, DOI 10.1007/978-1-62703-414-2_13, © Springer Science+Business Media, LLC 2013

proteins in the extract. After denaturation, these proteins are separated by sodium dodecyl sulfate-polyacrylamide gel electrophoresis (SDS-PAGE) and the protein of interest can be detected by immunoblot, with an anti-FLAG antibody.

To perform a pull-down assay, two major issues concerning the "bait" and "prey" protein expression and detection should be considered. First, an expression and immobilization system of the "bait" protein has to be defined. Here, we use expression in *E. coli* because it is a rapid and simple method, but alternative eukaryotic expression systems, such as insect cells or *Xenopus* oocytes, might be required for some proteins that could misfold in prokaryotic systems or need posttranslational modifications. Although we use an MBP fusion system of the "bait" protein, other commercially available systems, such as hexahistidine (6× His), glutathione-*S*-transferase (GST), or mixed tag-combining systems, are commonly used [1–3].

Second, the expression and detection method of the "prey" protein has to be considered. When a "prey"-specific antibody is not available to directly detect the protein in plant extracts, "preys" (fused to specific tags) can be produced *in planta*, synthesized in vitro by coupled transcription and translation systems by using commercial kits, or expressed in heterologous systems, such as *E. coli*, *Xenopus* oocytes or insect cells [1, 2, 4–6]. We recommend transgenic expression *in planta* to avoid misfolding or misregulation that could occur in vitro or in heterologous systems. In this case, "preys" can be directly used as a protein extract or be preimmunopurified.

For detection, we used a FLAG tag [7] that allows immunodetection with a commercial antibody, but other suitable tags for transgenic expression are hemagglutinin (HA), green fluorescent protein (GFP), or c-Myc [6, 8, 9], against which many commercial antibodies are available. For expression *in planta* it is indispensable to have either a specific antibody for the "prey" protein or an *Arabidopsis* transgenic line [10]. When the "prey" is synthesized in vitro by coupled transcription and translation, it can be radiolabeled (^{35}S-Met) during the synthesis and readily be detected by autoradiography [4, 11], whereas purified proteins from heterologous expression systems might be detected by Coomassie Brilliant Blue or Silver staining (only when large protein amounts can be recovered).

Over the last 5 years, pull-down in combination with yeast two-hybrid assays shed light on the molecular interactions among genetically identified components of the jasmonic acid (JA) signaling pathway. Pull-down assays were a powerful tool to disclose the hormone dose-dependent interaction between the JASMONATE ZIM-DOMAIN (JAZ) repressors and the F-box protein CORONATINE INSENSITIVE 1 (COI1), suggesting JAZ-COI1 as a JA-Ile (jasmonate-Isoleucine) coreceptor [2, 3, 8, 12, 13].

The mechanism of JA-Ile perception turned out to be highly similar to that of auxin perception, which was also analyzed by pull-down experiments and served as inspiration for our experimental design [1, 4, 5, 14]. Regarding the JA signaling core, pull-down assays were important for almost all discoveries and provided evidence for molecular interactions between JAZ proteins and basic helix-loop-helix (bHLH) transcription factors (such as MYC2, MYC3, and MYC4), supporting the role of JAZs as repressors of the JA signaling pathway by inhibiting the MYC activity [4, 8, 9, 11]. In combination with yeast two-hybrid assays, pull-down experiments demonstrated that JAZ proteins could associate to form homodimers and heterodimers and are part of a repressor complex by interacting with the NOVEL INTERACTOR OF JAZ (NINJA) [6, 8]. Thanks to the possibility of easy "bait" expression, a number of truncated JAZ proteins were generated, allowing the identification of the JAZ domains that interact with the above-described interactors. Paradoxically, upon discovery of the JA core signaling module, a series of pull-down assays were a key tool to demonstrate that the real hormone that activates the pathway was not JA itself, but the (+)-7-iso-JA-Ile molecule [2, 13, 14]. Mostly, full-length or truncated JAZ proteins were expressed as MBP fusions and functioned as efficient "bait" proteins, whereas COI1, JAZ, MYC, and NINJA proteins, expressed in *Arabidopsis* and fused to different tags, were commonly used as "prey" proteins. The protocol presented here uses JAZ as the "bait" and COI1 as the "prey," not only because this was the first pair of proteins for which a pull-down assay was optimized in our laboratory, but also because they have the fascinating particularity to interact depending on the presence of a hormone.

2 Materials

All solutions should be prepared with ultrapure water and stored at room temperature, unless otherwise stated.

2.1 *E. coli* Protein Expression and Purification

1. pDESTH1 vector [15], a Gateway derivative of pMALC (New England Biolabs, Ipswich, MA, USA) (*see* **Note 1**).
2. *Escherichia coli* BL21 (DE3) pLysS (Invitrogen) competent cells.
3. Luria Bertani (LB) medium (BD, Franklin Lakes, NJ, USA): Dissolve 20 g in 1 L of ddH_2O, autoclave at 120 °C for 15 min.
4. Antibiotics: Ampicillin: Prepare a 100 mg/mL solution in water, sterilize by filtration, aliquot, and store at −20 °C; chloramphenicol: Prepare a 25 mg/mL solution in ethanol, aliquot, and store at −20 °C.

5. Isopropyl-β-D-thiogalactopyranoside (IPTG) (Sigma-Aldrich, St. Louis, MO, USA): Prepare 0.1 M solution in ddH_2O, sterilize by filtration, aliquot, and store at −20 °C.
6. Lysis buffer: 20 mM Tris–HCl (pH 7.4), 200 mM NaCl, 10 % (v/v) glycerol, 1 mM ethylenediaminetetraacetic acid (EDTA), and 1 mM phenylmethanesulfonyl fluoride (PMSF).
7. 1 M stock solution Tris–HCl (pH 7.4): Dissolve 157.6 g Tris–HCl into 800 mL ddH_2O, adjust pH to 7.4 by adding NaOH droplets while stirring with a magnet, fill up to 1 L, and autoclave.
8. 2 M stock solution NaCl: Dissolve 116.88 g NaCl in 800 mL ddH_2O, transfer to a 1 L graduated cylinder, fill up to 1 L, and autoclave.
9. 0.5 M stock solution EDTA: Add 73 g EDTA to 400 mL ddH_2O, adjust to pH 8.0 with NaOH droplets, and fill up to 500 mL with ddH_2O.
10. PMSF (Sigma-Aldrich): Prepare a 0.1 M stock solution in ethanol, aliquot, and store at −20 °C.
11. Triton X-100 (Sigma-Aldrich).
12. Poly-prep chromatography columns, 10 mL (Bio-Rad) or similar.
13. Amylose resin (New England Biolabs) stored at 4 °C.
14. 2× protein loading buffer: Prepare 100 mL by mixing 10 mL glycerol, 30 mL of 10 % (w/v) SDS, 5 mL of 0.5 M Tris–HCl (pH 6.8), and 10 mg bromophenol blue.

2.2 Pull-Down Assay

1. Plant material: 10- to 12-day-old wild-type and transgenic seedlings grown in Murashige and Skoog agar plates [16] in a growth chamber at 21 °C under a 16-h light/8-h dark cycle (*see* **Note 2**).
2. Purified MBP-"bait" fusion protein expressed in *E. coli* and MBP for negative control.
3. Pull-down buffer: 50 mM Tris–HCl (pH 7.4), 80 mM NaCl, 10 % (v/v) glycerol, 0.1 % (v/v) Tween-20, 1 mM PMSF, 50 μM MG132 (optional, *see* **Note 3**), 1 Complete protease inhibitor EDTA-free cocktail tablet (Roche Applied Science, Indianapolis, IN, USA), and 1 mM dithiothreitol (DTT) (optional, *see* **Note 3**). Prepare fresh from stock solutions and keep on ice. Calculate 1 mL per sample.
4. Washing buffer: 50 mM Tris–HCl (pH 7.4), 80 mM NaCl, 10 % (v/v) glycerol, 0.1 % (v/v) Tween-20, 1 mM PMSF, and 1 Complete protease inhibitor EDTA-free cocktail tablet. Prepare fresh from stock solutions and keep on ice. Calculate 5 mL per sample (depending on the number of washes).

5. Proteasome inhibitor MG132 (Sigma-Aldrich): Dissolve in dimethyl sulfoxide (DMSO) to obtain a 10 mM stock solution, aliquot, and store at −20 °C.
6. Coronatine (Sigma-Aldrich) (a bacterial mimetic of the hormone JA-Ile): Prepare a 10-mM stock solution by dissolving in ethanol, aliquot, and store at −20 °C.
7. Homogenizer (IKA-Werke, Eurostar Power-b, or similar), mortar and pestle.
8. Syringes: 1 mL syringe with a 0.3 mm diameter needle (*see* **Note 4**); 1 mL syringe with a 0.9 mm diameter needle.
9. Anti-FLAG antibody (Sigma-Aldrich), or other tag-specific or "prey" protein-specific antibody.

3 Methods

3.1 E. coli Protein Expression and Purification

1. Clone the cDNA coding the "bait" protein in pDESTH1 (or an equivalent plasmid) (*see* **Note 1**).
2. Transform BL21-competent cells and select transformants in an LB agar plate containing 100 μg/μL ampicillin and 25 μg/μL chloramphenicol (for plasmid and *E. coli* strain selection, respectively).
3. Select two to five colonies and grow them for plasmid miniprep.
4. Isolate the plasmid and check the insertion by sequencing.
5. Streak a single colony with the "bait" protein cloned in pDESTH1 on a Petri dish containing LB solid medium and antibiotics.
6. Incubate overnight at 37 °C.
7. Inoculate 3 mL liquid LB medium supplemented with antibiotics and incubate at 37 °C overnight at 250 rpm.
8. Add 2 mL of the saturated liquid culture to 200 mL LB in a 1-L flask (*see* **Note 5**).
9. Incubate at 37 °C in a shaker at 250 rpm until the cells reach an OD (600 nm) between 0.4 and 0.6 (approximately 2 h) (*see* **Note 6**).
10. Induce the protein expression by adding 0.1 mM IPTG and incubate for 4 h at room temperature with shaking at 250 rpm (*see* **Note 7**).
11. Transfer the culture to 250-mL centrifuge flasks and centrifuge at 4,000 × *g* for 10 min at 4 °C to pellet the cells.
12. Discard the supernatant and freeze the pellet for at least 1 h at −80 °C.

13. Add 10 mL of cold lysis buffer (*see* **Note 8**) and resuspend the pellet. From now on, work on ice or in a 4 °C chamber.
14. Sonicate the cells for 15 s four to six times, resting 30 s in between, until the solution is homogeneous. Always keep the tubes on ice.
15. As optional step, add Triton X-100 to a final concentration of 1 % (v/v) and agitate gently for 20 min (increases protein solubility, but might decrease binding to the column).
16. Centrifuge at 13,000 × *g* for 30 min at 4 °C.
17. Recover the supernatant and keep on ice. It should be clean of debris; otherwise repeat the centrifugation step (*see* **Note 9**).
18. To prepare the amylose column, gently mix the amylose resin and transfer by pipetting 400 μL of the beads in suspension to a 10-mL column to obtain a final bead volume of 200 μL.
19. Fill up the column with 10 mL of cold lysis buffer and wash the beads by gravity.
20. Transfer the column to the 4 °C room and repeat the procedure.
21. Add the bacterial protein extract to the column, seal it, and incubate under vertical rotation for 2 h at 4 °C.
22. Place the column in a vertical position, open it, and allow the protein extract to flow by gravity.
23. Wash extensively with cold washing buffer 3 to 5 times the full column volume (approximately 50 mL in total).
24. Leave the column with 400 μL of washing buffer (twice the bead volume). Do not elute the column.
25. Gently resuspend the beads and transfer 5 μL into a 1.5-mL tube. Store the remaining resin in the column with 0.1 % (w/v) sodium azide, at 4 °C, until use.
26. Add 2× protein loading buffer to the sample, incubate for 5 min at 96 °C, and analyze by SDS-PAGE.
27. Check MBP-fused protein purity and estimate protein concentration by Coomassie staining (*see* **Notes 10** and **11**).
28. After quantification of the "bait" fusion protein, proceed to the pull-down assay.

3.2 Pull-Down Assay

Work in a 4 °C chamber or at room temperature always keeping the samples on ice. Centrifugation and incubation steps should be carried out at 4 °C.

1. Quickly collect and freeze plant material in liquid nitrogen.
2. Grind with mortar and pestle. Approximately 0.5–2 mL powder tissue will be sufficient for one reaction (*see* **Note 12**).

3. Homogenize with ¼ powder volume of pull-down buffer with a homogenizer.
4. Centrifuge for 15 min at 16,000 × *g* at 4 °C.
5. Transfer the supernatant to a new tube and centrifuge again at 16,000 × *g* for 15 min.
6. Transfer the supernatant to a new tube.
7. Check there is no material in suspension; otherwise, if there is cell debris in the suspension, centrifuge again or filter the supernatant with Miracloth (Millipore, Bedford, MA, USA).
8. Keep extracts as a bulk on ice. Always use freshly prepared extracts (*see* **Note 13**).
9. Keep 40 μL of each bulk extract to load in SDS-PAGE (*see* Subheading 3.2, **step 25**).
10. Label 1.5-mL tubes (one for each pull-down reaction) and pipette 5–30 μL of amylose resin slurry from the column prepared (*see* Subheading 3.1, **step 24**) to have approximately 6 μg resin-bound MBP-fusion protein (*see* **Note 14**).
11. Wash the resin by adding 1 mL of washing buffer, mix well, and centrifuge at 500 × *g* for 1 min (*see* **Note 15**).
12. Remove the washing buffer with the 0.9-mm needle syringe without disturbing the beads.
13. Add 1 mL of plant extract and mix by inverting the tube.
14. Add 10 μM coronatine (exceptional for JAZ-COI1 interactions).
15. Incubate at 4 °C under orbital rotation for 1 h (*see* **Note 16**).
16. Centrifuge at 500 × *g* for 2 min.
17. Carefully remove the plant extract, first with the 0.9-mm needle syringe without disturbing the beads and then with the 0.3-mm needle syringe (beads can be disturbed with this needle to remove all the extract). It is important to remove as much extract as possible (*see* **Note 4**).
18. Add 1 mL of washing buffer and keep the tubes under orbital rotation for 2 min.
19. Centrifuge at 500 × *g* for 1 min and remove as much buffer as possible with the syringes.
20. Repeat twice (*see* Subheading 3.2, **step 19**) (*see* **Note 17**).
21. Resuspend the resin in 43 μL of washing buffer and mix well (*see* **Note 18**).
22. Transfer 3 μL of the resuspended resin to a new tube. Add 2× protein loading buffer, and incubate for 5 min at 96 °C. Load this sample on an SDS-PAGE gel stained with Coomassie solution to control the amount of *E. coli* protein used in each reaction.

23. Add 8 μL of 5× protein loading buffer to the pull-down reaction tube.
24. Incubate for 5 min at 96 °C and keep on ice.
25. Load the pull-down samples on an SDS-PAGE gel and perform the immunoblot with a primary antibody that recognizes the tag of the "prey" protein. See the results obtained in a typical hormone-dependent pull-down assay in Fig. 2.

4 Notes

1. To select the cloning vector for the "bait" protein to use in the pull-down assay, (a) choose the type of fusion that might work better for the protein, either an N- or a C-terminal fusion and (b) select a tag, such as MBP, GST, and 6× His, which are some of the most common. Although small tags are preferable (such as 6× His) because they will interfere less with protein folding and properties, GST and MBP tend to increase the protein solubility, making them easy to express and to work with. Keep in mind that *E. coli*-expressed proteins can be used for many different applications and that in many cases the tag should be eliminated. We strongly recommend the use of vectors with a protease cleavage sequence between the tag and the "bait" protein. The vector of choice, without insert, should be processed in parallel throughout the protocol to obtain the expressed epitope (for instance, MBP here) that will serve as the "bait" negative control in the pull-down assay.
2. We use transgenic and wild-type seedlings grown in Murashige and Skoog (MS) [16] agar as the source of material for "prey" protein expression and negative control, respectively. Other types of tissues might be used. In any case, it is important that before the protein pull-down assay is started, the proper "prey" protein accumulation in the transgenic line has been confirmed by immunoblotting. A clean, reliable pull-down can only be achieved with good protein expression. As a control, wild-type plant extracts will allow the identification of unspecific bands.
3. The use of proteasome inhibitors, such as MG132, is needed only when one of the proteins of interest is a target for proteasome degradation. The reducing agent DTT might enhance the activity/interaction of some proteins by keeping them in a reduced state.
4. The syringe needle diameter is crucial. The 0.3-mm needle will allow the removal of the buffer without resin losses.
5. Before starting a medium- to large-scale protein induction, as described here, we recommend to do a small-scale induction in 2 mL of LB to test whether the "bait" protein is expressed

in *E. coli*. To optimize the protein induction, test different IPTG concentrations and induction times from 1 to several hours, or decrease the induction temperature (for instance, 18 °C should increase the protein solubility by reducing the formation of inclusion bodies).

6. When expressing a protein for the first time, keep 20 μL of the liquid culture, as non-induced control (NIC), for loading in SDS-PAGE gels (*see* Fig. 1).
7. When expressing a protein for the first time, keep 20 μL of the liquid culture as induced control (IC) and load in the SDS-PAGE gel together with the NIC. Meanwhile, the procedure can be kept in standby at **step 12** (in Subheading 3.1). After Coomassie staining, an extra, sometimes faint, band at the estimated mass ("bait" protein + MBP) should be present in the IC sample compared to the NIC sample (*see* Fig. 1).
8. Protein solubility may be enhanced by increasing the salt concentration of the lysis buffer. In our hands, 400 mM final concentration of NaCl was fine.
9. Keep 10 μL of the soluble fraction (SF). Before discarding the pellet, resuspend it on 10 mL of water and save 10 μL of insoluble fraction (IF). To assess the solubility of the fusion protein and, before proceeding with the protocol, run both fractions in the SDS-PAGE gels. The MBP-"bait" band should be detected in the soluble fraction (*see* Fig. 1). When large amounts of the protein are in the IF, optimize the protocol accordingly to **Notes 5** and **8**.
10. As the MBP-"bait" fusion protein expressed in *E. coli* is not eluted from the column, the easiest way to estimate the protein concentration is to load increasing dilutions of a bovine serum albumin stock and MBP-fusion protein in the SDS-PAGE gel. The MBP-"bait" protein should be visible by Coomassie staining, allowing the assessment of protein purity and quantity. Although some faint unspecific bands might be present, a strongly predominant band is to be expected at the molecular mass of the "bait" protein + MBP (approximately 40 kDa).
11. To confirm the strong band obtained after the purification procedure corresponds to the MBP-"bait" fusion protein use 2 μL of resin slurry to perform immunoblot with an anti-MBP antibody (New England Biolabs). Use antibody dilutions above 1/5,000 (in some cases we use 1/10,000 due to high unspecific background).
12. The amount of starting material to use will depend on the expression of the "prey" protein in transgenic plants and the detection efficiency; it does not directly depend on total protein concentration. For COI1-FLAG samples, 1.5 mg of total protein per sample works fine. This was achieved by using

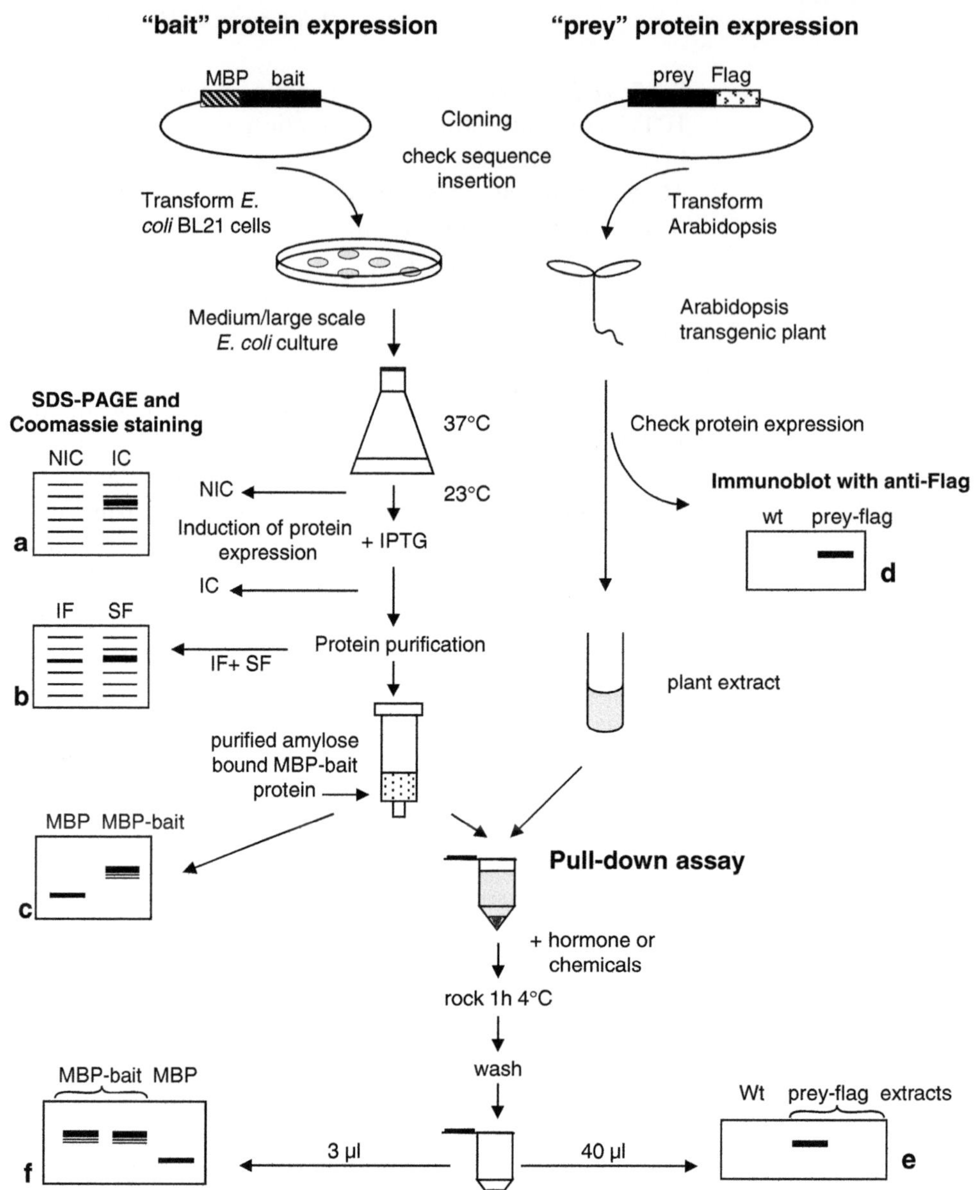

Fig. 1 Basic scheme of the "bait" and "prey" expression strategies and pull-down (PD) protocol. Coding sequences of both "prey" and "bait" are cloned in suitable vectors. Upon selection, *E. coli* containing the "bait" protein is grown in liquid medium at 37 °C and transferred to room temperature. A sample serving as non-induced control (NIC) is taken and IPTG added to the culture to induce expression. Four hours later, an induced control (IC) sample is reserved and loaded into an SDS-PAGE gel together with the NIC. A differential band should be visible in the IC lane, corresponding to the MBP+ "bait" mass (**a**). The bacterial culture is then lysed and the soluble protein (soluble fraction, SF) is separated from the pelleted cell material (insoluble fraction, IF). The MBP-"bait" fusion protein should be mostly present in the soluble fraction (**b**). The soluble protein is loaded into a column containing the previously packed amylose resin. After extensive washing, the purity of the expressed protein is assessed. For most JAZ proteins, full-length and truncated versions are visible (**c**). Transgenic *Arabidopsis* plants are selected for "prey" protein expression by immunoblot with an anti-FLAG antibody. A differential band should be visible in this extract versus the wild-type (wt) plant control extract (**d**). In the pull-down assay, the plant extract expressing the "prey" protein is incubated with the resin-bound "bait" protein. After washing, the presence of the "prey" protein is detected by immunoblot (**e**) and the "bait" protein is visualized by Coomassie staining to ensure that similar protein amounts were used in all samples (**f**)

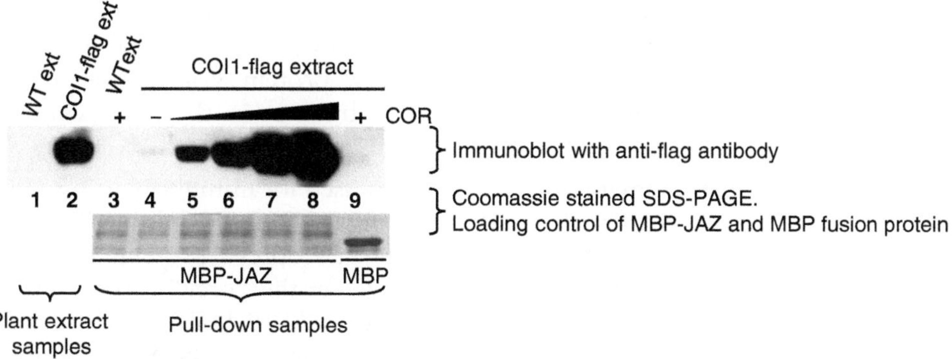

Fig. 2 Example of a COI1-JAZ pull-down (PD) assay in the presence of coronatine (COR). *Top gel*, immunoblot with anti-FLAG antibody. In *lanes 1* and *2*, 20 μL of 1.5 mg/mL of total protein of the wild-type (wt) and COI1-FLAG extracts were loaded. This sample of total protein extract is useful to indicate the size of the "prey" protein and to check that no unspecific bands exist in the wt extract (*lane 1*). *Lanes 3–9*, PD samples. *Lane 3* shows a control sample in which the MBP-JAZ protein was incubated with the wt extract to reveal that no protein was pulled down. In *lane 4*, only a residual interaction between COI1 and JAZ was detected in the absence of the hormone. In *lanes 5–8*, increasing COR concentrations were added to the PD, showing that the COI1–JAZ interaction is hormone dependent. *Lane 9* corresponds to the "bait" control, in which MBP alone is not sufficient to pull down the COI1 protein, even in the presence of high COR concentrations. *Bottom gel*, Coomassie-stained SDS-PAGE gel in which 3 μL of the PD reactions were loaded to ensure that equal amounts of bait proteins were used

2 mL powder tissue according to this protocol. Use Fig. 2 for comparison and scale-up or scale-down depending on the expression levels of the protein of interest.

13. At this stage, the extract to be pulled down is ready to use. If the extraction was made in several tubes, mix the extracts expressing the same protein together. It is important to dispense to each reaction extract from a bulk to ensure that each pull-down sample contains the same amount of protein extract. There is no need to quantify the total protein, except when two types of extracts (wild-type versus COI1-FLAG for example) will be used. In this case, measure the total protein extract with a Bradford assay [17]. This is also the step to perform pre-incubations with distinct conditions or chemicals [2].
14. MBP alone should be used as control. When working with different *E. coli*-expressed proteins, it is important to equilibrate the sample volumes by adding empty prewashed (in washing buffer) amylose resin to the same final volume.
15. Centrifugation velocity depends on the resin used. It should be high enough to recover all the resin and to allow a good resuspension of the beads. Follow the manufacturer's instructions.
16. Incubation time might range from 30 min to 4 h, depending on the proteins of interest.

17. The number of washes should be optimized. Normally, the optimal number of washes varies between 2 and 6. If the interaction is positive, the "prey" protein should be detected in the lane of the bait protein, but not in the MBP control after immunoblotting (*see* Fig. 2).
18. Sometimes the antibodies used for immunoblotting can cross-react with MBP, giving rise to faint, but visible, bands. If the sizes of the MBP-"bait" and "prey" proteins are similar, it might be impossible to detect the "prey" protein. As COI1-FLAG had the same size as MBP-JAZ3, we overcame this major drawback by cutting in between the MBP and the JAZ3 proteins with a protease (factor Xa; New England Biolabs). Note that this method can only be used when the protease does not cleave the "prey" protein.

Acknowledgments

We thank the members of the laboratory for critical reading of the manuscript. Research in R.S lab was supported by grants from the Ministry of Science and Innovation (BIO2007-66935, BIO2010-21739, CSD2007-00057-B, and EUI2008-03666) to R.S. S.F. was the recipient of a postdoctoral fellowship from the Portuguese Foundation for Science and Technology and a JAE-Doc contract from the Consejo Superior de Investigaciones Científicas.

References

1. Dharmasiri N, Dharmasiri S, Estelle M (2005) The F-box protein TIR1 is an auxin receptor. Nature 435:441–445
2. Fonseca S, Chini A, Hamberg M, Adie B, Porzel A, Kramell R, Miersch O, Wasternack C, Solano R (2009) (+)-7-*iso*-Jasmonoyl-L-isoleucine is the endogenous bioactive jasmonate. Nat Chem Biol 5:344–350
3. Katsir L, Schilmiller AL, Staswick PE, He SY, Howe GA (2008) COI1 is a critical component of a receptor for jasmonate and the bacterial virulence factor coronatine. Proc Natl Acad Sci USA 105:7100–7105
4. Chini A, Fonseca S, Fernández G, Adie B, Chico JM, Lorenzo O, García-Casado G, López-Vidriero I, Lozano FM, Ponce MR, Micol JL, Solano R (2007) The JAZ family of repressors is the missing link in jasmonate signalling. Nature 448:666–671
5. Kepinski S, Leyser O (2005) The *Arabidopsis* F-box protein TIR1 is an auxin receptor. Nature 435:446–451
6. Pauwels L, Fernández Barbero G, Geerinck J, Tilleman S, Grunewald W, Cuéllar Pérez A, Chico JM, Vanden Bossche R, Sewell J, Gil E, García-Casado G, Witters E, Inzé D, Long JA, De Jaeger G, Solano R, Goossens A (2010) NINJA connects the co-repressor TOPLESS to jasmonate signalling. Nature 464:788–791
7. Feng S, Ma L, Wang X, Xie D, Dinesh-Kumar SP, Wei N, Deng XW (2003) The COP9 signalosome interacts physically with SCF^{COI1} and modulates jasmonate responses. Plant Cell 15: 1083–1094
8. Chini A, Fonseca S, Chico JM, Fernández-Calvo P, Solano R (2009) The ZIM domain mediates homo- and heteromeric interactions between Arabidopsis JAZ proteins. Plant J 59: 77–87
9. Fernández-Calvo P, Chini A, Fernández-Barbero G, Chico J-M, Gimenez-Ibanez S, Geerinck J, Eeckhout D, Schweizer F, Godoy M, Franco-Zorrilla JM, Pauwels L, Witters E, Puga MI, Paz-Ares J, Goossens A, Reymond P,

De Jaeger G, Solano R (2011) The *Arabidopsis* bHLH transcription factors MYC3 and MYC4 are targets of JAZ repressors and act additively with MYC2 in the activation of JA responses. Plant Cell 23:701–715

10. Clough SJ, Bent AF (1998) Floral dip: a simplified method for *Agrobacterium*-mediated transformation of *Arabidopsis thaliana*. Plant J 16:735–743
11. Niu Y, Figueroa P, Browse J (2011) Characterization of JAZ-interacting bHLH transcription factors that regulate jasmonate responses in *Arabidopsis*. J Exp Bot 62: 2143–2154
12. Melotto M, Mecey C, Niu Y, Chung HS, Katsir L, Yao J, Zeng W, Thines B, Staswick P, Browse J, Howe GA, He SY (2008) A critical role of two positively charged amino acids in the Jas motif of Arabidopsis JAZ proteins in mediating coronatine- and jasmonoyl isoleucine-dependent interactions with the COI1 F-box protein. Plant J 55:979–988
13. Yan J, Zhang C, Gu M, Bai Z, Zhang W, Qi T, Cheng Z, Peng W, Luo H, Nan F, Wang Z, Xie D (2009) The *Arabidopsis* CORONATINE INSENSITIVE1 protein is a jasmonate receptor. Plant Cell 21:2220–2236
14. Thines B, Katsir L, Melotto M, Niu Y, Mandaokar A, Liu G, Nomura K, He SY, Howe GA, Browse J (2007) JAZ repressor proteins are targets of the SCF^{COI1} complex during jasmonate signalling. Nature 448:661–665
15. Hammarström M, Hellgren N, Van Den Berg S, Berglund H, Härd T (2002) Rapid screening for improved solubility of small human proteins produced as fusion proteins in *Escherichia coli*. Protein Sci 11:313–321
16. Murashige T, Skoog F (1962) A revised medium for rapid growth and bio assays with tobacco tissue cultures. Physiol Plant 15:473–497
17. Bradford MM (1976) A rapid and sensitive method for the quantitation of microgram quantities of protein utilizing the principle of protein-dye binding. Anal Biochem 72:248–254

Chapter 14

Yeast Two-Hybrid Analysis of Jasmonate Signaling Proteins

Amparo Pérez Cuéllar, Laurens Pauwels, Rebecca De Clercq, and Alain Goossens

Abstract

Protein–protein interaction studies are crucial to unravel how jasmonate (JA) signals are transduced. Among the different techniques available, yeast two-hybrid (Y2H) is commonly used within the JA research community to identify proteins belonging to the core JA signaling module. The technique is based on the reconstitution of a transcriptional activator that drives the reporter gene expression upon protein–protein interactions. The method is sensitive and straightforward and can be adapted for different approaches. In this chapter, we provide a detailed protocol to perform targeted Y2H assays to test known proteins and/or protein domains for direct interaction in a pairwise manner and present the possibility to study ternary protein complexes through Y3H.

Key words Jasmonate signaling, Yeast two-hybrid, Yeast three-hybrid, Protein–protein interaction, Protein complexes, Protein domain

1 Introduction

Over the last decade, a core module of jasmonate (JA) signaling has been characterized in which hormone perception and consequent signal transduction are tightly regulated by protein complexes formed by different protein types. So far, the main players identified include transcription factors regulating JA-responsive genes (such as MYC2-like basic helix-loop-helix), the JASMONATE-ZIM DOMAIN (JAZ) repressor proteins, corepressors such as the Novel Interactor of JAZ (NINJA) and TOPLESS, and the F-box protein CORONATINE INSENSITIVE1 (COI1) that forms part of the Skp1/Cullin/F-box (SCF)^{COI1} ubiquitin E3 ligase complex [1]. The protein complexes formed are determined by the absence or the presence of bioactive JAs, demonstrating the importance of protein–protein interaction studies in this field.

Among the different protein–protein interaction techniques available [2–4], yeast (*Saccharomyces cerevisiae*) two-hybrid (Y2H) is widely applied within the JA research community. For instance,

Alain Goossens and Laurens Pauwels (eds.), *Jasmonate Signaling: Methods and Protocols*, Methods in Molecular Biology, vol. 1011, DOI 10.1007/978-1-62703-414-2_14,

Table 1
Characteristics of the different yeast strains commonly used in Y2H assays

Strain	Markers	Other genotype characteristics	Reporter	Available marker	Reference
PJ69-4	*trp1-901*	*gal4Δ*	HIS3	TRP	[22]
	leu2-3, 112	*gal80Δ*	ADE2	LEU	
	ura3-52	LYS2::GAL1-HIS3	LacZ	URA	
	his3-200	GAL2-ADE2			
		met2::GAL7-LacZ			
AH109	*trp1-901*	*gal4Δ*	HIS3	TRP	Clontech
	leu2-3, 112	*gal80Δ*	ADE2	LEU	
	ura3-52	LYS2::$GAL1_{UAS}$-$GAL1_{TATA}$-HIS3	LacZ		
	his3-200	$GAL2_{UAS}$-$GAL2_{TATA}$-ADE2			
		URA3::$MEL1_{UAS}$-$MEL1_{TATA}$-LacZ			
EGY48	*ura3*	$LexA_{op(x6)}$-LEU2	LEU	HIS	[14]
(p8opLacZ)	*his3*	p8opLacZ (Ura)	LacZ	TRP	
	trp1			URA	

by means of Y2H assays, the JAZ proteins were confirmed to be the COI1 targets [5] and to interact with many different transcription factors (for an overview, ref. 1, Table 1). Moreover, Y2H not only can be used as an alternative but is also often used to validate other protein–protein interaction techniques [2, 6, 7]. For example, parallel tandem affinity purification and Y2H screens had originally identified NINJA as a member of the JA signaling core module [8] and Y2H assays confirmed that both the JAZ-NINJA and the NINJA-TOPLESS interactions were direct. These findings extended the current model for JA signaling and demonstrated the power of combining multiple techniques in unraveling protein complexes.

The Y2H technique as described [9] is based on the reconstitution of a functional transcriptional activator (such as the yeast galactosidase 4 [GAL4]) to promote the expression of (a) reporter gene(s). This transcriptional activator is split into its two functional protein domains: a DNA-binding domain and an activation domain. These domains are fused to a protein of interest, generating two "hybrid" proteins (bait and prey) that are transformed in a compatible yeast strain. In the case of bait–prey interaction, the transcriptional activator is reconstituted and the reporter gene(s) is (are) expressed (*see* Fig. 1) [2, 6, 7, 9, 10].

The Y2H technique has many advantages: it is sensitive, easy to perform, relatively cheap, and can be widely used thanks to its

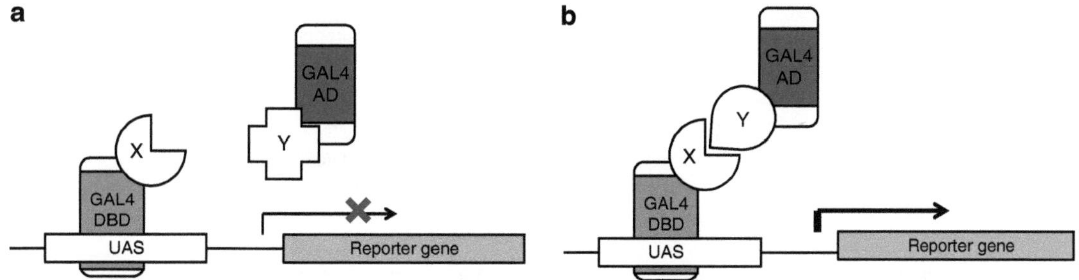

Fig. 1 Molecular basis of the Y2H technique. A transcriptional activator (such as the yeast GAL4) is split into its DNA-binding domain (DBD) and activation domain (AD) that are fused to the proteins of interest (X and Y), generating two "hybrid" proteins, the DBD-X (bait) and AD-Y (prey). (**a**) Lack of reporter gene expression, when the proteins do not interact. (**b**) Reconstitution of the transcriptional activator upon bait and prey interaction, allowing that the reporter gene expression is driven by a promoter containing the upstream activating sequence (UAS)

versatility, because it can be adapted to different approaches, such as cDNA-library screens to identify protein interactors. In the JA field, TOPLESS has been used as bait [11]. In matrix-based Y2H assays, a large set of proteins is tested for interaction in a pairwise manner. Such assays have been carried out with cloned ORFeomes in several model species and recently also in *Arabidopsis thaliana* [12]. Finally, targeted Y2H assays are chosen when the objective is to confirm interactions between two proteins observed through other methods and/or map the domain(s) necessary and sufficient for the interaction.

Together with Y2H, targeted yeast three-hybrid (Y3H) assays can also be performed to study the formation of ternary protein complexes by cotransformation with a third protein of interest acting as a bridge between bait and prey. Within the JA signaling field, it has recently been shown the formation of a ternary JAZ3–NINJA–TPL complex [13].

In this chapter, we focus on targeted Y2H and Y3H assays and we describe a comprehensive protocol that covers all the steps needed (*see* Fig. 2). We also give an overview of the variations in the experimental setup regarding the use of different transcriptional activator systems (i.e., the GAL4 and LexA systems), yeast strains, vectors, and reporter genes that are generally used in the JA signaling research field.

2 Materials

2.1 Equipment

1. 250-mL Erlenmeyer flasks.
2. Incubator shaker set at 30 °C and 300–400 rpm.
3. Spectrophotometer set for reading optical densities (OD) at $\lambda = 600$ nm (OD_{600}).

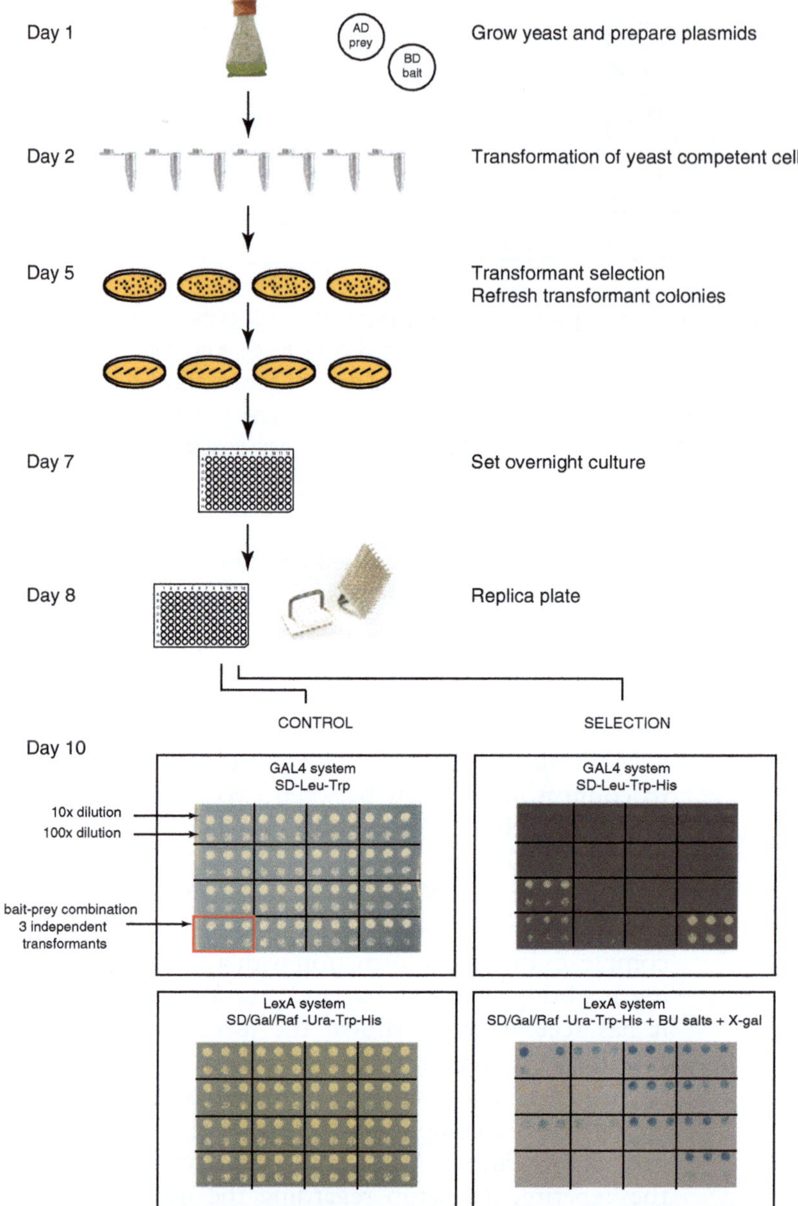

Fig. 2 Overview of the procedure for targeted Y2H assays. (Day 1) Inoculation of a yeast culture overnight in 2× YPDA medium and preparation of bait and prey plasmid constructs. (Day 2) Preparation of competent yeast cells and cotransformation with bait and prey constructs, followed by plating on selective media and incubation for 2 days at 30 °C. For the GAL4 and LexA systems, SD-Leu-Trp and SD Gal/Raf-Ura-Trp-His are used, respectively. (Day 5) Picking up and transfer of independent transformants to a new selective plate, followed by incubation for 2 days at 30 °C. (Day 7) Inoculation and incubation of three independent transformant cultures per each bait–prey combination overnight in liquid-selective media in 96-well plates. (Day 8) Dilution of the overnight cultures to 1/10 and 1/100 in sterile water. The 16 bait–prey combinations and their corresponding dilutions are set up in one 96-well plate by means of a replica plater to drop yeast cultures on the desired control and selective media. (Day 10) After incubation for 2 days at 30 °C, positive protein interactions are scored based on growth or staining on selective media

4. Tabletop centrifuge (both for conical centrifugation and 1.5-mL microcentrifuge tubes).
5. Thermoblock.
6. Steel bacterial cell spreader or glass beads (3 mm).
7. Sterile Petri dishes: round (9 mm diameter) and square (120 × 120 mm).
8. Microporous tape sheets for covering 96-well plates.
9. Steel replica plater for 96-well plates (Sigma-Aldrich, St. Louis, MO, USA).
10. Both U-shaped and flat-bottom 96-well plates.
11. Vortex.
12. Sterile toothpicks.
13. Laminar flow.

2.2 Media, Buffers, and Solutions

All solutions are prepared with purified water. There is no need to adjust the pH in the yeast growth media and, once sterile, they can be kept at room temperature.

2.2.1 Yeast Growth

1. 2× yeast peptone dextrose (YPD) adenine (YPDA) liquid medium: Add 100 g of yeast YPD (Clontech, Mountain View, CA, USA) and 73 mg of adenine (Sigma-Aldrich) for 1 L of medium (*see* **Note 1**).
2. YPDA solid medium: Add 50 g of YPD (Clontech), 73 mg of adenine (Sigma-Aldrich), and 20 g of agar for 1 L of medium. Autoclave and pour in sterile round Petri dishes.
3. Yeast synthetic defined (SD) media, both liquid and solid: Yeast nitrogen base ammonium sulfate and a carbon source, either dextrose (Minimal SD Base) or galactose and raffinose (Minimal SD Base Gal/Raf) and Drop Out (DO) supplements (Clontech). For the GAL4 system, add 26.7 g/L of minimal SD Base (Clontech); for the LexA system, add 37 g/L of minimal SD Base Gal/Raf (Clontech). Amounts required for the DO supplements (Clontech) depend on their composition. If preparing media to pour plates, add agar to 2 % (20 g/L). Autoclave and pour in sterile round Petri dishes for transformant selection or in sterile square Petri dishes for replica plating.

2.2.2 Yeast Transformation

All solutions are prepared with purified water.

1. 1 M lithium acetate (LiAc). Autoclave and keep at 4 °C.
2. 50 % (w/v) Polyethylene glycol (PEG) 3350: Add the PEG 3350 to less water than the final volume, dissolve by heating to 60 °C, add water to adjust to the correct volume, and cool down on ice.

3. 10× tris(hydroxymethyl)aminomethane (Tris)-ethylenediaminetetraacetic acid (EDTA) (TE): 100 mM Tris–HCl, 10 mM EDTA (pH 8.0). Autoclave and keep at 4 °C.
4. Carrier DNA: 10 mg/mL salmon sperm DNA.

2.2.3 5-Bromo-4-Chloro-Indolyl-β-D-Galactopyranoside (X-Gal) Assay for LexA-Based Y2H Systems

1. X-gal liquid solution: Prepare a 20 mg/mL stock solution by dissolving the X-gal in dimethyl sulfoxide (DMSO). Store in the dark at −20 °C.
2. 10× buffered (BU) salt solution: For 1 L, add 70 g of $Na_2HPO_4.7H_2O$ and 30 g of NaH_2PO4. Adjust pH to 7. Autoclave and store at room temperature.
3. X-gal plates: For 1 L, mix the SD Gal/Raf medium, DO supplements, and 20 g/L agar in 900 mL water. Autoclave and cool down to 55 °C. Add 100 mL of 10× stock BU salts for a final 1× BU salts concentration and add 4 mL of the 20 mg/mL X-gal stock, for a final 80 μg/L concentration. Mix and pour plates.

2.3 Selection of the Y2H System

The two predominant Y2H systems that have been used are based on the GAL4 and LexA transcriptional activators, respectively [2, 6]. In the GAL4 system, reconstitution of the yeast transcriptional activator GAL4 allows reporter gene expression. The LexA system is based on the *Escherichia coli* repressor DNA-binding domain-providing protein LexA that is combined with the *E. coli* B42 activation domain. The stringency of the latter system is based on the number of LexA operator elements present in the promoter of the reporter genes. Very sensitive reporters include up to eight LexA-binding sites, such as the p8opLacZ plasmid (*see* **Note 2**) [14, 15]. In both systems bait and prey proteins are targeted to the nucleus either by the fusion with GAL4 or LexA DNA-binding domains or the nuclear localization signal (NLS) which is added to GAL4 and B42 activation domains in the corresponding vectors.

2.4 Yeast Strains

Several yeast strains have been developed for Y2H (*see* Table 1 and **Note 3**). These strains contain auxotrophic markers, used to select plasmids or report protein interactions, and reporters, such as the gene encoding the enzyme β-galactosidase (*LacZ*), allowing interaction to be reported by yeast staining.

2.5 Vectors

Plasmid-assisted yeast complementation is used for transformant selection. Table 2 lists the properties of the most commonly used Y2H vectors (*see* **Note 4**). In the case of Y3H, a destination vector expressing the bridging protein fused to a NLS and an epitope tag is desired. The use of the PJ69-4A strain together with the GAL4 system allows the use of a vector with Ura as auxotrophic marker. This vector can be easily obtained with MultiSite Gateway using pMG426 as a destination vector [13].

Table 2
Y2H vectors frequently used in JA signaling research in *Arabidopsis*

Vector	Fusion	Yeast marker	*Escherichia coli* marker	Promoter	Epitope	Replication mechanism	Reference
pGADT7	GAL4AD	LEU2	Amp	ADH1	HA	2 μ	Clontech
pGBKT7	GAL4BD	TRP1	Km	ADH1	c-Myc	2 μ	Clontech
pGAD424	GAL4AD	LEU2	Amp	ADH1	–	2 μ	Clontech
pGBT9	GAL4BD	TRP1	Amp	ADH1	–	2 μ	Clontech
pDEST22	GAL4AD	TRP1	Amp	ADH1	–	CEN	Invitrogen
pDEST32	GAL4DB	LEU2	Gm	ADH1	–	CEN	Invitrogen
pGILDA	LexA BD	HIS3	Amp	GAL1	–	CEN	Clontech
pB42AD (pJG4-5)	B42AD	TRP1	Amp	GAL1	HA	2 μ	Clontech; [15]

Amp ampicillin, *Gm* gentamicin, *Km* kanamycin

3 Methods

In this section, we describe the methods to perform directed Y2H assays, designed to study the interaction between two known proteins and/or their protein domains. In this adapted yeast transformation [16], yeast is cotransformed with bait and prey constructs as an alternative to yeast mating. The complete procedure takes approximately 10–11 days (*see* Fig. 2).

3.1 Cloning Bait and Prey

1. To obtain bait and prey constructs, clone the genes encoding the proteins of interest with attention to the reading frame and generate an entry clone by BP reaction with pDONR207 or pDONR223 (*see* **Note 5**) that confer gentamicin and spectinomycin resistance in *E. coli*, respectively, and are compatible with the Y2H Gateway® destination vectors (*see* **Note 6**).
2. Verify the entry clone by sequencing.
3. Generate an expression clone by LR reaction with the entry clone and the chosen destination vector(s) (Table 2, *see* **Note 6**).

3.2 Competent Yeast Cells (Days 1 and 2)

1. Inoculate yeast (from plate or cryostock) in 5 mL of 2× YPDA.
2. Incubate overnight at 200 rpm and 30 °C (*see* **Note 7**).
3. Preincubate 50 mL of 2× YPDA at 30 °C overnight (*see* **Note 8**).
4. Determine the OD_{600} of the overnight culture (*see* **Note 9**).
5. Inoculate the preincubated 2× YPDA medium with the yeast culture grown overnight to yield an $OD_{600}=0.25$.

6. Incubate for two cell divisions (approximately 4 h) until $OD_{600} = 1$.
7. Record the OD of the culture ($OD_{600} = 1$ is optimal).
8. When the optimal OD_{600} is reached, centrifuge at 800 × *g* for 5 min in conical centrifugation tubes in a tabletop centrifuge.
9. Discard the supernatant.
10. Wash cells in 25 mL of precooled sterile water.
11. Centrifuge at 800 × *g* for 5 min.
12. Discard the supernatant.
13. Resuspend the yeast cells in 1 mL of sterile water.
14. Transfer resuspension to a sterile 1.5-mL microcentrifuge tube.
15. Centrifuge at 21,000 × *g* for 30 s.
16. Discard the supernatant.
17. Bring the yeast cell suspension back to $OD_{600} = 1$ by adding a volume of water equal to the OD_{600} recorded in **step 6** (for instance, 0.9 mL if OD_{600} was 0.9), or use to prepare stocks of frozen yeast-competent cells (*see* **Note 10**).
18. Vortex vigorously to resuspend the cells.
19. Transfer by pipetting 100-μL aliquots to sterile 1.5-mL microcentrifuge tubes.
20. Centrifuge at 21,000 × *g* for 30 s (*see* **Note 11**).
21. Remove supernatant.
22. Hold the cells on ice.

3.3 Yeast Transformation (Day 2)

1. For the preparation of DNA mixes, add 1 μg of both bait and prey constructs to a final volume of 34 μL of water in a 1.5-mL microcentrifuge tube (*see* **Note 12**).
2. In the meantime, boil just the needed aliquot of the carrier DNA for 10 min at 95 °C.
3. For the transformation mix, add 240 μL of 50 % (w/v) PEG, 36 μL of 1 M LiAc, 10 μL of 10 mg/mL carrier DNA, 34 μL of 30 ng/μL plasmid, 35 μL of sterile water, and 5 μL of 10 × TE (*see* **Note 13**).
4. Keep on ice.
5. Transfer the transformation mix to a microcentrifuge tube containing the yeast-competent cells.
6. Mix by pipetting up and down or vortexing shortly.
7. Incubate in a thermoblock set at 42 °C and shake at 300–400 rpm for 40 min (*see* **Note 14**).
8. Centrifuge at 21,000 × *g* for 30 s.

9. Remove the supernatant.
10. Resuspend the cells in 200 μL of sterile water.
11. Plate the cells on the appropriate selective SD medium either with a steel bacterial cell spreader or with sterile glass beads (*see* **Note 15**).
12. Incubate plates at 30 °C (*see* **Note 16**).

3.4 Isolation of Yeast Transformants (Day 5)

1. Pick a number (five is usually sufficient) of transformant colonies with sterile toothpicks for each bait–prey combination.
2. Streak on a fresh selective plate.
3. Grow for 2 days at 30 °C.

3.5 Verification of the Bait–Prey Interaction by Replica Plating (See Note 17)

3.5.1 Liquid Culture Overnight (Day 7)

1. Take a U-bottom 96-well plate (with lid) and fill it with 200 μL of the appropriate selective liquid SD medium per well (*see* **Note 18**).
2. Select three transformants per bait–prey combination.
3. Inoculate the yeast to the corresponding well with sterile toothpicks (*see* **Note 19**).
4. Seal the plate with a microporous tape sheet.
5. Incubate overnight at 30 °C and shake at 200 rpm.

3.5.2 Preparation of 96-Well Plates for Dilution Series (Day 8) (See Note 20)

1. Use a sterile flat-bottom, 96-well plate and fill it with 180 μL of sterile water per well.
2. For each well from the overnight culture, transfer the culture by pipetting up and down and transfer 20 μL to another (corresponding) well in the fresh flat-bottom plate filled with water.
3. Mix by pipetting up and down to generate a well-mixed tenfold dilution.
4. Repeat **steps 1–3** to generate a 100-fold dilution (*see* **Note 21**).

3.5.3 Replica Plate (Day 8)

1. Sterilize the replica plater (RP) twice by flaming it using ethanol.
2. Let it cool down.
3. Insert the RP in the flat-bottom 96-well plate with the diluted cultures.
4. Lift the RP to the middle of the culture, keeping it perfectly horizontal.
5. Mix cells by rotating the RP horizontally by hand.
6. Lift the RP out the 96-well plate (*see* **Notes 22** and **23**).
7. Put the RP perfectly horizontal on the selective plate (*see* **Note 22**).

Table 3
Drop out media for the different systems

System	Control media	Selective media
GAL4 (Y2H)	SD Base-Leu-Trp	SD Base-Leu-Trp-His
GAL4 (Y3H)	SD Base-Leu-Trp-Ura	SD Base-Leu-Trp-Ura-His
LexA	SD Base Gal/Raf-Ura-Trp-His	SD Base Gal/Raf-Ura-Trp-His + 1 × BU salts + X-gal

8. Lift the RP from the medium (*see* **Notes 22** and **24**).
9. Let the droplets dry in the laminar flow cabinet for 5 to 10 min.
10. Repeat **steps 3–9** for replica plating of the transformants on the desired number of control and selective media (*see* Table 3).
11. Incubate the plates at 30 °C for 2 days (day 10).
12. Once colonies start appearing, score growth or staining (*see* **Note 20**).
13. Store plates at 4 °C.

4 Notes

1. YPD medium is a rich medium used to grow yeast without plasmid selection. Yeast with mutations in adenylosuccinate synthetase *(Ade2)* accumulate red pigment and are slowed down in growth. Adding extra adenine to the media is required (for instance PJ69-4 and AH109 strains) [17, 18].
2. Different results can be obtained depending on the chosen system, most prominently illustrated by the reported JAZ homodimerization and heterodimerization. The LexA system yielded 47 out of 132 possible interactions versus 14 for the GAL4 system. Although the LexA system might be more sensitive, four interactions observed with GAL4 were not detected with it [19, 20]. Moreover, a GAL4-based system with different setups can yield different interactions [21].
3. For the GAL4 system, the most commonly used strains are PJ69-4 [22] and the derivative AH109 (Clontech), in which the endogenous *GAL4* and the inhibitory *GAL80* genes are knocked out. For the LexA system, the strain EGY48 (p8opLacZ) is widely used [14, 15] and allows the use of the *GAL1*-inducible promoter, because it does not require the knocked out *GAL4*.

4. Some of the vectors carry epitope tags (i.e., Myc and hemagglutinin) that allow the verification of the fusion protein production by immunoblot analysis. The vectors pGAD424 and pGBT9 are very similar to pGADT7 and pGBKT7, but lack epitopes and T7. In the Gateway®-compatible pDEST22 and pDEST32 vectors (Invitrogen), the auxotrophic markers are swapped compared to the vectors above, so they cannot be combined.
5. We focus on the use of Gateway®-compatible vectors and we refer to the manufacturer's guidelines (Invitrogen) for detailed cloning protocols. The entry clones in the pDONR223 (pENTR223) can be obtained for most of the *Arabidopsis* genes from the Arabidopsis Biological Resource Center (http://www.abrc.osu.edu/).
6. When using the Multisite Gateway vectors for bridging protein expression, the ORF must be cloned without stop codon to allow C-terminal fusion. In a MultiSite Gateway reaction this ORF can then be combined with any desired yeast promoter and tag. Available tags include NLS-3xV5, NLS-3xc-myc, and NLS-3xFLAG-6xHIS. These tags ensure nuclear localization and detection by immunoblot complementary to tags used in the Y2H vectors. For details about MultiSite Gateway cloning *see* ref. 13.
7. When the yeast culture is started from a fresh plate, resuspend the cells first in 1 mL of YPDA liquid medium in a microcentrifuge tube and vortex vigorously during 5 min to break yeast clumps. Transfer this 1-mL aliquot to set up the overnight culture.
8. This 50-mL volume is sufficient for ten transformations and should be adapted according to the required transformation number.
9. To monitor the OD600 of the overnight culture, prepare 1/10 dilutions of the culture with the YPDA medium, because the overnight culture will have an OD600 > 1.
10. To prepare frozen yeast-competent cells, resuspend the cells in a volume corresponding to the recorded OD_{600} with a 5 % (v/v) glycerol and 10 % (v/v) DMSO. Transfer to microcentrifuge tubes and freeze the cells gradually and slowly to guarantee cell survival (for instance with a MrFrosty container; Nalgene, Rochester, NY, USA). Cells stored at −80 °C can be used for up to 1 year [23].
11. If frozen yeast-competent cells are used for transformation, thaw the cell samples at 37 °C. For each transformation, aliquot 100 μL in microcentrifuge tubes, spin down at 17,000 × *g* for 2 min, and proceed to **step 20** of **step 3**, Subheading 3.3.
12. In the case of Y3H assays, mix 1 μg of each of the three constructs.

13. The volumes are given for one transformation reaction. Preparation of a master mix is recommended, including all components except the plasmid DNA, if multiple transformations are done simultaneously.
14. Incubation period and temperature might differ depending on the yeast strain used [16]. For the yeast strains used here, incubation at 42 °C for 40 min is efficient.
15. For the GAL4 and LexA Y2H systems, use SD Base Leu-Trp and SD Base Gal/Raf-Ura-Trp-His, respectively. In the Y3H presented here, use SD Base Leu-Trp-Ura.
16. Transformants are visible after 3–4 days.
17. This method gives the greatest reproducibility and nicest pictures (*see* Fig. 2). Alternatively, independent transformant cultures can be streaked or dropped manually on the appropriate selective plates to assess bait–prey interactions.
18. One well needs to be filled per independent transformant to be tested.
19. Normalizing ODs at this point is not necessary.
20. By using the 96-well replica plate system, up to 48 different bait–prey combinations (in two dilutions each) can be tested in one plate. However, because it is highly recommended to test three biological replicates per bait–prey combination, we normally test up to 16 combinations, including negative controls (for instance, bait and prey with the corresponding empty vectors) to detect false positives derived from the autoactivation of either bait and prey constructs, and eventually also positive controls (known interactors). Sometimes, false negatives appear among the three replicates. Western blotting is then recommended to confirm that both bait and prey fusion proteins are expressed.
21. A 1,000-fold dilution is optional. For a typical experimental setup, *see* Fig. 2.
22. It is very important that **steps 5–7** of Subheading 3.5.3 are performed in a fluent, swift manner, because hesitation will lead to unequal loading or deformed droplets. Also make sure that the replica plater does not move horizontally when in contact with the plate medium.
23. On every pin, an equal amount (approximately 2–3 μL) of yeast culture will be loaded.
24. Droplets are formed and an equal volume per well is expected to be plated.

References

1. Pauwels L, Goossens A (2011) The JAZ proteins: a crucial interface in the jasmonate signaling cascade. Plant Cell 23:3089–3100
2. Brückner A, Polge C, Lentze N, Auerbach D, Schlattner U (2009) Yeast two-hybrid, a powerful tool for systems biology. Int J Mol Sci 10:2763–2788
3. Shoemaker BA, Panchenko AR (2007) Deciphering protein–protein interactions. Part I. experimental techniques and databases. PLoS Comput Biol 3:e42
4. Berggård T, Linse S, James P (2007) Methods for the detection and analysis of protein-protein interactions. Proteomics 7:2833–2842
5. Thines B, Katsir L, Melotto M, Niu Y, Mandaokar A, Liu G, Nomura K, He SY, Howe GA, Browse J (2007) JAZ repressor proteins are targets of the SCF^{COI1} complex during jasmonate signalling. Nature 448:661–665
6. Causier B (2004) Studying the interactome with the yeast two-hybrid system and mass spectrometry. Mass Spectrom Rev 23:350–367
7. Cusick ME, Klitgord N, Vidal M, Hill DE (2005) Interactome: gateway into systems biology. Hum Mol Genet 14:R171–R181
8. Pauwels L, Fernández Barbero G, Geerinck J, Tilleman S, Grunewald W, Cuéllar Pérez A, Chico JM, Vanden Bossche R, Sewell J, Gil E, García-Casado G, Witters E, Inzé D, Long JA, De Jaeger G, Solano R, Goossens A (2010) NINJA connects the co-repressor TOPLESS to jasmonate signalling. Nature 464:788–791
9. Fields S, O-k S (1989) A novel genetic system to detect protein-protein interactions. Nature 340:245–246
10. Causier B, Davies B (2002) Analysing protein-protein interactions with the yeast two-hybrid system. Plant Mol Biol 50:855–870
11. Causier B, Ashworth M, Guo W, Davies B (2012) The TOPLESS interactome: a framework for gene repression in Arabidopsis. Plant Physiol 158:423–438
12. Arabidopsis Interactome Mapping Consortium (2011) Evidence for network evolution in an *Arabidopsis* interactome map. Science 333:601–607
13. Nagels Durand A, Moses T, De Clercq R, Goossens A, Pauwels L (2012) A MultiSite Gateway™ vector set for the functional analysis of genes in the model *Saccharomyces cerevisiae*. BMC Mol Biol 13:30
14. Estojak J, Brent R, Golemis EA (1995) Correlation of two-hybrid affinity data with in vitro measurements. Mol Cell Biol 15:5820–5829
15. Gyuris J, Golemis E, Chertkov H, Brent R (1993) Cdi1, a human G1 and S phase protein phosphatase that associates with Cdk2. Cell 75:791–803
16. Gietz RD, Schiestl RH (2007) High-efficiency yeast transformation using the LiAc/SS carrier DNA/PEG method. Nat Protoc 2:31–34
17. Bergman L (2001) Growth and maintenance of yeast. Methods Mol Biol 177:9–14
18. Saghbini M, Hoekstra D, Gautsch J (2001) Media formulations for various two-hybrid systems. Methods Mol Biol 177:15–39
19. Chini A, Fonseca S, Chico JM, Fernández-Calvo P, Solano R (2009) The ZIM domain mediates homo- and heteromeric interactions between Arabidopsis JAZ proteins. Plant J 59:77–87
20. Chung HS, Howe GA (2009) A critical role for the TIFY motif in repression of jasmonate signaling by a stabilized splice variant of the JASMONATE ZIM-domain protein JAZ10 in *Arabidopsis*. Plant Cell 21:131–145
21. Rajagopala SV, Hughes KT, Uetz P (2009) Benchmarking yeast two-hybrid systems using the interactions of bacterial motility proteins. Proteomics 9:5296–5302
22. James P, Halladay J, Craig EA (1996) Genomic libraries and a host strain designed for highly efficient two-hybrid selection in yeast. Genetics 144:1425–1436
23. Gietz RD, Schiestl RH (2007) Frozen competent yeast cells that can be transformed with high efficiency using the LiAc/SS carrier DNA/PEG method. Nat Protoc 2:1–4

Chapter 15

Modified Bimolecular Fluorescence Complementation Assay to Study the Inhibition of Transcription Complex Formation by JAZ Proteins

Tiancong Qi, Susheng Song, and Daoxin Xie

Abstract

The jasmonate (JA) ZIM-domain (JAZ) proteins of *Arabidopsis thaliana* repress JA signaling and negatively regulate the JA responses. Recently, JAZ proteins have been found to inhibit the transcriptional function of several transcription factors, among which the basic helix-loop-helix (bHLH) (GLABRA3 [GL3], ENHANCER OF GLABRA3 [EGL3], and TRANSPARENT TESTA8 [TT8]) and R2R3-MYB (GL1 and MYB75) that can interact with each other to form bHLH–MYB complexes and further control gene expression. The bimolecular fluorescence complementation (BiFC) assay is a widely used technique to study protein–protein interactions in living cells. Here we describe a modified BiFC experimental procedure to study the inhibition of the formation of the bHLH (GL3)–MYB (GL1) complex by JAZ proteins.

Key words JAZ protein, Bimolecular fluorescence complementation, Transcription factor, bHLH, R2R3-MYB

1 Introduction

Jasmonates (JAs), including jasmonic acid and its oxylipin derivatives, are plant hormones that play essential roles in multiple plant developmental processes, including stamen development, root growth, anthocyanin accumulation, and trichome initiation, and that also function in plant defense responses against insect attack and pathogen infection [1–4]. In *Arabidopsis thaliana*, the JA ZIM-domain (JAZ) proteins consist of 12 members (JAZ1 to JAZ12) and negatively regulate JA responses by directly inhibiting diverse downstream transcription factors [5–13]. Bioactive JAs and coronatine, a JA mimic, can activate the recognition of JAZ proteins by the SCF^{COI1} perception complex and, subsequently, the JAZ degradation through the 26S proteasome [13–18]. The JAZ-repressed transcription factors include the basic helix-loop-helix (bHLH)-type factors GLABRA3 (GL3), ENHANCER OF GLABRA3 (EGL3)

Alain Goossens and Laurens Pauwels (eds.), *Jasmonate Signaling: Methods and Protocols*, Methods in Molecular Biology, vol. 1011, DOI 10.1007/978-1-62703-414-2_15, © Springer Science+Business Media, LLC 2013

and TRANSPARENT TESTA8 (TT8), MYC2/MYC3/MYC4, and R2R3-MYB-type factors GL1, MYB75, and MYB21/MYB24 [5–10]. The bHLH factors (GL3, EGL3, and TT8) were found to interact with the R2R3-MYB factors (GL1 and MYB75) and the WD-repeat protein TRANSPARENT TEST GLABRA1 (TTG1) to form the bHLH/MYB/TTG1 transcriptional complexes that regulate multiple plant responses, including anthocyanin accumulation and trichome initiation [8, 19, 20]. The bHLH–MYB interaction is essential for the formation and transcriptional function of the bHLH/MYB/TTG1 complex.

The bimolecular fluorescence complementation (BiFC) assay is widely used to explore protein–protein interactions in living cells. Modified BiFC assays are further used to study mechanisms of molecular proximity, such as competition between alternative interaction partners [21–24]. The BiFC technique is based on the reconstitution of a fluorescent protein from its two nonfluorescent fragments. Two candidate proteins are fused to the N-terminal and C-terminal fragments of a fluorescent protein. If the two proteins interact with each other, the two nonfluorescent fragments are drawn very closely and reconstitute a fluorescent protein. Here we describe a modified BiFC experimental procedure to study the effect of JAZ proteins on the interaction of GL3 with GL1 and the subsequent formation of the bHLH/MYB/TTG1 complex.

2 Materials

2.1 Equipment

1. Electroporation apparatus ECM 399 (BTX Instrument Division, Harvard Apparatus, Inc., Holliston, MA, USA).
2. Needleless syringe.
3. Fluorescent microscope LSM710 (Carl Zeiss, Göttingen, Germany) with the ZEN 2009 software.
4. Real-time PCR system (ABI7500) (Applied Biosystems, Foster City, CA, USA).

2.2 Plasmids and *Agrobacterium tumefaciens* Strain

1. The binary vectors nYFP and cYFP contain the N-terminal domain (amino acids 1–155) and the C-terminal fragment (amino acids 156–239) of the yellow fluorescent protein (YFP), respectively.
2. The full-length coding sequences of the *GL1* and *GL3* genes of *A. thaliana* fused with the nYFP or cYFP fragments to generate the GL1-nYFP and cYFP-GL3 vectors, respectively [8].
3. The coding sequence of the *Arabidopsis JAZ3* gene cloned into a modified binary vector pCAMBIA1300 (Cambia, Brisbane, QLD, Australia) with four JAZ3-driving cauliflower mosaic virus 35S promoters to generate Super1300-JAZ3.

4. The pBI121 binary vector that contains an intron-containing β-glucuronidase (GUS) reporter gene.
5. The *A. tumefaciens* strain GV3101, of which the competent cells are prepared and stored at −80 °C.
6. Antibiotics: Gentamycin, rifampicin, and another antibiotic conferred by the transformed plasmid (*see* **Note 1**).

2.3 Culture Media and Buffers

1. Murashige and Skoog (MS) medium: 4.3 g of MS basal salt mixture (Sigma-Aldrich, St. Louis, MO, USA), stored at 4 °C, supplemented with 20 g of sucrose (*see* **Note 2**), dissolved in approximately 800 mL of distilled water. Add 7 g agar and additional distilled water to obtain a total volume of 1 L and adjust pH (with KOH) to 5.85–5.95. Autoclave at 121 °C for 15–20 min.
2. Luria-Bertani (LB) medium: 10 g NaCl, 5 g yeast extract, and 10 g tryptone dissolved in approximately 800 mL of distilled water. Add 15 g agar (only for solid LB medium) and additional distilled water to obtain a total volume of 1 L. Adjust pH to 7.0 and autoclave at 121 °C for 15–20 min.
3. Infiltration buffer: 10 mM of 2-*mercapto*-ethane sulfonic acid (MES) (Amresco, Solon, OH, USA), 10 mM $MgCl_2$ and 0.2 mM acetosyringone (Sigma-Aldrich), dissolved with dimethylformamide to a 0.2 M concentration and stored at −20 °C.
4. Coronatine: The stock solution (1 mg/mL) (Sigma-Aldrich) dissolved with ethanol and diluted with infiltration buffer to the final concentration of 5 μM.
5. 4′,6-diamidino-2-phenylindole (DAPI) solution: Dissolve the solid powder (Roche Diagnostics, Shanghai, China) with double-distilled water (ddH_2O) to 10 mg/mL as stock solution and store at −20 °C. The final concentration is 10 μg/mL (diluted with infiltration buffer).

2.4 Plant Material

1. *Nicotiana benthamiana* seeds disinfected in 20 % bleach for 10–20 min, rinsed five times with distilled water, plated on MS medium, and cultivated for 10 days in a growth chamber under a 16-h light (26–28 °C)/8-h dark (21–23 °C) photoperiod (light intensity is approximately 300 μmol m^{-2} s^{-1}). Thereafter, the *N. benthamiana* seedlings are transferred to soil composed of a 1:1 volume ratio of loam and vermiculite for 40 days before use for the BiFC assay.

2.5 RNA Extraction and Quantitative Real-Time PCR

1. TRIzol reagent (Invitrogen, Carlsbad, CA, USA) stored at 4 °C.
2. RNA precipitation solution: 0.8 M sodium citrate, 1.2 M sodium chloride, 0.1 % diethyl pyrocarbonate (DEPC) is added, incubated at room temperature overnight, autoclaved at 121 °C for 20–30 min, and stored at 4 °C.

3. Reverse Transcriptase M-MLV (RNase H⁻) kit (TaKaRa, Otsu, Japan) stored at −20 °C.
4. SYBR Premix Ex Taq™ (Perfect Real Time) kit (TaKaRa), stored in the dark at 4 °C.
5. SYBR Green Real-time PCR kit (TaKaRa).
6. ROX reference dye II (TaKaRa).
7. DNase (New England Biolabs, Ipswich, MA, USA) and DNase buffer.

3 Methods

3.1 *Agrobacterium*-Mediated Electric Transformation

1. Place GV3101-competent cells (*see* **Note 3**), plasmid DNA (nYFP, cYFP, GL1-nYFP, cYFP-GL3, Super1300-JAZ3, and pBI121), and 0.2-cm-gap electroporation cuvette (*see* **Note 4**) on ice.
2. Add 1–3 μL of plasmid DNA to the competent cells and mix well.
3. Transfer the mixture into a chilled cuvette (*see* **Note 5**).
4. Set the voltage of the electroporation apparatus to 2.0–2.2 kV.
5. Wipe and dry the outer surface of the cuvette and put it into the holder.
6. Press the button to deliver the pulse for approximately 4–6 ms.
7. Immediately, add 1 mL of LB medium into the cuvette and mix well.
8. Transfer the mixture into a 1.5-mL centrifuge tube and incubate at 28 °C with shaking at 150–180 rpm for approximately 1 h (*see* **Note 6**).
9. Spray 100 μL of cells on the LB medium containing the appropriate antibiotic(s) and incubate at 28 °C for 2–3 days.

3.2 BiFC Assay

The BiFC assay is applied to test whether the vectors nYFP, cYFP, GL1-nYFP, and cYFP-GL3 are functional.

1. Inoculate separately the successfully transformed *A. tumefaciens* strains containing nYFP, cYFP, GL1-nYFP, and cYFP-GL3 in 3 mL of antibiotic-containing LB liquid medium.
2. Incubate at 28 °C with shaking at 230 rpm for 1–2 days (*see* **Note 7**).
3. Harvest the *A. tumefaciens* cells with a 1.5-mL centrifuge tube and centrifuge at 1,500 × *g* at room temperature for 4 min.
4. Remove the supernatant (*see* **Note 8**).
5. Resuspend the cells in the infiltration buffer to an ultimate concentration of $OD_{600} = 0.5$.

6. Mix together equal volumes of different *A. tumefaciens* resuspension solutions in 2-mL centrifuge tubes to generate the combinations GL1-nYFP/cYFP-GL3, GL1-nYFP/cYFP, and nYFP/cYFP-GL3 (*see* **Note 9**).
7. Keep at room temperature for 3 h.
8. By means of a needleless syringe, infiltrate different areas of the same *N. benthamiana* leaf with equal volumes of the different combinations of *A. tumefaciens* strains.
9. Place the *N. benthamiana* plants at 24 °C for approximately 50 h before the YFP fluorescence is detected by the fluorescent microscope.
10. Infiltrate the *N. benthamiana* leaves with 10 μg/mL of DAPI solution 2 h before the YFP fluorescence detection (*see* **Note 10**).

3.3 Modified BiFC Assay

1. Inoculate the *A. tumefaciens* strains containing GL1-nYFP, cYFP-GL3, Super1300-JAZ3, and pBI121 in 3 mL of LB liquid medium supplemented with the appropriate antibiotics (*see* Subheading 3.2, **step 1**).
2. Harvest cells and resuspend in the infiltration buffer to an ultimate concentration of $OD_{600} = 0.5$ (*see* Subheading 3.1, **step 3**).
3. Mix together equal volumes of different *A. tumefaciens* resuspension solutions to generate the combinations GL1-nYFP/cYFP-GL3, GL1-nYFP/cYFP-GL3/Super1300-JAZ3, and GL1-nYFP/cYFP-GL3/pBI121.
4. Incubate at room temperature for 3 h (*see* **Note 11**).
5. With a needleless syringe, infiltrate different parts of *N. benthamiana* leaves (Fig. 1a, b) with equal volumes of different combinations of *A. tumefaciens* strains.
6. Grow the *N. benthamiana* plants infiltrated with *A. tumefaciens* in a growth chamber under a 16-h light (22–24 °C)/8-h dark (17–19 °C) photoperiod for approximately 50 h, where after the YFP fluorescence it is detected by fluorescent microscopy (Fig. 1c).
7. Infiltrate the *N. benthamiana* leaves with 5 μM coronatine (or solvent buffer as control) (Fig. 1b) for 10 h before the YFP fluorescence detection (*see* **Note 12**).
8. Analyze the YFP fluorescence intensity quantitatively with the ZEN 2009 software and draw bar graphs (Fig. 1d, e) (*see* **Note 13**).

3.4 Quantitative Real-Time PCR to Verify Gene Expression

The transiently expressed genes in *N. benthamiana* leaves are detected by a quantitative real-time PCR assay. The leaves coinfiltrated with the various combined constructs (Fig. 1a, b) are further used for RNA extraction and real-time PCR analysis (*see* **Note 14**).

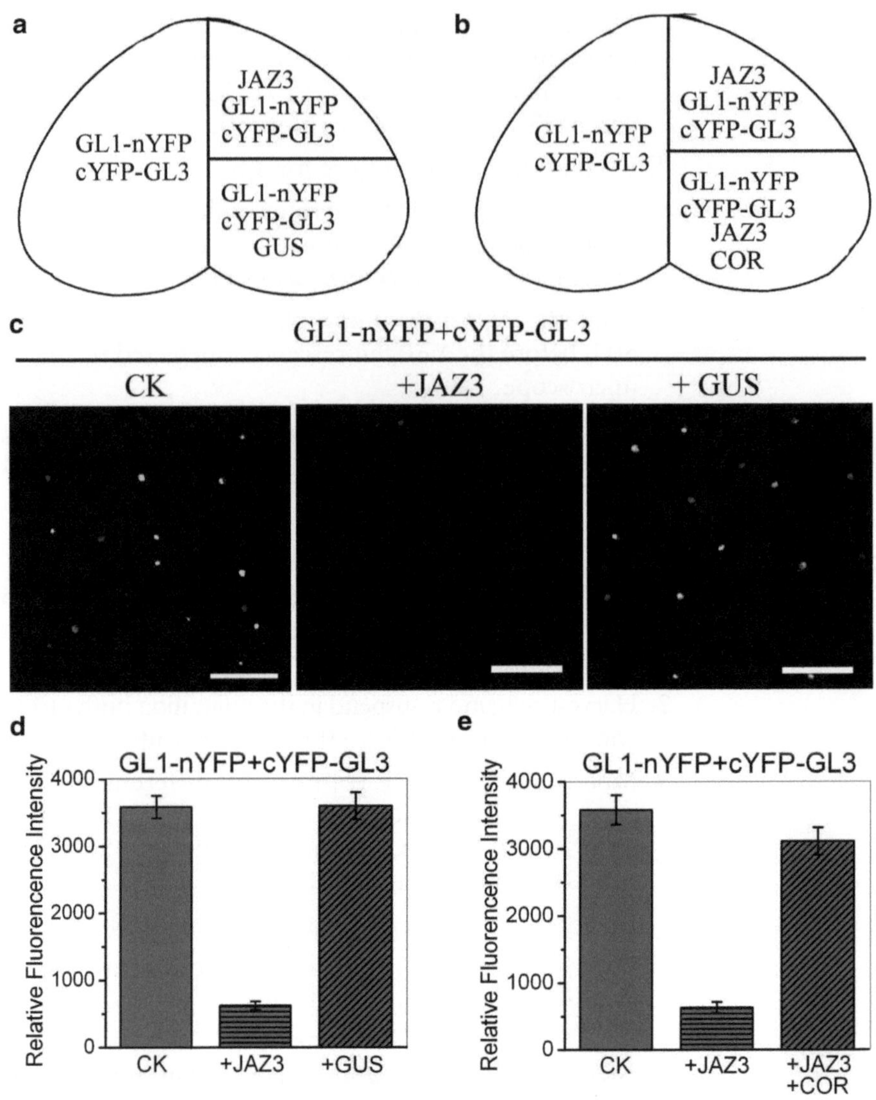

Fig. 1 Modified BiFC assay to study the inhibition of the interaction between GL3 and GL1 by JAZ3. (**a**, **b**) Schematic diagrams for the coinfiltration of *Agrobacterium tumefaciens* strains in *Nicotiana benthamiana* leaves. (**c**) Modified BiFC assay showing that JAZ3 attenuates the interaction of GL3 with GL1. The YFP fluorescence was detected 50 h after coproduction of GL1-nYFP/cYFP-GL3 (CK), JAZ3/GL1-nYFP/cYFP-GL3 (+JAZ3), or GUS/GL1-nYFP/cYFP-GL3 (+GUS). The scale bar represents 100 μm. (**d**) Quantitative data of YFP fluorescence intensity. Fifty independent fluorescent spots were assessed for average fluorescence intensity. Error bars represent SE. (**e**) Quantitative data of YFP fluorescence intensity revealing that coronatine attenuates the JAZ3-mediated inhibition of the interaction of GL3 with GL1. Fifty independent fluorescent spots were assessed for the quantitative data. Error bars represent SE

3.4.1 RNA Extraction

1. Place a proper amount (approximately 0.1 g) of *N. benthamiana* leaves in a mortar with enough liquid nitrogen and grind quickly before complete volatilization (*see* **Note 15**).
2. Quickly transfer the homogenized power into a cold, previously soaked in liquid nitrogen, 1.5-mL RNase-free tube.

3. Add 1 mL of TRIzol reagent.
4. Mix the mixture thoroughly and incubate at room temperature for 5 min.
5. Add 0.2 mL of chloroform.
6. Shake vigorously for 15 s.
7. Incubate at room temperature for 5 min.
8. Centrifuge the samples at 12,000 × *g* for 15 min at 2–8 °C (*see* **Note 16**).
9. Transfer the aqueous phase to a new DEPC-treated tube.
10. Add 1/2 volume of isopropyl and 1/2 volume of high-salt RNA precipitation solution to the aqueous phase.
11. Mix well and incubate at room temperature for 10 min.
12. Centrifuge at 12,000 × *g* for 10 min at 2–8 °C.
13. Remove the supernatant.
14. Very gently wash the pellet with 1 mL of 75 % ethanol.
15. Centrifuge at no more than 7,500 × *g* for 5 min at 2–8 °C (*see* **Note 17**).
16. Remove the supernatant and air-dry the RNA pellet at room temperature (*see* **Note 18**).
17. Dissolve the RNA pellet in RNase-free water and incubate at 55–60 °C for 10 min.
18. Store at −80 °C.

3.4.2 Reverse Transcription

1. Add 2–3 μg total RNA, 0.5 μL of DNase, 1 μL of DNase buffer, and additional DEPC-treated ddH_2O to obtain the ultimate volume to 10 μL.
2. Incubate at 37 °C for 10 min.
3. Add 0.1 μL of 0.5 M EDTA.
4. Inactivate at 75 °C for 10 min.
5. Add 5 μL of 10 μM oligo(dT) and mix thoroughly.
6. Incubate at 75 °C for 10 min.
7. Immediately, put on ice.
8. Add 0.6 μL reverse transcriptase M-MLV (RNase H^-), 4 μL of 5 × M-MLV buffer, 2 μL 2.5 mM of dNTP mixture, and 0.4 μL of RNase inhibitor.
9. Incubate at 42 °C for 1 h and 70 °C for another 15 min, and then store at −20 °C (*see* **Note 19**).

3.4.3 Quantitative Real-Time PCR

1. Perform the real-time PCR analyses with the SYBR Green Real-time PCR kit and the ABi7500 real-time PCR system.
2. For each reaction system, add 1 μL of cDNA, 10 μL of 2 × SYBR Premix Ex Taq, 0.4 μL of 5′ forward primer, 0.4 μL of 3′

reverse primer, 0.4 μL of 50×ROX Reference Dye II, and 7.8 μL of ddH_2O to obtain the total volume to 20 μL.

3. Proceed with the real-time PCR amplification with a holding stage at 95 °C for 30 s, cycling stages at 95 °C for 5 s and at 60 °C for 34 s (repeated for 40 cycles), and a melt curve stage at 95 °C for 15 s, at 60 °C for 1 min and increased to 95 °C at 1 % speed, and at 95 °C for 15 s (*see* **Note 20**).

4 Notes

1. The *A. tumefaciens* strain GV3101 is resistant to gentamycin and rifampicin, so positive clones are selected with three antibiotics: gentamycin, rifampicin, and another antibiotic that is conferred by the transformed plasmid. The transformants are verified by PCR.
2. Also 3 % (w/v) sucrose can be used because it is not very important if not carbon-related biological processes are studied.
3. It is always better to use freshly prepared competent cells, because the transformation efficiency decreases with increasing storage time. For each plasmid transformation, the used volume of competent cells should be more than 20 μL. Usually, we use 40–50 μL.
4. The electroporation cuvette is kept in absolute ethanol at room temperature. Before electroporation, it is placed in a ventilation cabinet to make sure that the remaining ethanol is evaporated completely.
5. Be careful not to produce bubbles during the transfer process. The appropriate concentration of plasmid DNA is approximately 10–100 ng/μL.
6. The rotational speed should be less than 200 rpm.
7. The GV3101 cells should be fresh before inoculation. They are usually harvested at the logarithmic growth phase.
8. It is not necessary to work under aseptic conditions. Normally, a vacuum pump is used to completely remove the supernatant.
9. The total volume of each combination needed depends on the experimental design. Generally, 500 μL of different *A. tumefaciens* resuspension solutions are added in each combination. To obtain high transformation efficiency, the *A. tumefaciens* combinations are kept at room temperature for approximately 3–5 h before inoculation.
10. The YFP signal is visible 45–60 h after the *Agrobacterium* infiltration. The duration is influenced by the interaction strength

of the two proteins. The stronger the interaction, the shorter the time needed. The experiment is repeated at least three times.

11. Usually, we add 500 μL of cells of each strain for each combination. The ultimate volume for each combination must be the same, meaning that in the GL1-nYFP/cYFP-GL3 combination, 500 μL additional infiltration buffer is required.
12. As the coronatine inoculation volume should be equal or larger than the inoculation volume of the *Agrobacterium* strains, we inoculate 50 μL of *Agrobacterium* combination solution and 70 μL of coronatine solution. The different combinations (Fig. 1a, b) are detected under the microscope with identical gain settings. The experiment is repeated at least three times and, generally, five *N. benthamiana* leaves are inoculated for each repetition.
13. Fifty independent fluorescent spots are assessed for fluorescence intensity to generate the quantitative data.
14. Protein production can also be detected by immunoblot assay, if suitable antibodies are available or through tags (for example, the myc-, flag-, or HA-tag).
15. Liquid nitrogen is added in the mortar 3–4 times just before its complete volatilization.
16. After centrifugation, the mixture separates into a bottom phenol–chloroform phase, an interphase, and a top aqueous phase, in which the total RNA is mainly dissolved. The volume of the aqueous phase is approximately 60 % of that of the TRIzol reagent used for homogenization.
17. Do not disturb the pellet because the small RNAs are partially soluble in the 75 % ethanol without salt.
18. Do not dry the RNA pellet completely; otherwise the solubility decreases.
19. Usually, these steps are done with the PCR equipment and the hot-lid procedure is started to avoid liquid evaporation.
20. Primers used for real-time PCR analysis are designed by the Roche software on https://www.roche-applied-science.com/sis/rtpcr/index.jsp. *Nicotiana Actin* (JQ256516) is used as the internal control. The real-time PCR data are analyzed with the 7500 Software v2.0.6.

Acknowledgments

This research was supported by the National Science Foundation of China (91017012) and grants from the National Basic Research Program of China (973 Program 2011CB915404).

References

1. Browse J (2009) Jasmonate passes muster: a receptor and targets for the defense hormone. Annu Rev Plant Biol 60:183–205
2. Howe GA, Jander G (2008) Plant immunity to insect herbivores. Annu Rev Plant Biol 59:41–66
3. Hause B, Wasternack C, Strack D (2009) Jasmonates in stress responses and development. Phytochemistry 70:1483–1484
4. Creelman RA, Mullet JE (1997) Biosynthesis and action of jasmonates in plants. Annu Rev Plant Physiol Plant Mol Biol 48:355–381
5. Cheng Z, Sun L, Qi T, Zhang B, Peng W, Liu Y, Xie D (2011) The bHLH transcription factor MYC3 interacts with the jasmonate ZIM-domain proteins to mediate jasmonate response in *Arabidopsis*. Mol Plant 4:279–288
6. Niu Y, Figueroa P, Browse J (2011) Characterization of JAZ-interacting bHLH transcription factors that regulate jasmonate responses in *Arabidopsis*. J Exp Bot 62: 2143–2154
7. Fernández-Calvo P, Chini A, Fernández-Barbero G, Chico J-M, Gimenez-Ibanez S, Geerinck J, Eeckhout D, Schweizer F, Godoy M, Franco-Zorrilla JM, Pauwels L, Witters E, Puga MI, Paz-Ares J, Goossens A, Reymond P, De Jaeger G, Solano R (2011) The *Arabidopsis* bHLH transcription factors MYC3 and MYC4 are targets of JAZ repressors and act additively with MYC2 in the activation of JA responses. Plant Cell 23:701–715
8. Qi T, Song S, Ren Q, Wu D, Huang H, Chen Y, Fan M, Peng W, Ren C, Xie D (2011) The jasmonate-ZIM-domain proteins interact with the WD-repeat/bHLH/MYB complexes to regulate jasmonate-mediated anthocyanin accumulation and trichome initiation in *Arabidopsis thaliana*. Plant Cell 23:1795–1814
9. Song S, Qi T, Huang H, Ren Q, Wu D, Chang C, Peng W, Liu Y, Peng J, Xie D (2011) The jasmonate-ZIM domain proteins interact with the R2R3-MYB transcription factors MYB21 and MYB24 to affect jasmonate-regulated stamen development in *Arabidopsis*. Plant Cell 23:1000–1013
10. Zhu Z, An F, Feng Y, Li P, Xue L, A M, Jiang Z, Kim J-M, To TK, Li W, Zhang X, Yu Q, Dong Z, Chen W-Q, Seki M, Zhou J-M, Guo H (2011) Derepression of ethylene-stabilized transcription factors (EIN3/EIL1) mediates jasmonate and ethylene signaling synergy in *Arabidopsis*. Proc Natl Acad Sci USA 108:12539–12544
11. Pauwels L, Fernández Barbero G, Geerinck J, Tilleman S, Grunewald W, Cuéllar Pérez A, Chico JM, Vanden Bossche R, Sewell J, Gil E, García-Casado G, Witters E, Inzé D, Long JA, De Jaeger G, Solano R, Goossens A (2010) NINJA connects the co-repressor TOPLESS to jasmonate signalling. Nature 464:788–791
12. Santner A, Estelle M (2007) The JAZ proteins link jasmonate perception with transcriptional changes. Plant Cell 19:3839–3842
13. Thines B, Katsir L, Melotto M, Niu Y, Mandaokar A, Liu G, Nomura K, He SY, Howe GA, Browse J (2007) JAZ repressor proteins are targets of the SCF^{COI1} complex during jasmonate signalling. Nature 448: 661–665
14. Sheard LB, Tan X, Mao H, Withers J, Ben-Nissan G, Hinds TR, Kobayashi Y, Hsu F-F, Sharon M, Browse J, He SY, Rizo J, Howe GA, Zheng N (2010) Jasmonate perception by inositol-phosphate-potentiated COI1-JAZ co-receptor. Nature 468:400–405
15. Fonseca S, Chini A, Hamberg M, Adie B, Porzel A, Kramell R, Miersch O, Wasternack C, Solano R (2009) (+)-7-*iso*-Jasmonoyl-l-isoleucine is the endogenous bioactive jasmonate. Nat Chem Biol 5:344–350
16. Yan J, Zhang C, Gu M, Bai Z, Zhang W, Qi T, Cheng Z, Peng W, Luo H, Nan F, Wang Z, Xie D (2009) The *Arabidopsis* CORONATINE INSENSITIVE1 protein is a jasmonate receptor. Plant Cell 21:2220–2236
17. Katsir L, Schilmiller AL, Staswick PE, He SY, Howe GA (2008) COI1 is a critical component of a receptor for jasmonate and the bacterial virulence factor coronatine. Proc Natl Acad Sci USA 105:7100–7105
18. Chini A, Fonseca S, Fernández G, Adie B, Chico JM, Lorenzo O, García-Casado G, López-Vidriero I, Lozano FM, Ponce MR, Micol JL, Solano R (2007) The JAZ family of repressors is the missing link in jasmonate signalling. Nature 448:666–671
19. Zhang F, Gonzalez A, Zhao M, Payne CT, Lloyd A (2003) A network of redundant bHLH proteins functions in all TTG1-dependent pathways of *Arabidopsis*. Development 130:4859–4869
20. Pesch M, Hülskamp M (2009) One, two, three ... models for trichome patterning in *Arabidopsis*? Curr Opin Plant Biol 12: 587–592
21. Hu C-D, Chinenov Y, Kerppola TK (2002) Visualization of interactions among bZIP and

Rel family proteins in living cells using bimolecular fluorescence complementation. Mol Cell 9:789–798

22. Magliery TJ, Wilson CGM, Pan W, Mishler D, Ghosh I, Hamilton AD, Regan L (2005) Detecting protein–protein interactions with a green fluorescent protein fragment reassembly trap: scope and mechanism. J Am Chem Soc 127:146–157

23. Hu C-D, Kerppola TK (2003) Simultaneous visualization of multiple protein interactions in living cells using multicolor fluorescence complementation analysis. Nat Biotechnol 21: 539–545

24. Kerppola TK (2008) Bimolecular fluorescence complementation (BiFC) analysis as a probe of protein interactions in living cells. Annu Rev Biophys 37:465–487

Chapter 16

Agroinfiltration of *Nicotiana benthamiana* Leaves for Co-localization of Regulatory Proteins Involved in Jasmonate Signaling

Volkan Çevik and Kemal Kazan

Abstract

Protein–protein interactions play important roles in many cellular processes, including the regulation of phytohormone signaling pathways. Identification of interacting partners of key proteins involved in the cellular signaling control can provide potentially unexpected insights into the molecular events occurring in any signaling pathway. Over the years, various techniques have been developed to examine protein–protein interactions, but, besides certain advantages, most of them have various pitfalls, such as yielding nonspecific interactions. Therefore, additional information obtained through different methods may be needed to substantiate protein–protein interaction data. One of these techniques involves the co-localization of proteins suspected to interact in the same subcellular compartment. In this chapter, we describe a method for co-expression of proteins associated with jasmonate signaling in *Nicotiana benthamiana* for studies such as co-localization.

Key words *Nicotiana benthamiana*, Jasmonate signaling, Mediator complex, Yellow fluorescent protein (YFP), Cyan fluorescent protein (CFP), Red fluorescent protein (RFP), Agrobacterium, Agroinfiltration

1 Introduction

Identification of interacting partner(s) of a protein can potentially reveal new insights into its function. A number of techniques are available to identify protein–protein interactions in plant cells. Some of these techniques, such as yeast two-hybrid (Y2H), tandem affinity purification, mass spectroscopy, and protein microarrays, are among the most commonly used and particularly suitable for high-throughput analyses [1, 2]. However, a positive protein–protein interaction detected with some of these techniques, particularly those operating in vitro, is only an indication of an interaction between the two proteins. In addition, as some methods are notorious

Alain Goossens and Laurens Pauwels (eds.), *Jasmonate Signaling: Methods and Protocols*, Methods in Molecular Biology, vol. 1011, DOI 10.1007/978-1-62703-414-2_16, © Springer Science+Business Media, LLC 2013

in producing false positives (such as nonspecific interactions), the data obtained should preferably be supported by independent analyses that employ alternative techniques, such as fluorescence resonance energy transfer, co-immunoprecipitation, and bimolecular fluorescence complementation (for a review, *see* ref. 3). One of the methods that provides supporting evidence for in vitro protein–protein interaction analysis, such as Y2H, is protein co-localization in which the subcellular localizations are investigated of two assumed interacting proteins [4]. If candidate proteins co-localized in the same subcellular compartment, the reliability would be confirmed of the initial interaction detected before the functional studies had been undertaken. In addition, such studies could also provide information about the spatial and temporal distribution of the interacting proteins.

Briefly, in protein co-localization studies, proteins are fused to different fluorescent tags cloned into different expression plasmids and introduced into plant cells through transformation mediated by *Agrobacterium tumefaciens* or microprojectiles. To monitor the intracellular localization of fluorescently tagged proteins, transformed cells are examined under a confocal microscope. Because different fluorescent tags display different emission maxima, the localization of these proteins can be detected individually under different wavelengths. When two proteins are produced in the same subcellular compartment, the fluorescent tags should be co-localized as well [4, 5].

In this chapter, we provide specific details of this co-localization method by using selected regulatory proteins with relatively well-characterized roles in jasmonate (JA) signaling that regulates plant defense and development (for reviews, *see* refs. 6, 7). The selected proteins are MYC2, ETHYLENE RESPONSE FACTOR 1 (ERF1), and OCTADECANOID-RESPONSIVE *ARABIDOPSIS* APETALA 2/ERF 59 (ORA59), all of which are key transcription factors regulating the JA signaling pathway [8–13] and PHYTOCHROME FLOWERING TIME 1/MEDIATOR 25 (PFT1/MED25), a subunit of the plant Mediator complex [14–16] implicated in JA signaling [17]. Previously, mutant analysis had suggested that PFT1/MED25 and the Mediator complex functioned as a bridge between DNA-bound transcription factors and RNA polymerase II transcriptional machinery in the JA signaling control [17, 18]. Recent Y2H and immunoprecipitation experiments have revealed physical interactions between the MED25 protein and ERF1, ORA59, and MYC2, suggesting that MED25 is required for the specific functions of these key transcriptional regulators in the JA signaling [18].

Here, we describe a method for co-expression in *N. benthamiana* of proteins fused to fluorescent tags for studying co-localization. We discuss (1) production of binary vectors for fusions to red fluorescent protein (RFP) and cyan fluorescent protein (CFP), (2) introduction into leaf epidermal cells of tobacco (*N. benthamiana*) via *Agrobacterium tumefaciens*-mediated transformation, and (3) confocal microscopy of the transformed cells.

2 Materials

2.1 Equipment

1. Laminar flow.
2. Shaker.
3. Sterile Petri plates.
4. 1-mL syringe without needle.
5. Laser scanning microscope (LSM710; Carl Zeiss, Jena, Germany) and image analysis software, such as the ZEN software (Carl Zeiss) and ImageJ (http://rsbweb.nih.gov/ij/) [19].

2.2 Plant Material and Vectors

1. Tobacco (*Nicotiana benthamiana*) seeds.
2. Gateway-compatible binary vectors pEarlygate102 [20] and pVGWmRFP (V. Çevik, unpublished).
3. 35S:mRFP and 35S:CFP control vectors (V. Çevik, unpublished).

2.3 Media, Buffers, and Solutions

1. 150 mM acetosyringone stock solution: 0.59 g of 4′-hydroxy-3′,5′-dimethoxyacetophenone (acetosyringone; Sigma-Aldrich, St. Louis, MO, USA) dissolved in 20 mL dimethyl sulfoxide (DMSO). Aliquot into 1.5-mL Eppendorf tubes and store at −20 °C.
2. Kanamycin (100 mg/mL) and gentamicin (25 mg/mL) stock solutions. Dissolve 1 g kanamycin sulfate (Sigma-Aldrich) or 0.25 g gentamicin sulfate (Sigma-Aldrich) in 10 mL sterile deionized water and filter-sterilize. Store in 1 mL aliquots at −20 °C.
3. Rifampicin stock solution (25 mg/mL). Dissolve 0.25 g rifampicin (Sigma-Aldrich) in DMSO (*see* **Note 1**). Store in 1 mL aliquots at −20 °C.
4. *Agrobacterium* liquid and solid growth medium: Dissolve 5 g yeast extract, 10 g NaCl, and 10 g peptone in a final volume of 1 L deionized water and adjust pH to 7.2. Store the sterile liquid medium at room temperature, and add sterile antibiotics (e.g., for *Agrobacterium* strain GV3101 (pMP90) add rifampicin, kanamycin (*see* **Note 2**), and gentamicin to give a final concentration of 50, 50, and 25 μg/mL, respectively). For solid growth medium, add 10 g of agar per liter of liquid medium and sterilize by autoclaving at 121 °C for 15 min. Store at room temperature and before use, melt the medium in a microwave oven, cool it to ~50–60°C, and add sterile antibiotics. Briefly and gently shake the media and

dispense (approximately 25 mL) into sterile Petri plates under sterile conditions (e.g., in a laminar flow hood) (*see* **Note 3**).

5. YEB medium: 5 g/L beef extract, 1 g/L yeast extract, 5 g/L peptone, 5 g/L sucrose, and 2 mM $MgSO_4$.
6. Infiltration medium: Filter-sterilized 10 mM of 2-(*N*-morpholino) ethanesulfonic acid (MES) (pH 5.6), 10 mM $MgCl_2$, and 150 μM acetosyringone. The infiltration medium without acetosyringone can be stored at RT for several months. Add acetosyringone just before use.

3 Methods

3.1 Plant Culture

1. Use seeds that are stored under suitable conditions for uniform germination.
2. Store dried seeds in a sealed container at low humidly at 4 °C.
3. Grow *N. benthamiana* plants from seeds in pots containing compost or any other medium that supports plant growth in a greenhouse or a plant growth chamber under 16-h light photoperiod at 25–28 °C.
4. Keep the plants free from pests and pathogens by following hygienic growth conditions.

3.2 Agrobacterium and E. coli Cultures

1. Grow *E. coli* and *A. tumefaciens* carrying binary vector plasmids in liquid media by shaking at 37 °C overnight for *E. coli* and at 28 °C for 2 days (or until the culture becomes cloudy) for *A. tumefaciens.*
2. Add up to 50 % (w/v) glycerol and store at −80 °C.
3. When required, start a new culture by streaking the bacteria onto a solid medium containing appropriate antibiotic(s).
4. Incubate the plates at 28 °C for *A. tumefaciens* and 37 °C for *E. coli* cultures.
5. Pick up a single colony using a sterile toothpick and start a liquid culture as described above.

3.3 Vector Construction

For *Agrobacterium*-mediated transient expression of proteins in tobacco epidermal cells, fluorescently tagged proteins should be cloned into binary expression vectors, such as those reported [20], according to a molecular biology laboratory manual using standard cloning procedures [21].

1. Amplify the full-length coding sequences of your genes of interest from cDNA without stop codon, attaching Gateway-compatible sites according to the manufacturer.

2. Subclone into the binary vector, e.g., pEarleyGate102 [22] for C-terminal fusions to CFP or, e.g., pVGWmFR for C-terminal fusions to mRFP (*see* **Note 4**).

3.4 Transformation of Agrobacterium

Agrobacterium can be transformed by a variety of methods as described [23].

1. Dilute an overnight culture of *Agrobacterium* in YEB medium containing the appropriate antibiotics.
2. Grow *Agrobacterium* at 28 °C by shaking until the culture reaches an OD_{550} value of 0.5–0.8.
3. Pellet the cells by centrifugation at 3,000 × *g* for 20 min at room temperature.
4. Wash the pellet with sterile distilled water and resuspend it in a smaller volume of YEB (1/10th of the initial volume of YEB used to grow the bacteria should be sufficient for this purpose).
5. Freeze 0.2-mL aliquots in liquid nitrogen.
6. Store at –80 °C.
7. For transformation, thaw the cells on ice.
8. Add the purified vector DNA through miniprep.
9. Incubate on ice for 5 min.
10. Transfer the mixture into liquid nitrogen for 5 min.
11. Incubate at 37 °C for 5 min.
12. Add 1 mL of liquid growth medium into the tube.
13. Shake at 28 °C for 2–4 h.
14. Working in a laminar flow, spread small aliquots of the culture into Petri plates containing the appropriate antibiotics.
15. Incubate the plates in a laminar flow until they are completely dry.
16. Incubate in the dark at 28 °C for 2 days or until distinct colonies emerge.
17. Grow a small culture from one of the colonies.
18. Confirm the presence of the vector either by polymerase chain reaction or restriction digest of miniprep DNA.
19. Prepare the glycerol stock of the confirmed culture.
20. Keep at –80 °C, until further use.

3.5 Transient Expression in N. benthamiana Epidermal Cells

1. Grow each *Agrobacterium* strain harboring a binary vector in YEB and liquid growth media by shaking (220 rpm) at 28 °C for 24 h. Grow next to the mRFP- and CFP-tagged proteins of interest also *Agrobacteria* containing the 35S:mRFP and 35S:CFP control vectors.

2. Harvest the cells by centrifugation at 3,000 × *g* for 20 min at 20 °C.
3. Resuspend the bacterial pellet in the infiltration medium.
4. Adjust the bacterial density to give a final OD_{550} of 0.1–0.2.
5. Mix equal volumes of *Agrobacterium* cultures containing mRFP and CFP fusions.
6. Prepare control mixes containing the 35S:CFP or 35S:mRFP in addition to mRFP or CFP fusions, respectively.
7. Infiltrate into leaves of approximately 4-week-old *N. benthamiana* plants with a needless syringe (e.g., 1 mL).
8. Carefully and slowly inject the bacteria into the abaxial side (backside) of the leaf (*see* **Note 5**).
9. Label infiltrated leaf with mixture composition. Multiple leaves can be infiltrated with different mixtures on one plant.

3.6 Visualization of Co-localization with Confocal Microscopy

1. Prepare samples 2–3 days after infiltration with *Agrobacterium*. Cut approximately 0.25 cm^2 leaf sample from the infiltrated area avoiding the veins and place the leaf sample onto a microscope slide. Add a drop of distilled water onto the leaf sample and then place the cover slide. Make sure that the abaxial side of the leaf sample faces the microscope objective.
2. Microscope settings are dependent on the laser scanning confocal microscopy being used. Laser power, gain, and pinhole sizes should be adjusted. Generally, 40× or 60× objective lens is used to resolve individual cell nuclei. Lasers at 458 nm and 543 nm are used to excite CFP and mRFP, respectively.
3. Images from multiple epidermal cells coexpressing both proteins should be examined. In addition, the proportion of the cells showing co-localization pattern should be reported (*see* **Note 6**).

In many cases, co-localization alone is not a sufficient evidence for protein–protein interaction and therefore one should be cautious about the interpretation of results from these analyses alone. In addition, potential pitfalls associated with image acquisition techniques have been identified. We therefore recommend various guidelines published elsewhere (e.g., 24, 25) be consulted to find out what these pitfalls are and how they can affect the interpretation of the data.

2–3 days after infiltration, images (Fig. 1) are obtained with a laser scanning microscope and processed with the ZEN software and ImageJ [19]. MED25 co-localizes to the nucleus of *N. benthamiana* cells, when co-expressed with the transcription factors ERF1, MYC2, and ORA59 (Fig. 1, *see* **Note 7**). In control experiments, co-expression of *MED25-RFP* with CFP or *MYC2-CFP* or *ERF1-CFP* with RFP did not induce speckle formation (Fig. 2), suggesting that speckle formation requires co-expression of *MED25* and the interacting transcription factors.

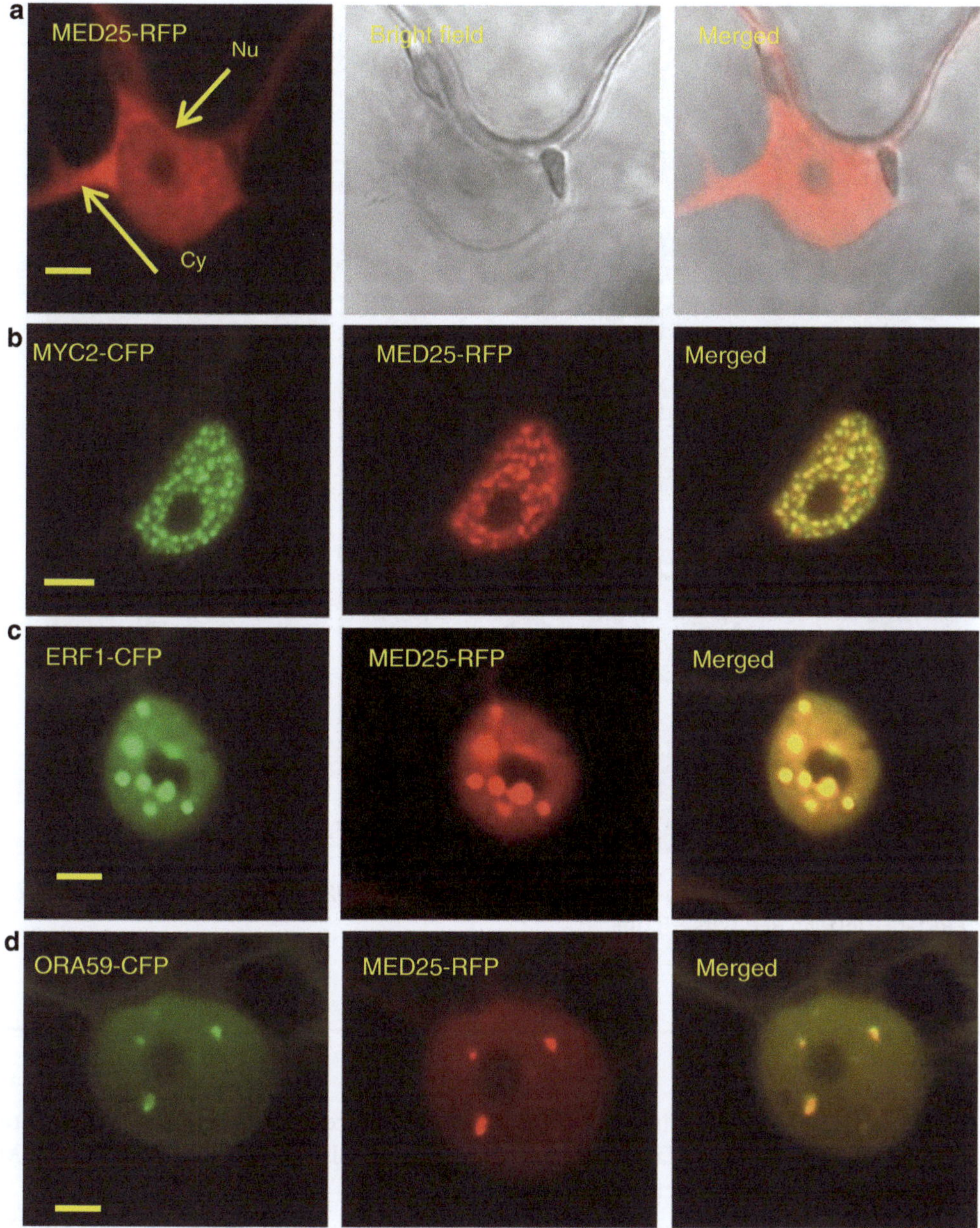

Fig. 1 Co-localization of *MED25-RFP* to the nucleus and cytoplasm of *N. benthamiana* cells, forming nuclear speckles when co-expressed with *ERF1-CFP*, *ORA59-CFP*, and *MYC2-CFP*. Confocal images were taken 48 h after *Agrobacterium*-mediated transient expression in the epidermal cells of *N. benthamiana* leaves. The expression of *MED25-RFP* was seen both in the nucleus and in the cytoplasm when transformed on its own (**a**). Nuclear speckles were observed when *N. benthamiana* leaves were co-transformed with MYC2-CFP with MED25-RFP (**b**), ERF1-CFP with MED25-RFP (**c**), and ORA59-CFP with MED25-RFP (**d**). Cy, cytoplasm; Nu, nucleus. Bars = 5 μm

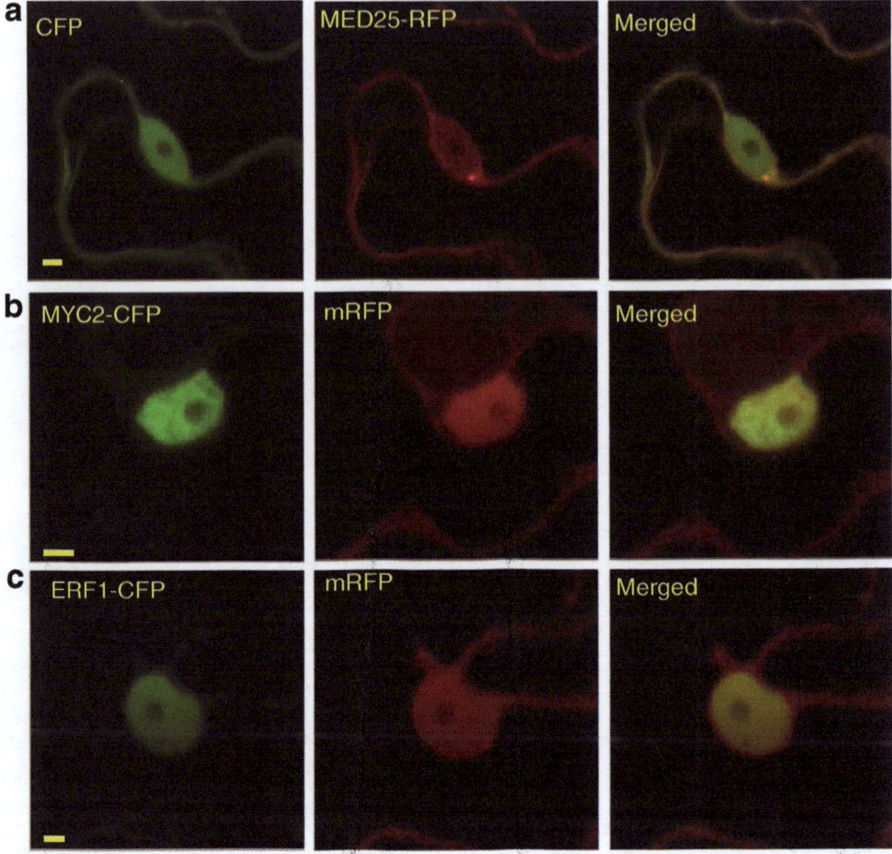

Fig. 2 Confocal images of control co-expression experiments of *MED25-RFP* with CFP (**a**) and of *MYC2-CFP* (**b**) or *ERF1-CFP* (**c**) with RFP. Bars = 5 μm

4 Notes

1. Filter sterilization is not required when antibiotics (for instance, rifampicin) are dissolved in solvents, such as EtOH or DMSO, before use.
2. For example, for the *Agrobacterium* strain GV3101 (pMP90), add rifampicin, kanamycin, and gentamicin to give a final concentration of 50, 50, and 25 μg/mL, respectively. Here, kanamycin was used in the selection, because the binary vectors confer kanamycin resistance.
3. Plates can be stored at 4 °C, wrapped in aluminum foil for up to 2 weeks.
4. The fluorescent tags can be fused to either the N-terminus or the C-terminus of the protein of which the co-localization is to be examined. Here, MED25 and the transcription factor

proteins were C-terminally tagged with the fluorescent proteins, but similar co-localization patterns are observed when they are N-terminally tagged. The decision on which terminus to fuse the candidate protein with a fluorescent tag depends on whether the fusion could affect the stability and/or activity of the candidate protein; for instance, if the stability of the protein is affected by the C-terminus fusion, then the N-terminus fusion or vice versa can be tested. It is also possible that a functional fusion protein cannot be produced for some proteins.

5. Although tobacco cells are larger and the transient expression is more efficient, *Arabidopsis* epidermal cells from cotyledons and root cells have also been reported as a suitable alternative [5, 26]. In addition, some *Arabidopsis* fusion proteins are seemingly not expressed in *N. benthamiana* and, therefore, the use of *Arabidopsis* cells is required. Epidermal cells of onion (*Allium cepa*) and carrot (*Daucus carota*) could be utilized as material for transient co-localization studies as well.
6. In some instances, MED25-RFP protein was also found in the cytoplasm (Fig. 1). Transient over-expression and/or tagging of proteins may sometimes alter their correct subcellular localization. Weaker promoters may be tested in transient expression studies if artifacts are observed. Again, fluorescent tags fused to the N-terminus or the C-terminus of the protein may be tested in order to eliminate potential artifacts.
7. Nuclear speckles are known to serve as hubs of enhanced mRNA metabolic activity and/or involved in active transcription sites (*see* ref. 27 for a review).

Acknowledgment

We thank Brendan Kidd for useful discussions.

References

1. Shoemaker BA, Panchenko AR (2007) Deciphering protein–protein interactions. Part I. Experimental techniques and databases. PLoS Comput Biol 3:e42
2. Berggård T, Linse S, James P (2007) Methods for the detection and analysis of protein–protein interactions. Proteomics 7:2833–2842
3. Fukao Y (2012) Protein–protein interactions in plants. Plant Cell Physiol 53:617–625
4. Nelson BK, Cai X, Nebenführ A (2007) A multicolored set of in vivo organelle markers for co-localization studies in Arabidopsis and other plants. Plant J 51:1126–1136
5. Marion J, Bach L, Bellec Y, Meyer C, Gissot L, Faure J-D (2008) Systematic analysis of protein subcellular localization and interaction using high-throughput transient transformation of Arabidopsis seedlings. Plant J 56:169–179
6. Kazan K, Manners JM (2012) JAZ repressors and the orchestration of phytohormone crosstalk. Trends Plant Sci 17:22–31
7. Pauwels L, Goossens A (2011) The JAZ proteins: a crucial interface in the jasmonate signaling cascade. Plant Cell 23:3089–3100
8. Lorenzo O, Piqueras R, Sánchez-Serrano JJ, Solano R (2003) ETHYLENE RESPONSE

FACTOR1 integrates signals from ethylene and jasmonate pathways in plant defense. Plant Cell 15:165–178

9. Lorenzo O, Chico JM, Sánchez-Serrano JJ, Solano R (2004) *JASMONATE-INSENSITIVE1* encodes a MYC transcription factor essential to discriminate between different jasmonate-regulated defense responses in Arabidopsis. Plant Cell 16:1938–1950
10. Boter M, Ruíz-Rivero O, Abdeen A, Prat S (2004) Conserved MYC transcription factors play a key role in jasmonate signaling both in tomato and *Arabidopsis*. Genes Dev 18: 1577–1591
11. Anderson JP, Badruzsaufari E, Schenk PM, Manners JM, Desmond OJ, Ehlert C, Maclean DJ, Ebert PR, Kazan K (2004) Antagonistic interaction between abscisic acid and jasmonate-ethylene signaling pathways modulates defense gene expression and disease resistance in Arabidopsis. Plant Cell 16:3460–3479
12. Dombrecht B, Xue GP, Sprague SJ, Kirkegaard JA, Ross JJ, Reid JB, Fitt GP, Sewelam N, Schenk PM, Manners JM, Kazan K (2007) MYC2 differentially modulates diverse jasmonate-dependent functions in *Arabidopsis*. Plant Cell 19:2225–2245
13. Pré M, Atallah M, Champion A, De Vos M, Pieterse CMJ, Memelink J (2008) The AP2/ERF domain transcription factor ORA59 integrates jasmonic acid and ethylene signals in plant defense. Plant Physiol 147:1347–1357
14. Cerdán PD, Chory J (2003) Regulation of flowering time by light quality. Nature 423: 881–885
15. Bäckström S, Elfving N, Nilsson R, Wingsle G, Björklund S (2007) Purification of a plant Mediator from *Arabidopsis thaliana* identifies PFT1 as the Med25 subunit. Mol Cell 26: 717–729
16. Kidd BN, Cahill DM, Manners JM, Schenk PM, Kazan K (2011) Diverse roles of the Mediator complex in plants. Semin Cell Dev Biol 22:741–748
17. Kidd BN, Edgar CI, Kumar KK, Aitken EA, Schenk PM, Manners JM, Kazan K (2009) The Mediator complex subunit PFT1 is a key regulator of jasmonate-dependent defense in *Arabidopsis*. Plant Cell 21:2237–2252
18. Çevik V, Kidd BN, Zhang P, Hill C, Kiddle S, Denby KJ, Holub EB, Cahill DM, Manners JM, Schenk PM, Beynon J, Kazan K (2012) MEDIATOR25 acts as an integrative hub for the regulation of jasmonate-responsive gene expression in Arabidopsis. Plant Physiol 160: 541–555
19. Abramoff M, Magalhães P, Ram S (2004) Image processing with ImageJ. Biophotonics Int 11:36–42
20. Nakagawa T, Kurose T, Hino T, Tanaka K, Kawamukai M, Niwa Y, Toyooka K, Matsuoka K, Jinbo T, Kimura T (2007) Development of series of Gateway binary vectors, pGWBs, for realizing efficient construction of fusion genes for plant transformation. J Biosci Bioeng 104: 34–41
21. Ausubel FM, Brent R, Kingston RE, Moore DD, Seidman JG, Smith JA, Struhl K (1994) Current protocols in molecular biology, vol 1. Wiley, New York, p 1600
22. Earley KW, Haag JR, Pontes O, Opper K, Juehne T, Song K, Pikaard CS (2006) Gateway-compatible vectors for plant functional genomics and proteomics. Plant J 45:616–629
23. Weigel D, Glazebrook J (2002) Arabidopsis: a laboratory manual. Cold Spring Harbor Laboratory Press, Cold Spring Harbor, NY, p 354
24. French AP, Mills S, Swarup R, Bennett MJ, Pridmore TP (2008) Co-localization of fluorescent markers in confocal microscope images of plant cells. Nat Protoc 3:619–628
25. North AJ (2006) Seeing is believing? A beginners' guide to practical pitfalls in image acquisition. J Cell Biol 172:9–18
26. Van Loock B, Markakis MN, Verbelen J-P, Vissenberg K (2010) High-throughput transient transformation of Arabidopsis roots enables systematic co-localization analysis of GFP-tagged proteins. Plant Signal Behav 5: 261–263
27. Mao SY, Zhang B, Spector DL (2011) Biogenesis and function of nuclear bodies. Trends Genet 27:295–306

Chapter 17

Electrophoretic Mobility Shift Assay for the Analysis of Interactions of Jasmonic Acid-Responsive Transcription Factors with DNA

Johan Memelink

Abstract

The electrophoretic mobility shift assay based on nondenaturing polyacrylamide gel electrophoresis is a simple, rapid, and sensitive method for the study of the interaction of transcription factors with DNA in vitro. It relies on a change in the electrophoretic mobility of a DNA fragment when bound to an interacting protein. The assay can be used to test DNA binding of either purified or recombinant proteins or uncharacterized binding activities present in crude protein extracts from plant cells or nuclei. It allows the determination of the abundance, affinity, association rate constants, dissociation rate constants, and binding specificity of DNA-binding proteins.

Key words DNA-binding protein, EMSA, Gel retardation, Gel shift, Jasmonic acid

1 Introduction

In the plant research field, transcription factors have emerged in the past decade as central components in signal transduction pathways that are often intimately coupled to signal receptors, such as in the regulation of gene expression in response to auxin or jasmonic acid (JA) [1]. In JA signaling, the receptor for the bioactive jasmonate JA-isoleucine (JA-Ile) is the F-box protein CORONATINE INSENSITIVE 1 (COI1) that, upon JA-Ile binding, interacts with members of the Jasmonate ZIM-domain (JAZ) family of repressors. These repressors are thought to be ubiquitinated by the SCF^{COI1} complex and are degraded by the 26S proteasome. As a result their repression targets, which consist of a growing number of transcription factors, including the basic helix-loop-helix (bHLH) protein MYC2 [2], are liberated and trigger their specific set of target genes. The specificity of the response genes relies, among others, on the presence in the promoters of specific binding sites for transcription factors.

Alain Goossens and Laurens Pauwels (eds.), *Jasmonate Signaling: Methods and Protocols*, Methods in Molecular Biology, vol. 1011, DOI 10.1007/978-1-62703-414-2_17, © Springer Science+Business Media, LLC 2013

The electrophoretic mobility shift assay (EMSA), also known as gel shift assay or gel retardation assay, is a simple and widely used method to study protein–DNA interactions in vitro. The technique has been recently described in volume 543 entitled "DNA–Protein Interactions" of this "Methods in Molecular Biology" series [3]. Another excellent description of the method can be found in [4]. EMSA consists of four simple steps: (1) labeling and isolation of the DNA probe, (2) preparation of a nondenaturing gel, (3) preparation of the protein–DNA binding reaction, and (4) electrophoresis of protein–DNA complexes, followed by drying of the gel and autoradiography.

EMSA can be performed with crude cellular protein extracts, nuclear protein extracts, conventionally purified proteins, recombinant proteins expressed in *Escherichia coli*, or proteins synthesized in vitro in a wheat germ extract or in a rabbit reticulocyte lysate. The technique has been pioneered for plants in the lab of Nam-Hai Chua [5, 6]. This led among others to the discovery of nuclear factors GT-1 and ASF-1 binding to the light-regulatory box II from the promoter of the pea (*Pisum sativum*) gene encoding the small subunit of ribulose-1,5-bisphosphate carboxylase (*RbcS-3A*) and the *as-1* sequence from the cauliflower mosaic virus (CaMV) 35*S* promoter, respectively, whereafter the corresponding proteins were cloned [7, 8].

The EMSA technique became very popular in the plant research field during the 1990s for studies that often combined analysis of in vivo promoter activity and interaction of the promoter with DNA-binding proteins in vitro. Nowadays, the EMSA technique is less popular, probably because of an interest shift and the emergence of new technical hypes. Despite its decreasing popularity, EMSA has not been replaced by a superior technique and it remains the method of choice for obtaining information about the interaction of proteins with DNA in vitro.

It has been used to study several cloned transcription factors that are involved in JA signaling and/or that bind to JA-responsive promoters, including APETALA2/ETHYLENE RESPONSE FACTOR (AP2/ERF) domain [9–13], bHLH [14–20], MYB [21], zinc finger [22], AT-hook [23], and basic leucine zipper (bZIP) transcription factors [24]. The assay has also been applied to study the interactions between DNA fragments from JA-responsive promoters with binding activities from crude nuclear protein extracts [25–27].

1.1 Principle of the Method

EMSA is based on the principle that under native conditions the mobility of a DNA fragment differs, depending on whether it is free or bound by a DNA-binding protein. The protein-bound DNA fragment usually migrates more slowly with a mobility shift as a consequence. A protein that can bind to a DNA fragment will form a complex, eventually resulting in an equilibrium between the association rate k_a and the dissociation rate k_d. The association and dissociation rates depend on ion concentration, pH, and temperature.

1.2 Overview of the Procedure

Several variables influence the result of the procedure.

1.2.1 DNA-Binding Protein

Total cellular or nuclear protein extracts can be prepared [28] and used for EMSA. Kinased probes labeled with γ-^{32}P-ATP should be avoided, because crude protein preparations may contain phosphatase activity that will dephosphorylate the probe. Cloned DNA-binding proteins can be produced in vitro in a wheat germ extract or in a rabbit reticulocyte lysate. Alternatively, they can be produced with an affinity tag in *E. coli* and purified by affinity chromatography. In *E. coli*, the expression levels of transcription factors are usually much lower than those of other proteins, such as enzymes, and degradation is often extensive. In most cases, the majority of the transcription factor accumulates in insoluble protein bodies, which can be used to isolate the protein when the levels in the soluble fraction are too low [12]. Partially degraded protein preparations may give rise to multiple bands in EMSA, depending on the location of the affinity tag relative to the DNA-binding domain. Full-length proteins can be selected by expression with two different N-terminal and C-terminal tags and sequential affinity purification [12, 17, 18].

1.2.2 DNA Probe

Fragments generated by restriction enzyme digestion of cloned DNA or polymerase chain reaction (PCR) fragments of 50–300 bp, or synthetic double-stranded oligonucleotides of 20–50 bp, can be used for EMSA. Detection of DNA–protein complexes usually involves labeling of the DNA with a ^{32}P-labeled deoxynucleotide. Other nonradioactive labeling methods have been described and are commercially available, such as digoxygenin or biotin labeling. To simplify the interpretation of EMSA results, probes are preferably designed to contain a single binding site for the protein of interest. A probe with multiple binding sites may generate multiple complexes, depending on the added amount of a noncooperatively binding protein (Fig. 1, right panel). However, most DNA-binding proteins bind cooperatively, resulting in a single complex, regardless of the number of binding sites on the DNA probe (Fig. 1, left panel).

1.2.3 Salt Concentration

The salt concentration is critical for DNA binding and for the rate with which the binding equilibrium is reached. Certain plant bZIP proteins reach binding equilibrium after 30 min at 100 mM salt, but not at the commonly used concentration of 40 mM even

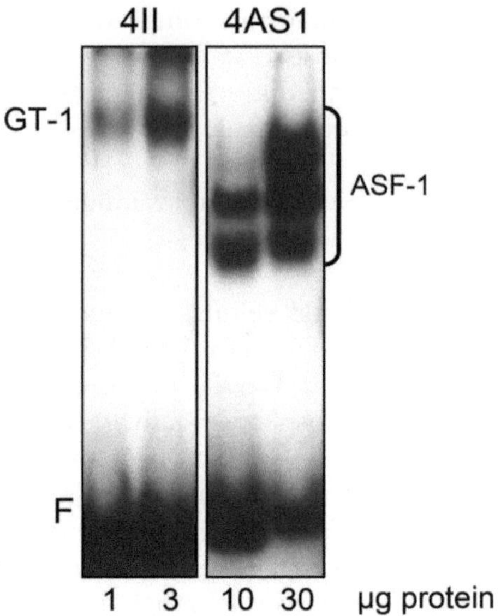

Fig. 1 Cooperative and noncooperative binding of the nuclear factors GT-1 and ASF-1, respectively. Labeled probes containing four copies of box II (4II) from the promoter of the pea *RbcS-3A* gene or four copies of the *as-1* sequence (4AS1) from the CaMV 35*S* promoter [29] were incubated with the indicated amounts of tobacco (*Nicotiana tabacum*) leaf nuclear protein extract. EMSA conditions were standard with 3 μg poly(dI-dC).poly(dI-dC) as nonspecific competitor per binding reaction. F indicates the free probe fragments

after 4 h of incubation [30]. For approximately ten different DNA-binding proteins tested, including GT-1, AP2/ERF, bHLH, and WRKY1, the binding was superior at 100 mM compared with 40 mM salt concentration.

1.2.4 Nonspecific Competitors

To ensure specificity of the DNA–protein interaction, a variety of nonspecific competitors may be used. This is particularly important with cellular or nuclear protein extracts, which contain a mixture of specific and nonspecific DNA-binding proteins. Competitor DNA is sonicated to obtain a length similar to that of the probe DNA. As nonspecific competitors, salmon or herring sperm DNA or calf thymus DNA may be used. To avoid the presence of specific binding sites for the protein of interest, synthetic DNA is utilized, such as poly(dI-dC).poly(dI-dC) or poly(dA-dT).poly(dA-dT). These synthetic competitor DNAs are not completely nonspecific, since they compete for GC-rich or AT-rich binding sites, respectively. EMSAs with crude extracts can yield different results when the two different competitors are used, because proteins binding either to AT-rich or to GC-rich sequences are preferentially visualized (Fig. 2). With recombinant proteins, nonspecific competitors may compete directly for binding. For example, poly(dI-dC).poly(dI-dC)

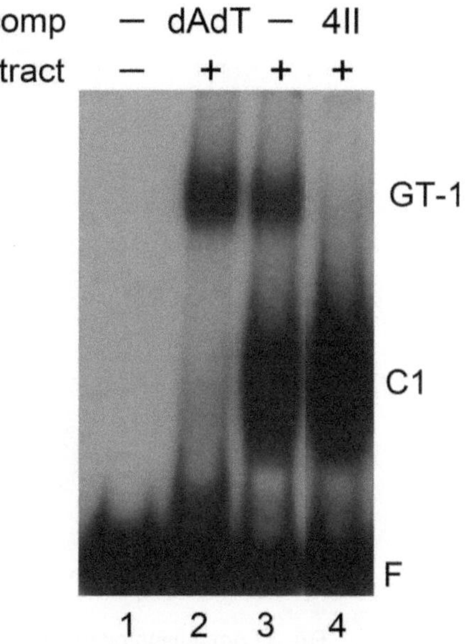

Fig. 2 Formation of different complexes depending on the nature of the nonspecific competitor. A labeled probe consisting of the −238 to −87 region of the tryptophan decarboxylase promoter from *Catharanthus roseus* (periwinkle) was incubated without (*lane 1*) or with (*lanes 2–4*) 2 μg of nuclear protein extract from suspension-cultured cells of *C. roseus*. All reactions contained 3 μg poly(dI-dC).poly(dI-dC). Reactions contained in addition 0.5 μg poly(dA-dT).poly(dA-dT) (*lane 2*) or 34 ng of 4II competitor corresponding to 350-fold molar excess (*lane 4*) as indicated. The upper complex is competed by the 4II fragment, consisting of four copies of box II from the promoter of the pea *RbcS-3A* gene, indicating that it is formed by the nuclear factor GT-1. The nonspecific competitor poly(dA-dT).poly(dA-dT) competes with the uncharacterized middle complex C1. EMSA conditions were standard. F indicates the free probe fragment

competes for the binding of AP2/ERF proteins to their target sequence GCCGCC or variants thereof (Fig. 3) and poly(dA-dT). poly(dA-dT) is the preferred nonspecific competitor for EMSAs with proteins belonging to this transcription factor class.

1.3 Common Misconceptions

One frequently sees misconceptions in manuscripts submitted for publication, in published articles, or in reviewer reports.

1.3.1 Competition with Unlabeled Probe Fragment Shows Binding Specificity

In a competition experiment the protein incubated with the labeled DNA fragment produces a mobility shift that disappears when the reaction is carried out with an excess of the same unlabeled fragment. This experimental setup does not demonstrate specificity, because both specific and nonspecific complexes are competed. In contrast, competition with an unlabeled DNA fragment without specific binding site, either due to a specific mutation or because it is unrelated, will reveal specificity, because under these conditions

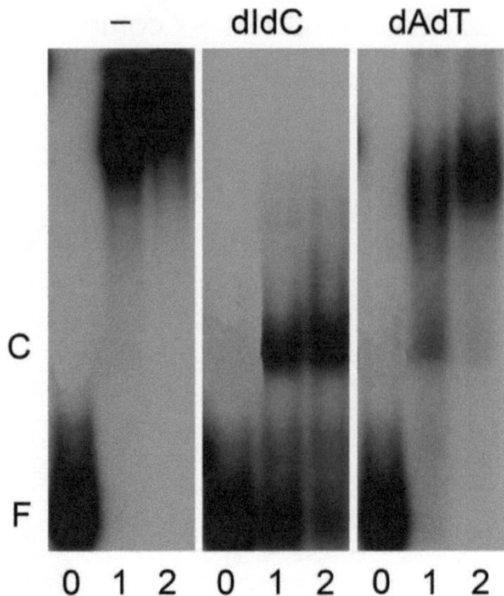

Fig. 3 Competition by the nonspecific competitor poly(dI-dC).poly(dI-dC) for binding by the AP2/ERF protein Octadecanoid-derivative Responsive *Catharanthus* AP2-domain protein 3 (ORCA3) to GC-rich sites. A labeled probe consisting of the −145 to +52 region of the promoter of the strictosidine synthase gene from *C. roseus* was incubated with 0, 1, or 2 μL recombinant His-tagged ORCA3 protein [10, 11]. Reactions did not contain nonspecific competitors, 100 ng poly(dI-dC).poly(dI-dC) or 500 ng poly(dA-dT).poly(dA-dT) as indicated. EMSA conditions were standard. F and C, the free probe fragment and the probe fragment bound by one ORCA3 molecule to its specific AGACCGCC binding site, respectively. Larger complexes are formed by less specific binding of additional ORCA3 molecules

a specific DNA–protein complex will not be competed, whereas a nonspecific DNA–protein complex will be competed.

1.3.2 A Mobility Shift Is Due to the Formation of a Larger DNA–Protein Complex

Frequently, the shifted complex is described as migrating more slowly because of its increased size. In fact, the DNA contribution to the mobility of the DNA–protein complex is marginal. The mobility is almost entirely due to the protein component of the complex and, in a nondenaturing gel, depends mainly on its charge, which, in turn, depends on its isoelectric point (pI) and the gel pH. In the buffered system described here, the pH is approximately 8, imposing a negative charge on most proteins and allowing them to migrate into the gel. Although most complexes run more slowly than the free DNA probe, complexes containing very acidic proteins may run faster. Proteins with a pI of 8 will not migrate at all (*see* **Note 1**), whereas those with a pI higher than 8 will be positively charged and migrate in the opposite direction into the upper buffer tank. In conclusion, there is little relationship between protein size and the mobility shift magnitude. Small proteins (such as GT-1) may cause large mobility shifts.

1.3.3 Competition Occurring Only with a Large Molar Excess of Unlabeled Fragment Indicates Low Specificity or Lack Thereof

When EMSA results are described in which competition is only effective with 500-fold or 1,000-fold molar excess of unlabeled fragments, the conclusion is often that the specificity of the DNA-binding protein is low. However, without knowledge about the kinetics of the reaction, statements on the specificity are impossible. A competition reaction is essentially a probe dilution experiment. The unlabeled fragment competes with the labeled fragment for binding. The total binding does not diminish, but the bound labeled fragment is replaced by bound unlabeled fragment. Complex formation between a protein (P) and a DNA fragment (D) with one single binding site is a simple bimolecular reaction according to

$$\mathrm{P} + \mathrm{D} \underset{k_\mathrm{d}}{\overset{k_\mathrm{a}}{\rightleftarrows}} \mathrm{PD}$$

At equilibrium, the amount of PD complex depends on the amounts of protein and DNA according to $[\mathrm{PD}] = k_\mathrm{a}/k_\mathrm{d}\ [\mathrm{P}][\mathrm{D}]$. Depending on the relative amounts, either P or D can be the limiting factor and the other component is the driving factor in complex formation. For effective competition to occur, the amount of the complex should not increase by addition of an unlabeled fragment. For example, if addition of a tenfold excess of unlabeled fragment would cause a tenfold increase in complex formation, then no competition would be visible. Therefore, effective competition with a low amount of competitor can only occur when the protein amount is the limiting factor. Under these conditions, total complex formation does not increase after DNA addition and the labeled DNA in the complex is replaced by unlabeled DNA. However, when the DNA is the limiting factor and the protein is in excess, addition of low amounts of a competitor will not result in competition because the total amount of the complex will increase. Competition will be visible only with large excesses of unlabeled DNA. In summary, the excess amount of unlabeled fragment necessary for competition depends on the exact experimental conditions. To avoid possible criticisms, the results could be presented without statement of the molar excess either by representing the increasing molar excess as a black wedge without numbers or by giving the absolute amount of DNA used for competition. To experimentally solve this issue, relatively large amounts of probe DNA and low amounts of protein should be used so that the protein amount is the limiting factor in the reaction and the DNA drives the complex formation.

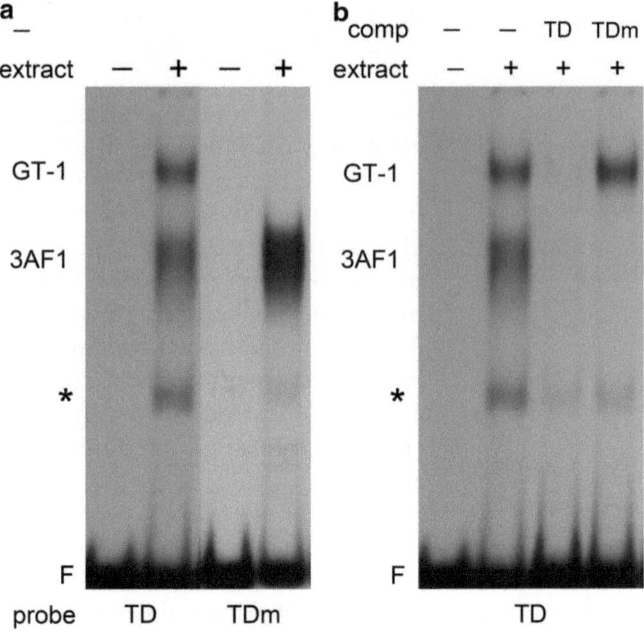

Fig. 4 Direct binding assay versus competition assay. The labeled TD probe consists of the −160 to −99 region of the promoter of the tryptophan decarboxylase gene. The TDm probe contains two nucleotide mutations that abolish the binding sites for the nuclear factor GT-1 [26]. (**a**) Direct binding assay. The probes were incubated with or without 8 μg of tobacco leaf nuclear protein extract. The free probe (F) and complexes formed by the nuclear factors GT-1 and 3AF1 are indicated and the *asterisk marks* a nonspecific complex. (**b**) Competition assay. The TD probe was incubated without or with 8 μg tobacco nuclear protein extract with a 200-fold molar excess of specific competitors as indicated (comp). EMSA conditions were standard with 3 μg poly(dI-dC).poly(dI-dC) as nonspecific competitor per binding reaction

1.3.4 Direct Protein-Binding Assays to Wild-Type and Mutant Fragments Are Not Sufficient for Determining the Binding Specificity, but Additional Competition Assays Are Required

When the results are presented of a series of direct protein-binding assays with labeled wild-type and mutant fragments (as in Fig. 4a), frequently additional competition assays (as in Fig. 4b) are requested to show the binding specificity. Competition assays are not superior to direct binding assays and provide no supplemental information. In fact, they are inferior because they have the shortcomings mentioned in Subheading 1.3.3 and they are not very sensitive to differences in affinity due to the fold excess used for competition. In contrast, direct binding assays clearly show whether a protein can bind to a DNA fragment and the amount of complex formation is a sensitive measure of the affinity. For example, binding affinities of DNA-binding proteins are determined via direct binding assays and not via competition assays [24, 30]. Actually, the only advantage of competition assays is the requirement of one single labeled probe.

2 Materials

2.1 Equipment

1. Standard vertical electrophoresis apparatus for polyacrylamide gels with 1.5-mm spacers and corresponding 10- or 15-well combs with teeth that are 8 or 6 mm wide, respectively.
2. Power supply.
3. UV transilluminator equipped with 302 nm UV-B lamps with UV-blocking spectacles and face shield (*see* **Note 2**). Optionally, a nanodrop microvolume spectrophotometer.
4. Silanized glass wool.
5. 2-mL glass Pasteur pipettes.
6. Sephadex G-75 gel filtration matrix (medium, not fine or superfine).
7. Geiger counter.
8. Whatman DE81 and 3MM paper.
9. Standard gel dryer.
10. X-ray films and autoradiography cassettes mounted with tungstate-intensifying screens.

2.2 Buffers, Solutions, and Media

2.2.1 DNA Isolation from Gel

1. Stock solution of 40 % (w/v) acrylamide containing a 19:1 ratio of acrylamide:*N*, *N′*-methylene bis-acrylamide; stored at 4 °C. Caution: Unpolymerized acrylamide is a neurotoxic agent (*see* **Note 3**).
2. Stock solution of 10× tris(hydroxymethyl)aminomethane (Tris)-borate-ethylenediaminetetraacetic acid (EDTA) (TBE): Per L add 108 g of Tris, 55 g of boric acid, and 40 mL of 0.5 M EDTA–NaOH (pH 8).
3. Stock of 10 % (w/v) ammonium persulfate (APS) in 0.5 mL aliquots and stored at −20 °C.
4. *N*,*N*,*N′*,*N′*-tetramethylethylenediamine (TEMED); stored at 4 °C.
5. Gel mix stock solution: 5 % (w/v) acrylamide mono:bis = 19:1, 1× TBE; stored at 4 °C.
6. Stock solution of loading mix: 12.5 % (w/v) Ficoll 400, 2 mM of EDTA–NaOH (pH 8), 0.25 % (w/v) bromophenol blue, 0.25 % (w/v) xylene cyanol FF. Omit dyes for estimation of fragment concentrations.
7. DNA size marker, such as pBluescript II SK(+) plasmid digested with *Hae*III, containing DNA fragments of 601, 458, 434, 290, 267, 254, 174, 142, 102, 80, 79, 50, 18, and 11 bp.

8. Ethidium bromide solution: 10 mg/L; stored at 4 °C. Caution: Ethidium bromide is a suspected mutagenic agent (*see* **Note 4**).
9. PAA elution buffer: 0.5 M NH_4-acetate, 1 mM EDTA, 0.1 % (w/v) sodium dodecyl sulfate (SDS).

2.2.2 Probe Labeling

1. Depending on the labeling method α-^{32}P-dCTP or γ-^{32}P-ATP with specific activities of 111 TBq/mmol. Caution: ^{32}P emits high-energy β radiation (*see* **Note 5**).
2. Approximately 50 ng of a double-stranded oligonucleotide with a typical length of 30 bp (*see* **Notes 6** and **7**) or approximately 20 ng of a double-stranded restriction fragment with a typical length of 100 bp (*see* **Note 7**). Adjust amounts for other fragment sizes.
3. Klenow fragment of *Escherichia coli* DNA polymerase I.
4. 10× Klenow buffer: 100 mM Tris–HCl (pH 8) and 50 mM $MgCl_2$; stored at −20 °C.
5. GAT mix: 1 mM dGTP, 1 mM dATP, 1 mM dTTP in H_2O; stored at −20 °C.
6. T_4 polynucleotide kinase.
7. 10× Kinase buffer: 500 mM Tris–HCl (pH 7.5), 100 mM $MgCl_2$, and 50 mM dithiothreitol (DTT); stored at −20 °C.
8. 0.5 M EDTA–NaOH (pH 8.0).
9. Sephadex G-75 slurry: Add 15 mL of 25 mM 4-(2-hydroxyethyl)-1-piperazineethanesulfonic acid (Hepes)–KOH (pH 7.2), 100 mM KCl, 1 mM EDTA per g of Sephadex G-75 powder and autoclave for 20 min at 120 °C.

2.2.3 EMSA

1. Stock solution of 40 % (w/v) acrylamide containing a 37.5:1 ratio of acrylamide:*N*, *N*′-methylene bis-acrylamide; stored at 4 °C. Caution: Unpolymerized acrylamide is a neurotoxic agent (*see* **Note 3**).
2. A stock solution of 10× TBE (*see* Subheading 2.2.1, **item 2**, and **Note 8**).
3. A stock of 10 % (w/v) APS (*see* Subheading 2.2.1, **item 3**).
4. DNA-binding protein preparation (*see* **Notes 9** and **10**).
5. Stock solution of 5× high-salt extraction buffer (HEB): 125 mM Hepes–KOH (pH 7.2), 500 mM KCl, 0.5 mM EDTA, and 50 % (v/v) glycerol (*see* **Note 11**).
6. Gel shift mix stock solution (5 % (w/v) acrylamide mono:bis = 37.5:1, 0.5× TBE); stored at 4 °C (*see* **Note 12**).
7. Double-stranded poly(dI-dC) or poly(dA-dT) sodium salt.
8. Stock solution of loading mix (*see* Subheading 2.2.1, **item 6**).

3 Methods

3.1 DNA Isolation from Gel

1. Clean the glass plates with water (and with soap, if necessary) and then with 96 % ethanol.
2. Assemble glass plates with 1.5-mm spacers.
3. Prepare a 5 % polyacrylamide (mono:bis = 19:1) gel by mixing 30 mL of gel mix (or as much as needed for the particular setup used) with 50 μL TEMED and 500 μL 10 % (w/v) APS.
4. Mix by swirling and pour between the plates.
5. Insert a 1.5-mm thick comb with 8-mm wide teeth.
6. Let the gel polymerize for 30 min and remove the comb.
7. Mount the gel in an electrophoresis apparatus and add 1× TBE to the electrophoresis tanks.
8. Digest 25–50 μg of plasmid DNA or 1–5 μg of a PCR fragment with appropriate enzymes (*see* **Note 7**).
9. To the digested DNA, add 0.25 volumes 5× loading mix and load in two separate wells.
10. Load DNA size marker, e.g., 1 μg of pBluescript II SK(+) digested with *Hae*III.
11. Run the gel at 100 V (*see* **Note 13**).
12. After running, stain the gel in H_2O containing 2 μg/mL ethidium bromide (*see* **Note 4**).
13. Cut out the bands of interest with a scalpel on a UV transilluminator equipped with 302 nm UV-B lamps (*see* **Note 2**).
14. Incubate the acrylamide slices with the fragments of interest in a 1.5-mL Eppendorf tube with 0.5 mL of PAA elution buffer with gentle shaking overnight.
15. Transfer the PAA elution buffer to a new Eppendorf tube.
16. Add 1 mL 96 % ethanol and centrifuge for 5 min at 15,000 × *g*.
17. Remove liquid and add 0.5 mL 70 % ethanol to the pellet.
18. Centrifuge for 5 min at 15,000 × *g*.
19. Remove the liquid.
20. Air-dry the pellet at room temperature or dry it in a speedvac apparatus without heating (*see* **Note 14**).
21. Resuspend the pellet in 20 μL 10 mM Tris–HCl (pH 7.5) and 1 mM EDTA.
22. Measure the DNA concentration with a Nanodrop microvolume spectrophotometer. Alternatively, estimate the concentration by running 1 and 2 μL of the purified fragment alongside 1 μg of pBluescript II SK(+) plasmid digested with *Hae*III and compare the band intensities after ethidium bromide staining (*see* **Note 15**).

3.2 Probe Labeling

Three different methods to obtain a labeled probe are described.

3.2.1 Labeling of Restriction Endonuclease Fragments

1. Label the purified restriction enzyme-digested fragment (*see* **Note 7**) by mixing 20 ng of the fragment (for a typical 100-bp probe) with 2 μL 10× Klenow buffer, 2 μL of 1 mM GAT mix, 0.5 μL of α-^{32}P-dCTP, 2 units of the Klenow fragment of DNA polymerase I, and H_2O to 20 μL.
2. Incubate for at least 30 min at room temperature (*see* **Note 16**).
3. Separate the labeled DNA from unincorporated nucleotides on a Sephadex G-75 column in a glass 2-mL Pasteur pipette.
4. To make the column, remove the narrow tip of the pipette with a glass knife, put in a silanized glass wool plug, and fill it to the constriction at the top with approximately 2 mL of packed Sephadex G-75 slurry.
5. Apply the labeling mixture to the column.
6. Elute the column with 25 mM Hepes–KOH pH 7.2, 100 mM KCl, and 1 mM EDTA as elution buffer, collecting fractions of 5 drops equaling approximately 200 μL in Eppendorf tubes (*see* **Note 17**).
7. Pool the first peak fractions by measuring the eluted radioactivity with a Geiger counter (*see* **Note 18**).

3.2.2 Labeling of Annealed Synthetic Oligonucleotides with α-^{32}P-dCTP

1. Mix 25 ng each of two complementary oligonucleotides (*see* **Notes 6** and 7) with 2 μL 10× Klenow buffer and H_2O in a final volume of 20 μL.
2. Heat at 5 °C over the melting temperature of the oligonucleotides for 5 min and let cool down at room temperature for 20 min.
3. Add 1 μL of 10× Klenow buffer, 2 μL of 1 mM GAT mix, 1 μL of α-^{32}P-dCTP, 2 units of the Klenow fragment of DNA polymerase I, and H_2O to 30 μL.
4. Incubate for at least 30 min at room temperature.
5. Separate the probe from the unincorporated nucleotides (*see* **step 2** in Subheading 3.2.1).

3.2.3 Labeling of Annealed Synthetic Oligonucleotides or a PCR Fragment with γ-^{32}P-ATP

1. Mix 25 ng each of two complementary oligonucleotides (*see* **Notes 6** and 7) with 2 μL 10× Kinase buffer and H_2O in a final volume of 20 μL.
2. Heat at 5 °C over the melting temperature of the oligonucleotides for 5 min and let cool down at room temperature for 20 min.
3. Add 1 μL of 10× Kinase buffer, 2 μL of γ-^{32}P-ATP, 20 units of T_4 polynucleotide kinase, and H_2O to 30 μL.
4. Incubate for 1 h at 37 °C.

5. Stop the reaction by adding 1 μL of 0.5 M EDTA–NaOH (pH 8.0).
6. In the case of a PCR fragment (purified from gel as described in Subheading 3.2.1), skip **steps 1** and **2** and perform the labeling in a 20-μL volume with 50 ng of fragment.
7. Separate the probe from unincorporated nucleotides (*see* **step 2** in Subheading 3.2.1).

3.3 EMSA

1. Clean the glass plates with water (and soap, if necessary) and then with 96 % ethanol.
2. Assemble them with 1.5-mm spacers.
3. Prepare a 5 % polyacrylamide (mono:bis = 37.5:1) gel by mixing 30 mL of gel shift mix (or as much as needed for the particular setup used) with 50 μL TEMED and 500 μL 10 % (w/v) APS.
4. Mix by swirling and pour between the plates.
5. Insert a 1.5-mm comb with 6 mm wide teeth.
6. Let the gel polymerize for 30 min and remove the comb.
7. Number the wells on the outer glass plate from 1 to n (n=number of samples) with a black marker pen.
8. Mount the gel in an electrophoresis apparatus and add 0.5× TBE to the electrophoresis tanks.
9. Prepare the n samples and an $n+1$ probe master mix containing per sample 0.1 ng probe (typical 100-bp probe; corresponding to 10,000–15,000 cpm), nonspecific competitor DNA (100 ng to 3 μg; poly(dI-dC).poly(dI-dC) or poly(dA-dT).poly(dA-dT) or a mixture), an appropriate amount of 5× HEB to adjust the concentration of the probe mixture to 1× HEB final concentration, and H_2O to adjust the volume to 4 μL.
10. For each sample tube numbered from 1 to n, add 4 μL of probe mix, 1× HEB to adjust the total final volume to 10 μL, and competitor DNA as needed.
11. Retrieve the DNA-binding protein preparation from the −80 °C freezer, thaw it, gently vortex it, and store on ice.
12. Add the desired amount of DNA-binding protein preparation to the sample tubes (*see* **Notes 8** and **19**).
13. Mix by gentle vortexing and incubate for 30 min at room temperature.
14. Freeze the DNA-binding protein preparation in liquid nitrogen and return to the −80 °C freezer (*see* **Note 9**).
15. Switch on the tension on the gel at 120 V.
16. Load the samples changing the pipette tip for each sample.

17. Run the gel at 120 V for 1.5 h (time can be adjusted for probe sizes differing from 100 bp).
18. Disassemble the gel.
19. Place a piece of Whatman DE81 paper cut to size on top of the gel and reverse the DE81 paper with the gel on a piece of Whatman 3MM paper placed on the surface of a gel dryer.
20. Cover the gel with a piece of Saran Wrap and dry at 80 °C for 40 min.
21. Place the dried gel in a plastic folder and expose to an X-ray film in an autoradiography cassette mounted with a tungstate-intensifying screen in a −80 °C freezer overnight.

4 Notes

1. Some protein preparations yield upon incubation with the probe DNA a complex that migrates into the gel, while at the same time a considerable amount of DNA–protein complex does not migrate at all. Performing the DNA binding reaction on ice sometimes solves this problem resulting in a negligible amount of complex remaining in the well.
2. UV radiation causes damage to unprotected eyes and skin. Always wear UV-blocking spectacles and face shield. Protect hands with gloves and other skin with a laboratory coat.
3. Unpolymerized acrylamide is a potent neurotoxic compound that is easily absorbed through the skin. It is recommended to order liquid stocks premixed at the desired mono:bis ratio to avoid handling the acrylamide powder. Avoid skin contact and wash immediately with water. Wearing gloves is an option, but be aware that contaminated gloves are a further source of spreading the chemical to other surfaces that are afterward handled without gloves. Acrylamide stock solutions should be stored at 4 °C in the dark.
4. Ethidium bromide is a suspected carcinogen. It is recommended to order a liquid stock to avoid handling the powder. Avoid contact with skin. Dispose of ethidium bromide waste according to the rules established at the local research facility.
5. Handling of α-^{32}P-dCTP or γ-^{32}P-ATP requires adequate protection of body, head, and eyes from the emitted high-energy β-particles. Follow standard procedures and guidelines on manipulation of radioactive material and on disposal of radioactive waste in effect at the local radioactive facility.
6. Order the oligonucleotides, including an additional purification step by high-performance liquid chromatography to remove shorter intermediate products.

7. For labeling with α-^{32}P-dCTP, the double-stranded oligonucleotide or the restriction fragment should have, at least, at one end a 5′ single-stranded G-containing overhang (as obtained after *Bam*HI, *Bgl*II, *Hin*dIII, *Sal*I, *Xba*I, or *Xho*I digestion). In the case of a PCR fragment, the restriction sites can be located at the ends of the amplified fragment or they can be introduced into the PCR primers. For labeling with γ-^{32}P-ATP, the double-stranded oligonucleotide should have, at least, one blunt end or one 5′ single-stranded overhang at one end. For example, order single-stranded oligonucleotides that, after annealing, will form sticky ends corresponding to an appropriate restriction enzyme (e.g., *Bam*HI, *Bgl*II, *Hin*dIII, *Sal*I, *Xba*I, or *Xho*I) that, additionally, will allow convenient cloning in a plasmid vector for other uses (such as construction of a synthetic promoter).
8. The most common buffers are Tris-borate-EDTA (TBE), Tris-acetate-EDTA (TAE) and Tris-glycine electrophoresis buffers at 0.25, 0.5× or 1× strength.
9. The optimal amount of DNA-binding protein added in binding reactions should be determined by titration with the wild-type probe. Optimally, 10–30 % of the probe is bound to protein.
10. Repeated freezing–thawing can inactivate the DNA-binding protein. Divide the protein preparation in working and stock amounts and discard the working solution when inactivation occurs.
11. Typical binding reaction conditions are 20–50 mM Hepes or Tris–HCl pH 7–8, 40–150 mM KCl or NaCl, 0–1 mM EDTA, 0–1 mM DTT, and 10–12 % glycerol.
12. DNA–protein complexes are typically analyzed on 4–5 % acrylamide gels with a ratio of mono to bis acrylamide of 37.5 to 1.
13. The bromophenol blue and xylene cyanol FF dyes comigrate with double-stranded DNAs of approximately 50 and 250 bp, respectively.
14. Heating of the purified probe fragment should be avoided at any stage during isolation or during use, because it may lead to partial denaturation resulting in additional minor bands on gel. After labeling of the fragment, repeated freezing/thawing should be avoided since it also can lead to partial denaturation. Store the radiolabeled probe at 4 °C.
15. In the latter case dyes will quench UV fluorescence when overlapping with the DNA bands; therefore use loading mix without dyes.
16. The labeling reaction can be left at room temperature overnight without negative effects on fragment integrity.

17. Different types of columns with enhanced flow rate using centrifugal force or air pressure are commercially available (e.g., illustra ProbeQuant G-50 Micro Columns; GE Healthcare Life Sciences). Columns with enhanced flow rate tend to retain up to 50 % of the probe DNA and the recovered probe usually contains up to 20 % unincorporated nucleotides. In contrast, gravity flow columns as described here completely recover the probe without free label contamination.
18. Usually most of the DNA is in a single fraction in tube 4 (if so, do not pool with other fractions to keep DNA concentration reasonably high). Specific activity can be closely approximated by Cerenkov counting of the (pooled) peak fractions, determining the total volume of the peak fractions and assuming that they contain all of the DNA and no unincorporated nucleotides. The specific activity should be around $1–2 \times 10^5$ cpm/ng. The DNA concentration in the selected peak fraction is usually between 0.05 and 0.1 ng/μL.
19. The control is a sample without DNA-binding protein. Instead add 1× HEB and 1 μL of loading mix.

Acknowledgments

We thank Pieter Ouwerkerk and Leslie van der Fits for providing some of the data.

References

1. Chini A, Boter M, Solano R (2009) Plant oxylipins: COI1/JAZs/MYC2 as the core jasmonic acid-signalling module. FEBS J 276: 4682–4692
2. De Geyter N, Gholami A, Goormachtig S, Goossens A (2012) Transcriptional machineries in jasmonate-elicited plant secondary metabolism. Trends Plant Sci 17:349–359
3. Gaudreault M, Gingras M-E, Lessard M, Leclerc S, Guérin SL (2009) Electrophoretic mobility shift assays for the analysis of DNA-protein interactions. Methods Mol Biol 543:15–35
4. Buratowski S, Chodosh LA (1996) Mobility shift DNA-binding assay using gel electrophoresis. In: Ausubel FM, Brent R, Kingston RE, Moore DD, Seidman JG, Smith JA, Struhl K (eds) Current protocols in molecular biology, vol 2, Supp 36. Wiley, New York, pp 12.2.1–12.2.11
5. Green PJ, Kay SA, Chua N-H (1987) Sequence-specific interactions of a pea nuclear factor with light-responsive elements upstream of the *rbc*S-3A gene. EMBO J 6:2543–2549
6. Green PJ, Yong M-H, Cuozzo M, Kano-Murakami Y, Silverstein P, Chua N-H (1988) Binding site requirements for pea nuclear protein factor GT-1 correlate with sequences required for light-dependent transcriptional activation of the *rbcS-3A* gene. EMBO J 7:4035–4044
7. Katagiri F, Lam E, Chua N-H (1989) Two tobacco DNA-binding proteins with homology to the nuclear factor CREB. Nature 340:727–730
8. Gilmartin PM, Memelink J, Hiratsuka K, Kay SA, Chua N-H (1992) Characterization of a gene encoding a DNA binding protein with specificity for a light-responsive element. Plant Cell 4:839–849
9. Menke FLH, Champion A, Kijne JW, Memelink J (1999) A novel jasmonate- and elicitor-responsive element in the periwinkle secondary metabolite biosynthetic gene *Str* interacts with a jasmonate- and elicitor-inducible AP2-domain transcription factor, ORCA2. EMBO J 18:4455–4463
10. van der Fits L, Memelink J (2000) ORCA3, a jasmonate-responsive transcriptional regulator

of plant primary and secondary metabolism. Science 289:295–297

11. van der Fits L, Memelink J (2001) The jasmonate-inducible AP2/ERF-domain transcription factor ORCA3 activates gene expression via interaction with a jasmonate-responsive promoter element. Plant J 25:43–53
12. Zarei A, Körbes AP, Younessi P, Montiel G, Champion A, Memelink J (2011) Two GCC boxes and AP2/ERF-domain transcription factor ORA59 in jasmonate/ethylene-mediated activation of the *PDF1.2* promoter in Arabidopsis. Plant Mol Biol 75:321–331
13. Shoji T, Kajikawa M, Hashimoto T (2010) Clustered transcription factor genes regulate nicotine biosynthesis in tobacco. Plant Cell 22:3390–3409
14. Boter M, Ruíz-Rivero O, Abdeen A, Prat S (2004) Conserved MYC transcription factors play a key role in jasmonate signaling both in tomato and *Arabidopsis*. Genes Dev 18: 1577–1591
15. Chini A, Fonseca S, Fernández G, Adie B, Chico JM, Lorenzo O, García-Casado G, López-Vidriero I, Lozano FM, Ponce MR, Micol JL, Solano R (2007) The JAZ family of repressors is the missing link in jasmonate signalling. Nature 448:666–671
16. Todd AT, Liu E, Polvi SL, Pammett RT, Page JE (2010) A functional genomics screen identifies diverse transcription factors that regulate alkaloid biosynthesis in *Nicotiana benthamiana*. Plant J 62:589–600
17. Montiel G, Zarei A, Körbes AP, Memelink J (2011) The jasmonate-responsive element from the *ORCA3* promoter from *Catharanthus roseus* is active in Arabidopsis and is controlled by the transcription factor AtMYC2. Plant Cell Physiol 52:578–587
18. Zhang H, Hedhili S, Montiel G, Zhang Y, Chatel G, Pré M, Gantet P, Memelink J (2011) The basic helix-loop-helix transcription factor CrMYC2 controls the jasmonate-responsive expression of the *ORCA* genes that regulate alkaloid biosynthesis in *Catharanthus roseus*. Plant J 67:61–71
19. Shoji T, Hashimoto T (2011) Tobacco MYC2 regulates jasmonate-inducible nicotine biosynthesis genes directly and by way of the *NIC2*-locus *ERF* genes. Plant Cell Physiol 52: 1117–1130
20. Zhang H-B, Bokowiec MT, Rushton PJ, Han S-C, Timko MP (2012) Tobacco transcription factors NtMYC2a and NtMYC2b form nuclear complexes with the NtJAZ1 repressor and regulate multiple jasmonate-inducible steps in nicotine biosynthesis. Mol Plant 5:73–84
21. van der Fits L, Zhang H, Menke FLH, Deneka M, Memelink J (2000) A *Catharanthus roseus* BPF-1 homologue interacts with an elicitor-responsive region of the secondary metabolite biosynthetic gene *Str* and is induced by elicitor via a JA-independent signal transduction pathway. Plant Mol Biol 44:675–685
22. Pauw B, Hilliou FAO, Sandonis Martin V, Chatel G, de Wolf CJF, Champion A, Pré M, van Duijn B, Kijne JW, van der Fits L, Memelink J (2004) Zinc finger proteins act as transcriptional repressors of alkaloid biosynthesis genes in *Catharanthus roseus*. J Biol Chem 279: 52940–52948
23. Vom Endt D, Soares e Silva M, Kijne JW, Pasquali G, Memelink J (2007) Identification of a bipartite jasmonate-responsive promoter element in the *Catharanthus roseus ORCA3* transcription factor gene that interacts specifically with AT-hook DNA-binding proteins. Plant Physiol 144:1680–1689
24. Sibéril Y, Benhamron S, Memelink J, Giglioli-Guivarc'h N, Thiersault M, Boisson B, Doireau P, Gantet P (2001) *Catharanthus roseus* G-box binding factors 1 and 2 act as repressors of strictosidine synthase gene expression in cell cultures. Plant Mol Biol 45:477–488
25. Lopes Cardoso MI, Meijer AH, Rueb S, Queiroz Machado J, Memelink J, Hoge JHC (1997) A promoter region that controls basal and elicitor-inducible expression levels of the NADPH:cytochrome P450 reductase gene (*Cpr*) from *Catharanthus roseus* binds nuclear factor GT-1. Mol Gen Genet 256:674–681
26. Ouwerkerk PBF, Trimborn TO, Hilliou F, Memelink J (1999) Nuclear factors GT-1 and 3AF1 interact with multiple sequences within the promoter of the *Tdc* gene from Madagascar periwinkle: GT-1 is involved in UV light-induced expression. Mol Gen Genet 261: 610–622
27. Pasquali G, Erven ASW, Ouwerkerk PBF, Menke FLH, Memelink J (1999) The promoter of the strictosidine synthase gene from periwinkle confers elicitor-inducible expression in transgenic tobacco and binds nuclear factors GT-1 and GBF. Plant Mol Biol 39: 1299–1310
28. Green PJ, Kay SA, Lam E, Chua N-H (1991) In vitro DNA footprinting. In: Gelvin SB, Schilperoort RA, Verma DPS (eds) Plant molecular biology manual, 1st edn. Kluwer, Dordrecht, pp B11.1–B11.22
29. Lam E, Chua N-H (1990) GT-1 binding site confers light responsive expression in transgenic tobacco. Science 248:471–474
30. de Pater S, Katagiri F, Kijne J, Chua N-H (1994) bZIP proteins bind to a palindromic sequence without an ACGT core located in a seed-specific element of the pea lectin promoter. Plant J 6:133–140

Chapter 18

Transient Expression Assays in Tobacco Protoplasts

Robin Vanden Bossche, Brecht Demedts, Rudy Vanderhaeghen, and Alain Goossens

Abstract

The sequence information generated through genome and transcriptome analysis from plant tissues has reached unprecedented sizes. Sequence homology-based annotations may provide hints for the possible function and roles of particular plant genes, but the functional annotation remains nonexistent or incomplete for many of them. To discover gene functions, transient expression assays are a valuable tool because they can be done more rapidly and at a higher scale than generating stably transformed tissues. Here, we describe a transient expression assay in protoplasts derived from suspension cells of tobacco (*Nicotiana tabacum*) for the study of the transactivation capacities of transcription factors. To enhance throughput and reproducibility, this method can be automated, allowing medium-throughput screening of interactions between large compendia of potential transcription factors and gene promoters.

Key words Protoplasts, Transient expression, Automation, Tobacco BY-2, Transactivation, Transcription factors, Promoters

1 Introduction

Transcription factors (TFs) are an important class of proteins that interact with specific DNA sequences to control DNA-to-RNA transcription, thereby influencing gene expression. TFs always contain one (or more) DNA-binding domain(s) that determine(s) to which elements in the gene promoters they will bind. Besides the DNA-binding domain, one or more other functional domains that are usually protein interaction sites are also present in TFs. Depending on the interaction partners, TFs can be considered activators or repressors. Typical activators are TFs that recruit RNA polymerases or other members of the transcription initiation complex, whereas repressors are TFs that block the formation of this complex. TFs can influence the transcription rate by modulating the chromatin density, or can contain motifs that recognize signaling molecules [1]. They can act independently or with other proteins in a complex.

Alain Goossens and Laurens Pauwels (eds.), *Jasmonate Signaling: Methods and Protocols*, Methods in Molecular Biology, vol. 1011, DOI 10.1007/978-1-62703-414-2_18, © Springer Science+Business Media, LLC 2013

The function of TFs has often been demonstrated by the introduction of transgenes into the chromosomes of stably transformed plants. Although this methodology can be successful, it suffers from several drawbacks, such as lethal, pleiotropic, or indirect effects of the genetic perturbations or, conversely, absence of a notable phenotype because the process regulated by the TF is too robust. Furthermore, generation and analysis of large amounts of stably transformed plants are often too laborious or impractical, impeding the parallel testing of a large number of TFs. Therefore, a transient expression assay (TEA) in plant protoplasts is a suitable assay to determine the effect of a certain TF on a given gene promoter.

In a TEA, optimally, minimally three different plasmids ("effector," "reporter," and "normalizer") are transfected into plant protoplasts, where they drive the transient expression of the TF (on the "effector" plasmid), a first reporter gene (on the "reporter" plasmid and under control of the target promoter), and a second reporter gene (on the "normalizer" plasmid and under control of a strong constitutive promoter that is presumably not bound by the TF) [2, 3] (*see* Fig. 1). The latter construct allows the normalization of the expression of the first reporter that may vary because of experimental factors that are unrelated to the TF activity, such as transfection efficiency. TEAs in plant protoplasts can be used to dissect transcriptional regulatory and other signaling networks and to determine whether a given TF acts as activator or repressor on a given promoter. TEAs have already been used successfully to identify the DNA motifs to which specific TFs can bind, to assess the effects of environmental signals, to unravel the assembly of TF complexes, or to evaluate the influence of other regulatory proteins (such as kinases) on the TF activity [4–7]. It should be noted that in protoplast assays direct and indirect effects of a TF may not be readily distinguishable because endogenous TFs may act downstream or concurrently with the target TF, whereas lack of indispensable interaction partners might provide a false negative result. Therefore, inclusion of complementary methods, such as yeast one-hybrid or the electrophoretic mobility shift assay, might be desirable in functional analysis of TFs. In this chapter, we describe a TEA protocol for protoplasts of tobacco Bright Yellow-2 (BY-2) cells in which transfections and readout measurements can be performed on robotic platforms.

2 Materials

2.1 Equipment

1. Innova 44 incubation shaker (New Brunswick Scientific, Enfield, CT, USA).
2. Eppendorf centrifuge 5810R (Eppendorf, Hamburg, Germany).

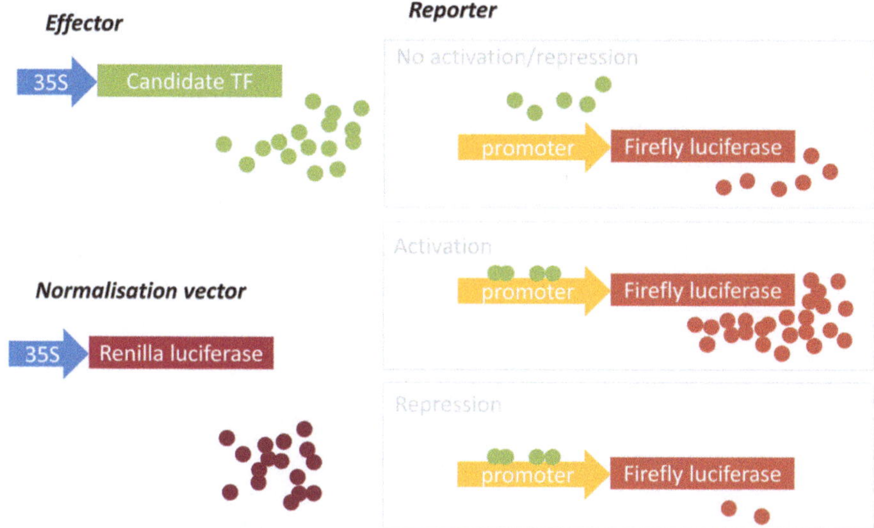

Fig. 1 Mechanism of the transient expression assay. Three different plasmids (effector, reporter, and "normalizer") are transfected into protoplasts. By the effector plasmid, a TF of choice is overexpressed through a strong constitutive promoter. This TF (*green dots*) can be examined for its transactivation activity on a promoter of choice (*yellow arrow*) that is cloned into the reporter plasmid and drives the expression of the firefly luciferase (fLUC) reporter. If the candidate TF does not bind to the promoter, no activation occurs and the basal reporter gene expression caused by endogenous TFs is measured. If the TF binds to the promoter and it is an activator, an increased fLUC activity is detected (as compared to transfection with a control construct), whereas if it is a repressor, a decreased fLUC activity is observed. By measuring the expression of a different reporter gene, in this case *Renilla* luciferase (rLUC), which is controlled by a strong constitutive promoter (such as the CaMV 35S promoter) that is not bound by the TF, the measured fLUC activities can be normalized for transfection variables by dividing the fLUC by the rLUC values

3. Sorvall RC 5plus centrifuge (Thermo Scientific, Waltham, MA, USA).
4. Light microscope.
5. Genesis® workstation 200 (Tecan, Männedorf, Germany).
6. 200-μL wide-bored tips (Molecular Bioproducts, San Diego, CA, USA).
7. 200-μL automation tips (Tecan).
8. 1,000-μL automation tips (Tecan).
9. LUMIstar Galaxy microplate luminometer (BMG LABTECH GmbH, Ortenberg, Germany).
10. Fuchs-Rosenthal counting chamber (LO-Laboroptik Ltd, Lancing, UK).
11. Orbital shaker (Bellco Glass,Vineland, NJ, USA).
12. Incubation oven (VWR, Radnor, PA, USA).
13. Tissue culture plate, 48-well, flat bottom with low evaporation lid (Becton Dickinson, Franklin Lakes, NJ, USA).

14. Modular reservoir quarter module, 40 mL capacity (Beckman Coulter, Brea, CA, USA).
15. Black FluoroNunc PolySorp plate (Thermo Scientific).
16. Plasmid maxi kit (Qiagen, Hilden, Germany).
17. Dual-Luciferase® Reporter system (Promega, Madison, WI, USA).
18. Falcon tubes (50 mL) and 100- and 250-mL flasks.
19. Petri dishes.

2.2 Suspensions, Buffers, and Media

All media should be prepared with purified deionized water at resistivity of 18 MΩ-cm at 25 °C (milliQ water) and adjusted to the indicated pH with 1 M KOH. Media can be stored at room temperature, unless stated otherwise.

2.2.1 Tobacco BY-2 Cell Suspension Cultures

1. 2,4-Dichlorophenoxyacetic acid (2,4-D) (Sigma-Aldrich, St. Louis, MO, USA), dissolved in EtOH at a concentration of 0.2 g/mL.
2. BY-2 vitamin mix: 18 mM 2,4-D, 3 mM thiamine, 0.6 M myo-inositol. Dissolve 1 mL of 2,4-D, 0.05 g of thiamine, and 5.4 g of myo-inositol in 49 mL of milliQ water. Filter sterilize and store in 1-mL aliquots at −20 °C.
3. BY-2 cell suspension culture medium: Murashige and Skoog (MS) medium (Duchefa, Haarlem, The Netherlands), 30 mM KH_2PO_4, 3 % (w/v) sucrose, pH 5.8. Weigh 4.302 g of MS, 0.2 g of KH_2PO_4, 30 g of sucrose and add 1 L milliQ water. Mix and adjust pH. Autoclave the medium and add 1 mL/L of the BY-2 vitamin mix when cooled (*see* **Note 1**).

2.2.2 Protoplast Preparation

1. Cell wall-degrading mixture: 1 % (v/v) cellulase Y-C (Kyowa Chemicals Products Co, Osaka, Japan), 0.1 % (v/v) pectolyase Y-23 (Kyowa Chemicals Products Co), 0.4 M mannitol (*see* **Note 2**), 5 mM 2-(*N*-morpholino) ethanesulfonic acid (MES). In a 100-mL flask, add 1 g of cellulase and 0.1 g of pectolyase with 99 mL of 0.4 M mannitol and 1 mL of 500 mM MES. Dissolve solution by gentle shaking or mixing on a vortex and heat at 60 °C in an oven for 10 min to remove (inactivate) proteases. When the solution is dissolved and cooled down, filter sterilize.
2. Wash buffer: 0.4 M mannitol, 2.5 mM $CaCl_2.2H_2O$, 1 mM MES, pH 5.7, filter sterilized. Weigh 36.44 g of mannitol, 0.184 g of $CaCl_2.2H_2O$, 0.10 g of MES and add 500 mL of milliQ water. Mix and adjust pH with 1 M KOH.
3. PEG solution: 40 % (w/v) polyethylene glycol (PEG) 3350 (Sigma-Aldrich), 0.4 M mannitol, 0.1 M $Ca(NO_3)_2.4H_2O$. Weigh 20 g of PEG 3350, 3.64 g of mannitol, 1.18 g of

$Ca(NO_3)_2.4H_2O$ and dissolve in 50 mL of milliQ water. Heat to dissolve all components and adjust pH to 6.0 with 0.1 M KOH. Filter sterilize immediately when the solution is still warm (*see* **Note 3**).

4. Mannitol magnesium solution (MaMg): 0.4 M mannitol, 15 mM $MgCl_2$, 5 mM MES, pH 5.6, filter sterilized. Weigh 36.4 g of mannitol, 0.714 g of $MgCl_2$, 0.533 g of MES and dissolve in 500 mL of milliQ water. Adjust the pH to 5.6 with 1 M KOH.
5. W5 medium: 0.4 M mannitol, 154 mM NaCl, 125 mM $CaCl_2.2H_2O$, 5 mM KCl, 5 mM glucose, 1.5 mM MES, pH 5.8, filter sterilized. Weigh 36.44 g of mannitol, 4.5 g of NaCl, 9.19 g of $CaCl_2.2H_2O$, 0.186 g of KCl, 0.5 g of glucose, and 0.15 g of MES and dissolve in 500 mL milliQ water. Adjust the pH to 5.8 with 1 M KOH.
6. Mannitol solution: 0.4 M mannitol, filter sterilized (*see* **Note 2**). Weigh 7.28 g of mannitol and add 100 mL milliQ water.
7. MES buffer: 500 mM MES, filter sterilized. Weigh 19.52 g of MES, add 200 mL milliQ water, and autoclave or filter sterilize.
8. Methyl jasmonate (MeJA) (Duchefa): Prepare a 100-mM stock solution in dimethyl sulfoxide (DMSO). Keep at 4 °C up to 1 month.

2.2.3 Dual-Luciferase Kit

All reagents are supplied within the Dual-Luciferase® Reporter kit (Promega), unless stated otherwise. The kit has to be stored at −20 °C until opened. Once opened, store individual components according to the manufacturer's instructions.

1. Luciferase assay reagent II (LARII): Once the luciferase assay substrate has been reconstituted according to the manufacturer's instructions, divide it into working aliquots and store at −70 °C up to 1 year. Defrost by placing the aliquot in a water-filled jar (*see* **Note 4**).
2. Stop&Glo® (Promega): Keep buffer and substrate at −20 °C and add 20 μL of substrate for every mL of buffer only just before use, because, once made, the reagent expires quickly even at −70 °C. For additional information, see the manufacturer's instructions.
3. Luciferase cell culture lysis 5× reagent (CCLR) (Promega): As this solution is not supplied in the Dual-Luciferase® Reporter system kit, add 6 mL milliQ water to 1.5 mL of CCLR. Equilibrate 1× lysis reagent to room temperature before use.

2.3 Vectors and Constructs

All vectors used (*see* Table 1) in this protoplast TEA are based on the Gateway™ technology (Invitrogen, Carlsbad, CA, USA) and are accessible through http://gateway.psb.ugent.be/ [8] or are available on request (*see* **Note 5**).

Table 1
Gateway vectors and clones used in this method

Name	Vector type	Cloning site	Composition	Size (kb)	Reference
pDONR221	Effector, donor	*att*P1-*ccdB*-*att*P2	–	4.8	–
pDONR207	Effector, donor	*att*P1-*ccdB*-*att*P2	–	5.7	–
p2GW7	Effector, destination	*att*R1-*ccdB*-*att*R2	35Spro::GCS::35Ster	5.9	[8]
pm41GWL7	Reporter, destination	*att*R4-*ccdB*-*att*R1	GCS::fLUC::35Ster	6.7	–
pGWL7	Reporter, destination	*att*R1-*ccdB*-*att*R2	GCS::fLUC::35Ster	6.6	–
p2B$_1$GUSB$_2$7	Internal control	–	35Spro::GUS::35Ster	6.1	[2]
p2rL7	Internal standard	–	35Spro::rLUC::35Ster	5.2	[2]
p2B$_1$rLB$_2$7	Internal standard	–	35Spro::rLUC::35Ster	5.2	[2]

35S CaMV 35S, *fLUC* firefly luciferase ORF, *GCS* Gateway cloning site, *rLUC Renilla* luciferase ORF, *pro* promoter, *ter* terminator

1. Effector vectors: Preferentially amplify open reading frames (ORFs) of the TFs from cDNA (libraries) with 5′ and 3′ gene-specific primers extended with the *attB1* and *attB2* recombination cassettes, respectively (Invitrogen). Entry clones are obtained by transferring the amplicons via BP clonase in pDONR221 (Invitrogen) and, finally, expression clones are generated via LR clonase in the p2GW7 destination vector (*see* **Note 6**). Negative controls (without effector plasmids) are compensated with p2B1_GUS_B27 to balance the amount of the cauliflower mosaic virus (CaMV) 35S promoter among the samples.
2. Reporter vectors: Similarly as the ORFs in the effector vectors, amplify the promoter sequences and transfer them to pDONR221. Recombine the resulting entry vectors with pGWL7 to create promoter::firefly luciferase (fLUC)::35Ster constructs. Alternatively, reporter constructs can be obtained by using destination vectors from the Multisite Gateway system (Invitrogen) with pDONRP4P1R to create entry vectors and pm41GWL7 as destination vector. In this case, the promoter amplicon should contain the *attB4* and *attB1R* recombination cassettes at their 5′ and 3′ ends, respectively (*see* **Note 6**).
3. Normalization vector: As internal standard, the ORF of *Renilla* luciferase (rLUC) is used under control of the CaMV 35S promoter as described [2].

3 Methods

The TEA can be done in a nonsterile environment and at room temperature. As protoplasts are fragile, they should be handled carefully and without sudden temperature shifts.

3.1 Plasmid Maxiprep and Sample Preparation

1. Prepare all plasmid vectors with plasmid maxiprep kits that give high-quality plasmids at sufficient yields (*see* **Note** 7).
2. Use 20 μL of plasmid mix for each sample, containing 2 μg of each plasmid construct, adjusted to a total volume of 20 μL with milliQ water (*see* **Note 8**).

3.2 BY-2 Cell Suspension Culture Maintenance and Preparation for TEA

1. Grow tobacco BY-2 cell suspension cultures in 250-mL flasks in a volume of 50 mL.
2. Weekly, make a new cell culture by diluting 1 mL of the 1-week-old saturated cell suspension culture into 49 mL of fresh BY-2 medium supplemented with the BY-2 vitamin mix.
3. Grow cells at 25 °C at 150 rpm in the dark.
4. Starting from a 6- to 10-day-old BY-2 culture, prepare a subculture by diluting 5 mL in 95 mL of fresh BY-2 medium supplemented with the BY-2 vitamin mix.
5. Leave the culture to grow for 3 days at 25 °C at 150 rpm in the dark (*see* **Note 9**).

3.3 Isolation of Protoplasts

1. Divide the 3-day-old cell suspension culture over two 50-mL Falcon tubes.
2. Spin down for 5 min at 145 ×*g* at room temperature.
3. Discard the supernatant by gently pouring it off.
4. Resuspend both the soft cell pellets in the freshly prepared cell wall-degrading enzyme mixture.
5. Leave to incubate in a large Petri dish at 25 °C under gentle shaking in the dark for 1–2 h depending on the batch of enzymes used.
6. Before continuing, check the cells by microscopy to make sure that the cell suspension culture is fully protoplasted (*see* Fig. 2).
7. Transfer the protoplasts into two sterile 50-mL Falcon tubes by carefully and slowly pipetting using a wide-bored pipette.
8. Centrifuge for 4 min at 145 ×*g*.
9. Carefully discard the supernatant by pipetting.
10. Resuspend the protoplasts in 40 mL of wash buffer.
11. Try to avoid pouring on the protoplast pellet, but rather on the sides of the Falcon tube.
12. Spin for 4 min at 145 ×*g*.

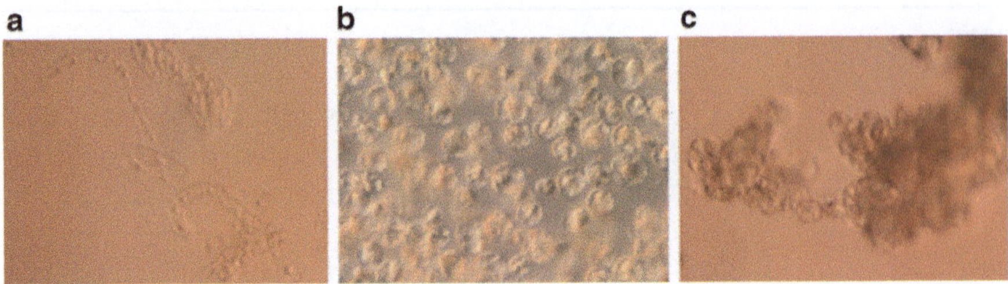

Fig. 2 Light microscopy of tobacco BY-2 cells before (**a**) and after (**b**) protoplasting and after PEG-Ca^{2+}-mediated transfection (**c**)

13. Gently remove the supernatant by pipetting.
14. Repeat **steps 10–13**.
15. Carefully add 15 mL of MaMg buffer to the Falcon tube.
16. Fully dissolve the pellet by gentle and repetitive tilting of the tube.
17. Then pour the protoplast suspension into another Falcon tube and similarly dissolve the pellet.
18. Verify that the protoplast transient expression is with 500 protoplasts per μL MaMg buffer, for instance with a Fuchs-Rosenthal counting chamber.
19. When the volume of one square is 0.0125 μL, count protoplasts in 16 squares (0.2 μL), because the protoplasts are large from BY-2 cell suspension cultures, and without taking into account the dead cells.
20. Adjust the volume of the protoplast suspension until the correct concentration of 500 protoplasts per μL (or 100 counts in 16 squares) is reached.

3.4 Automatization of the TEA

Here, the protoplast TEA is described for an automated assay in 48-well plates (*see* **Note 10**). Scripts for a Tecan Genesis Liquid Handling robot are available upon request [2]. Preferentially, the assay should be performed in eight biological replicates (*see* Fig. 3a), but could, minimally, be done with four biological repeats (*see* Fig. 3c). Optionally, the assay can be performed with eight biological repeats and MeJA elicitation (*see* Fig. 3b). Control samples should be included on each separate plate to limit technical variations. In such a control sample, the effector plasmid used is p2GW7-GUS for instance, thus lacking a true TF (*see* **Note 11**).

3.4.1 Protoplast Transfection

Here, the script ProTEA_T48_1.0 is used that drives the transfection procedure.

1. Distribute the DNA into a 48-well plate (*see* Fig. 3 and **Note 11**).

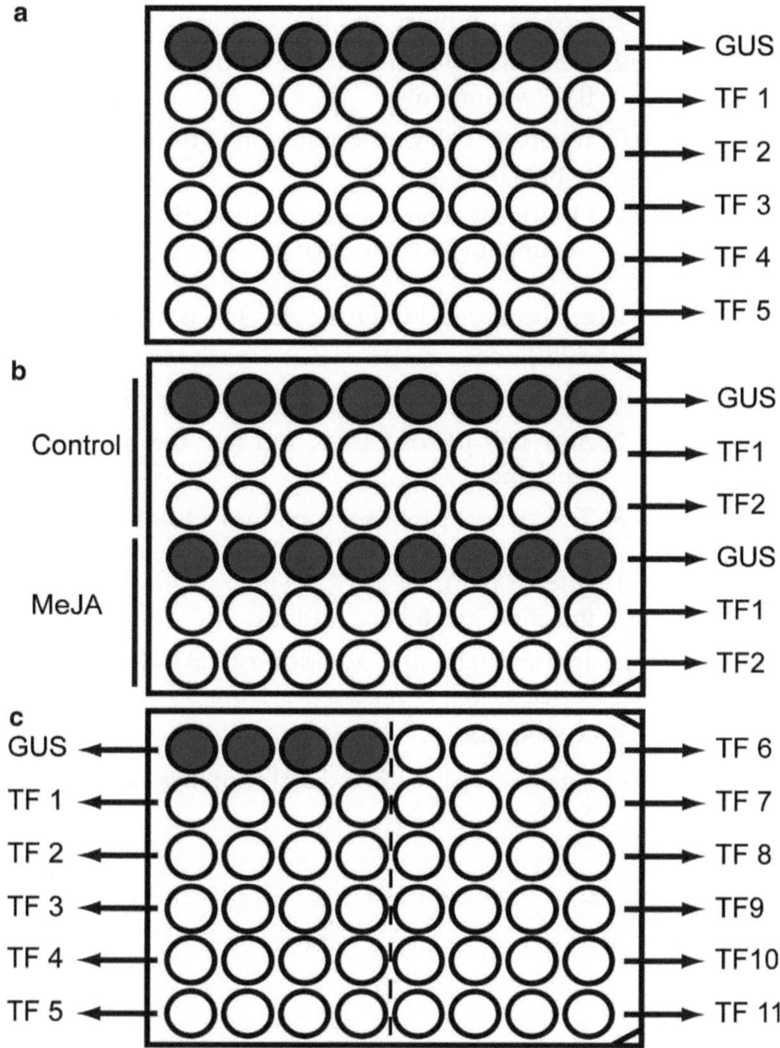

Fig. 3 Experimental setups of the TEA in 48-well plates. (**a**) Eight biological repeats. (**b**) Eight biological repeats and MeJA elicitation. (**c**) Four biological repeats

2. Carefully add 100 μL of protoplast suspension mix with wide-bore tips.
3. Leave the samples for 10 min.
4. Add 120 μL of PEG solution and mix gently with wide-bore tips.
5. Incubate for 10 min (*see* **Note 12**).
6. Add 600 μL of W5 medium to lower the viscosity of the mixture.
7. Wait for 10 min until protoplast clots stay at the bottom of the plate.

8. Carefully remove and discard the medium.
9. Immediately add 800 μL of the BY-2 medium without the BY-2 vitamin mix.
10. Incubate with gentle agitation in the dark for 16–24 h at room temperature.
11. Optionally, combine the transfection assay with a jasmonate treatment (*see* **Note 13**).

3.4.2 Protoplast Lysis

Here, the script ProTEA_L48_1.0 is used that drives the lysis procedure.

1. Carefully remove the medium by pipetting.
2. Add 125 μL of CCLR and lyse the protoplasts by vigorous pipetting.
3. Transfer 70 μL of solution into a black 96-well readout plate.

3.5 Dual-Luciferase Measurement

1. Add 100 μL of LARII either manually or by luminometer injection just before readout.
2. Immediately after addition of LARII, measure luminescence for each well every 2 s (*see* **Note 14** and Fig. 4).
3. Add Stop&Glo reagent.
4. Again, measure immediately luminescence (*see* **steps 1** and **3** in Subheading 3.5).

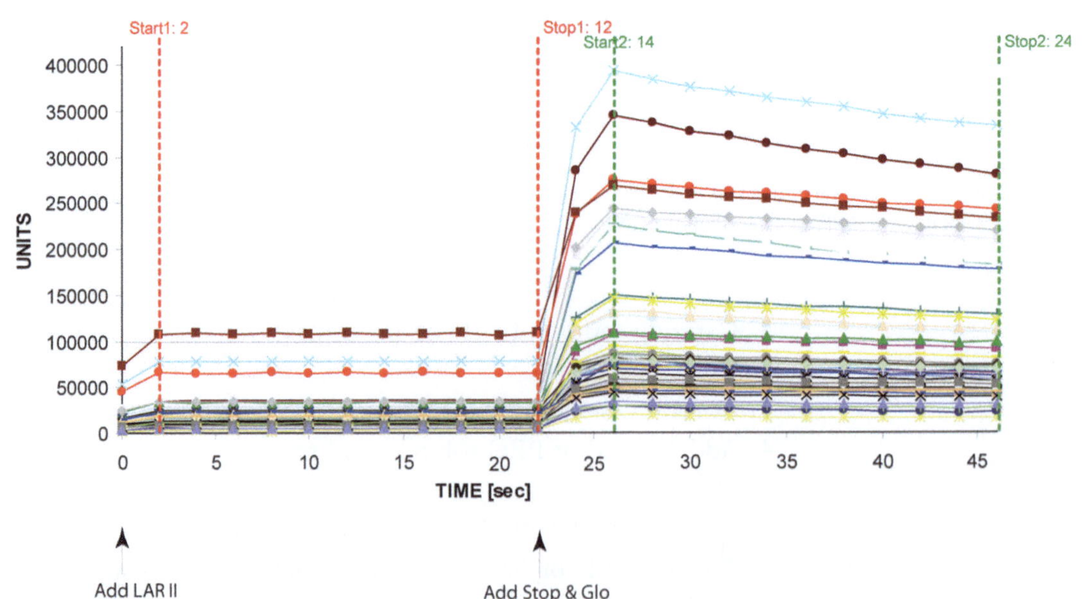

Fig. 4 Luminometer output. Luminescence for all 48 wells is plotted over time after addition of the fLUC substrate LARII and rLUC substrate Stop&Glo reagent (*arrows*). The intervals used for downstream analysis of data are indicated

3.6 Data Analysis

1. Calculate luminescence as the sum of all measurements (*see* **Note 15**).
2. Transfer the data to a spreadsheet software (such as Microsoft Excel).
3. Normalize the data by dividing the fLUC values with the rLUC values for each transfection sample (*see* **Note 16**).
4. Calculate the average for fLUC, rLUC, and the normalized values across all biological repeats (*see* **Note 17**).
5. Use the average of the normalized values to calculate the ratio relative to the control.
6. Determine the statistical significance of the differences by Student's *t*-test or other tests, depending on the experimental setup.

4 Notes

1. Medium without vitamin mix can be stored at room temperature, whereas when supplemented with the vitamin mix, it should be kept at 4 °C.
2. This solution must be made freshly before use.
3. When the solution cools down, it becomes too dense for filter sterilization due to the high PEG content.
4. This substance is sensitive to light degradation. Therefore, when thawed, keep it in the dark before use.
5. Any other type of cloning technology is suitable, provided the expression cassettes are cloned in minimal high-copy-number plasmids in which the experimental signals produced by identical expression cassettes were always higher than those in large binary T-DNA plasmids. This feature has been observed in tobacco BY-2 protoplasts as well as in *Arabidopsis thaliana* protoplasts [2].
6. A large collection of *Arabidopsis* TFs and promoter clones is available on The Arabidopsis Information Resource (www.arabidopsis.org).
7. Suitable kits are the NucleoBond™ PC500 (Macherey-Nagel, Düren, Germany) or the QIAfilter Plasmid Maxi Kit (Qiagen).
8. Minimally such a plasmid mix contains one effector, one reporter, and one normalizer construct, but multiple effector plasmids can be combined to assess additive or synergistic effects between effectors or to reconstitute TF complexes.
9. We use protoplasts prepared from tobacco BY-2 suspension cells, but TEAs can be done as well in protoplasts derived from other tissues (such as leaves) and/or other species (such as *Arabidopsis*)

or specific mutants thereof. For protoplast preparation and transfection protocols for such tissues and/or species, we refer to the literature.

10. The protoplast TEA can also be executed manually, but obviously with reduced throughput.
11. Taking into account that a minimum of four biological replicates should be measured per effector/reporter/treatment combination and that each plate should contain its own control, the activity of 11 effector/reporter combinations can be tested maximally on one 48-well plate. In the eight-biological repeat setup, maximally five effector/reporter combinations can be assessed on one 48-well plate. For an initial screen of large collections of TFs, the four-biological repeat setup is suitable, but for a detailed assessment of the transactivation capacity of a given TF (or combinations thereof), the eight-biological repeat setup is more appropriate.
12. During this step, protoplasts should agglutinate into clots visible to the eye (*see* Fig. 2).
13. Here MeJA was applied, but, in principle, any phytohormone, chemical, or environmental condition could be considered. For instance, for the MeJA treatment, instead of adding 800 μL of final BY-2 medium (**step 5** in Subheading 3.4.1), we first fill each well with 600 μL of the medium and, then, 200 μL of medium containing 200 μM MeJA or an equivalent amount of the solvent as a control (EtOH or DMSO) (*see* Fig. 3b).
14. The gain of the photomultiplier tube for a 2-s measurement interval is set at 255.
15. Of each measurement, the first repeat is excluded from calculations because the readout values do not reach the maximum yet (*see* Fig. 4).
16. At this stage, the eventual presence of proteases that cause degradation of the rLUC in the sample can occasionally be observed through a decrease in the rLUC values over time. In that case, the assay should be repeated because the normalization is incorrect.
17. Generally, the expression of the rLUC construct is not influenced by the potential TFs or other regulatory or environmental conditions, as reflected by the similar absolute mean rLUC values that are measured for the samples and of which the difference is usually less than twofold between the different effector/reporter combinations tested in a plate. Nonetheless, it is advised to verify the mean absolute rLUC values to detect possible artifacts.

References

1. Lee TI, Young RA (2000) Transcription of eukaryotic protein-coding genes. Annu Rev Genet 34:77–137
2. De Sutter V, Vanderhaeghen R, Tilleman S, Lammertyn F, Vanhoutte I, Karimi M, Inzé D, Goossens A, Hilson P (2005) Exploration of jasmonate signalling via automated and standardized transient expression assays in tobacco cells. Plant J 44:1065–1076
3. Wehner N, Hartmann L, Ehlert A, Böttner S, Oñate-Sánchez L, Dröge-Laser W (2011) High-throughput protoplast transactivation (PTA) system for the analysis of Arabidopsis transcription factor function. Plant J 68:560–569
4. Tiwari SB, Wang X-J, Hagen G, Guilfoyle TJ (2001) AUX/IAA proteins are active repressors, and their stability and activity are modulated by auxin. Plant Cell 13:2809–2822
5. Bharti K, von Koskull-Döring P, Bharti S, Kumar P, Tintschl-Körbitzer A, Treuter E, Nover L (2004) Tomato heat stress transcription factor HsfB1 represents a novel type of general transcription coactivator with a histone-like motif interacting with the plant CREB binding protein ortholog HAC1. Plant Cell 16: 1521–1535
6. Pape S, Thurow C, Gatz C (2010) The Arabidopsis *PR-1* promoter contains multiple integration sites for the coactivator NPR1 and the repressor SNI1. Plant Physiol 154: 1805–1818
7. De Boer K, Tilleman S, Pauwels L, Vanden Bossche R, De Sutter V, Vanderhaeghen R, Hilson P, Hamill JD, Goossens A (2011) APETALA2/ETHYLENE RESPONSE FACTOR and basic helix-loop-helix tobacco transcription factors cooperatively mediate jasmonate-elicited nicotine biosynthesis. Plant J 66:1053–1065
8. Karimi M, Inzé D, Depicker A (2002) GATEWAY™ vectors for *Agrobacterium*-mediated plant transformation. Trends Plant Sci 7:193–195

Chapter 19

Functional Analysis of Jasmonic Acid-Responsive Secondary Metabolite Transporters

Nobukazu Shitan, Akifumi Sugiyama, and Kazufumi Yazaki

Abstract

Jasmonic acid (JA) is a plant hormone that mediates a wide variety of plant developmental processes and defense responses. One of the major roles of JA is the versatile enhancement of the production of secondary metabolites that function as second messengers in plant defense responses. Recently, several genes have been identified as coding for JA-responsive transporters involved in the membrane transport of various secondary metabolites. Although in the literature such transport activities have been explored by a number of methods, only a few studies systematically provide a detailed technical basis of the transport assay. Here, we describe the established method to functionally analyze secondary metabolite transporters by means of a yeast cellular transport system. Moreover, the advantages and disadvantages of the method are summarized and the relevant technical points are noted.

Key words ABC transporter, Alkaloid, Cellular transport, Jasmonate-responsive, MATE transporter, Secondary metabolites, Transporter, Yeast

1 Introduction

Jasmonic acid (JA) is a plant-specific hormone that is involved in a number of developmental processes and defense responses [1]. One important function of JA is to induce the production of various secondary metabolites [2–5]. Many of these metabolites have a strong biological activity and function in defense responses against herbivores and pathogens [6]. To exhibit their biological activities in planta, these metabolites accumulate in a specific organelle of particular organs or are secreted from cells after their biosynthesis [7]. In the last decade, some proteins have been identified as metabolite transporters, of which most belong to the ATP-binding cassette (ABC) or multidrug and toxic compound extrusion (MATE) transporter families. These include for instance, among the ABC transporters, the berberine alkaloid transporters CjABCB1 [8] and CjABCB2 [9] of *Coptis japonica*, the anthocyanin

Alain Goossens and Laurens Pauwels (eds.), *Jasmonate Signaling: Methods and Protocols*, Methods in Molecular Biology, vol. 1011, DOI 10.1007/978-1-62703-414-2_19,

transporter multidrug resistance-associated protein 3 (ZmMRP3) of *Zea mays* [10], the auxin transporter AtABCB4 (AtPGP4) of *Arabidopsis thaliana* [11, 12], the inositol hexakisphosphate transporter AtABCC5 (AtMRP5) of *Arabidopsis* [13, 14], the sclareol diterpene transporter pleiotropic drug resistance NpPDR1 of *Nicotiana plumbaginifolia* [15, 16], SpTUR2 of *Spirodela polyrrhiza* [17], and AtPDR12 of *Arabidopsis* [18], and among the MATE transporters, the nicotine alkaloid transporters jasmonate-inducible alkaloid transporter 1 (NtJAT1) [19], NtMATE1 and NtMATE2 of *Nicotiana tabacum* [20], the anthocyanin transporters *antho*MATE1 and *antho*MATE3 of *Vitis vinifera* [21], the proanthocyanidin transporter TRANSPARENT TESTA 12 (TT12) of *Arabidopsis* [22], and MtMATE1 [23] and MtMATE2 of *Medicago truncatula* [24]. Recently, the *N. tabacum* gene *NtNUP1* encoding a nicotine uptake permease that belongs to the purine uptake permease family has been identified [25]. A number of these transporters were found to be inducible by JA [16, 18–20], suggesting that the genes encoding biosynthetic enzymes and transporters of the secondary metabolites might be regulated coordinately by JA in some systems. Hence, recent progress in proteome and transcriptome analyses of JA-treated cells might allow the identification of more candidate genes involved in the transport of secondary metabolites in the future. Regarding the currently utilized techniques, different methods are used by research groups to characterize transporters for such complicated organic metabolites, such as cellular transport systems with tobacco Bright Yellow-2 (BY-2) cells [12], HeLa cells [11], human embryonic kidney 293 (HEK293) cells [13], *Schizosaccharomyces pombe* (fission yeast) cells [25], *Saccharomyces cerevisiae* (budding yeast) cells [19, 20], *Xenopus laevis* (frog) oocytes [8], vesicular transport systems with plant cells [26, 27] or yeast cells [21–24], and a proteoliposome system [19]. All of these systems are known to have certain merits and drawbacks. Although it is very important to assay in several heterologous expression systems for the isolated transporters, it is a time-consuming and laborious task to learn these methods and test the many systems. Recently, we isolated the tobacco *NtJAT1* gene, of which the expression is co-regulated with that of alkaloid biosynthetic genes in JA-treated tobacco cells [19, 28]. We characterized NtJAT1 function in a cellular transport system with budding yeast. Yeast cells transformed with *NtJAT1* or an empty vector were incubated in a nicotine-containing synthetic dextrose (SD) medium. The intracellular nicotine content was quantitatively analyzed by high-performance liquid chromatography. The NtJAT1 transformants accumulated markedly less nicotine than the control yeast cells (Fig. 1a, b). The NtJAT1 protein fused to the green fluorescent protein (GFP) was mainly localized to the plasma membrane (Fig. 1c, d). These data suggested that NtJAT1 functions as a nicotine efflux transporter at the plasma

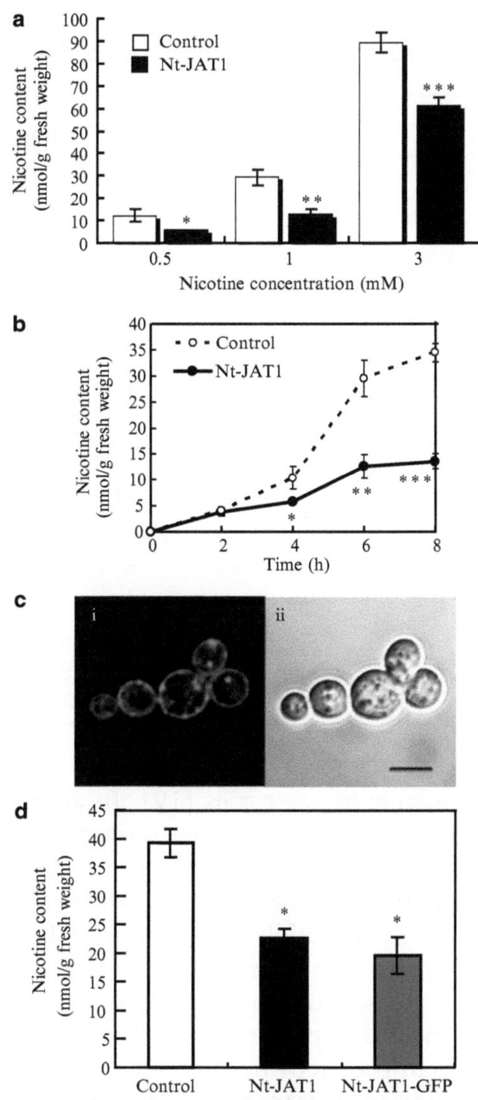

Fig. 1 NtJAT1-mediated nicotine transport in yeast cells. (Reproduced from ref. 19 with permission © the National Academy of Sciences; http://www.pnas.org/.) (**a**) Nicotine accumulation in yeast cells. Control (*open filled square*) or *NtJAT1*-expressing (*filled square*) yeast cells were incubated in half-strength SD medium, including nicotine at the indicated concentrations for 6 h. *Asterisks* indicate statistically significant difference compared to the control (Student's *t*-test; *$P<0.01$; **$P<0.001$; ***$P<0.0001$). (**b**) Time-course analysis of nicotine transport in *NtJAT1*-expressing yeast cells. Control (*open circle*) or *NtJAT1*-expressing (*filled circle*) yeast cells were incubated in half-strength SD medium supplemented with 1 mM nicotine, and sampled at the times indicated. Asterisks indicate statistically significant difference compared to control (Student's *t*-test; *$P<0.01$; **$P<0.001$; ***$P<0.0001$). (**c**) Localization of NtJAT1–GFP to the yeast plasma membranes. Yeast cells expressing *NtJAT1–GFP* were grown at 30 °C to the logarithmic growth phase and observed by fluorescence microscopy. (i) Fluorescence of yeast cells transformed with *NtJAT1–GFP*, (ii) bright-field contrast. Bar = 5 μm. (**d**) Functionality of NtJAT1–GFP in yeast cells. Nicotine content in control (*open filled square*), *NtJAT1*-expressing (*filled square*) or *NtJAT1–GFP*-expressing (*grey filled square*) yeast cells incubated in half-strength SD medium, including nicotine for 12 h. *Asterisks* indicate statistically significant differences (ANOVA Bonferroni test; *$P<0.01$)

Table 1
Advantages and disadvantages points of the yeast cellular transport method

Advantages	Disadvantages
Large inner volume size (advantageous for detection of transported substances)	Plasma membrane localization in yeast cells is necessary
No need for special equipment	Highly toxic compounds cannot be tested
Applicable for both primary and secondary transporters	Compounds impermeable at the plasma membrane cannot be evaluated in the export assay
Applicable for both uptake-type and efflux-type transporters	In contrast to patch clamp techniques for ion channels, direct transport events in cellular transport cannot be measured
	Due to the relatively long incubation time, the driving force performance cannot be characterized in detail
	Caution in interpretation of raw data, regarding the effects of inhibitors, such as ammonium ions

membrane in yeast cells [19]. In another study, by means of the same yeast system, we characterized CjABCB2, an ABC transporter that is involved in berberine (an isoquinoline alkaloid) accumulation in the rhizome of the medicinal part of *C. japonica* [9]. During the characterization processes of both NtJAT1 and CjABCB2, we found that this yeast cellular transport system is easier to apply to other transporters than alternative transport assay methods because (1) most transporters tested could be produced in yeast cells, (2) the transport activities of both primary and secondary transporters could be evaluated, and (3) the system can be applied in standard molecular biology laboratories without special equipment. Another advantage of this cellular transport assay is that the method is also applicable to both uptake-type and efflux-type transporters (*see* Table 1). Here, we describe the detailed protocol of this system, illustrated by the analysis of the nicotine transport activity of NtJAT1. This method may prove useful for the functional analysis of any JA-responsive secondary metabolite transporter.

2 Materials

2.1 Equipment

1. Acid-washed glass beads (425–600 μm) (Sigma-Aldrich, St. Louis, MO, USA) (*see* **Note 1**).
2. Shaker.

3. Vortex mixer, preferably with a platform head for multiple test tubes that is suitable for uniform extraction from many samples.
4. Confocal microscope.
5. 100-mL conical flasks.
6. Centrifuge.
7. Sterile microcentrifuge tubes.
8. Pipettes.

2.2 Solutions, Buffers, and Media

Prepare all solutions with ultrapure water and reagents of biochemical grade. Prepare and store all reagents at room temperature, unless otherwise stated.

1. Synthetic dextrose (SD) medium (-uracil): Dissolve 6.7 g of Difco Yeast Nitrogen Base without amino acids (Becton Dickinson, Franklin Lakes, NJ, USA), 20 g of glucose, and 1.92 g of Yeast Synthetic Dropout Medium supplement without uracil (Sigma-Aldrich) in 1 L water and autoclave.
2. Half-strength (1/2) SD medium (-uracil): Dissolve 3.35 g of Difco Yeast Nitrogen Base without amino acids (Becton Dickinson), 10 g of glucose, and 0.96 g of Yeast Synthetic Dropout Medium Supplement without uracil (Sigma-Aldrich) in 1 L water and autoclave.
3. Yeast AD12345678 strain (*see* **Note 2**) that contains yeast expression vector pDR196 (empty vector) (*see* **Note 3**), pDR196-NtJAT1, or pDR196-NtJAT1-GFP (*see* **Note 4**).
4. Nicotine (transport substrate) to be diluted to 630 mM with EtOH before use (*see* **Note 5**).
5. Extraction buffer: 50 % (v/v) ethanol, 50 % (v/v) methanol, and 0.5 % (v/v) acetic acid (*see* **Note 6**).

3 Methods

3.1 Incubation of Yeast Cells with Substrate

1. Incubate each yeast transformant containing a plasmid in 30 mL of SD medium (-uracil).
2. Grow overnight at 30 °C with vigorous shaking.
3. Measure OD_{600} of the overnight culture.
4. Take the inoculation volume necessary to initiate the culture of $OD_{600}=0.25$ in 40 mL of SD medium (-uracil) (*see* **Note 7**).
5. Inoculate the volume above to a fresh SD medium (-uracil) (total medium volume is 40 mL) in a new 100-mL conical flask.
6. Grow cells at 30 °C with shaking at 200 rpm (*see* **Note 8**).

7. At OD_{600} = 0.8–1.0 (logarithmic growth stage), spin down to collect the cell pellet at 3,000 × g for 5 min at room temperature.
8. Discard the supernatant.
9. Resuspend the cells in 40 mL of half-strength SD medium (-uracil).
10. Measure the OD_{600} of each transformant and control cells.
11. If the OD_{600} of the transformants and control cells differs, adjust to the same OD_{600} value by adding half-strength SD medium (-uracil) (*see* **Note 9**).
12. Add nicotine at the desired concentration (*see* **Note 10**).
13. Incubate the cells at 30 °C with shaking at 200 rpm for 6 h as a standard assay.
14. Harvest the cells by centrifugation at 3,000 × g for 5 min at 4 °C after 6 h or at a different time point for a time-course experiment (*see* **Note 11**).
15. Discard the supernatant by decantation, remove the remaining medium by pipetting, and resuspend the cells in 1 mL of cold water.
16. Transfer the cells to a sterile microcentrifuge tube (*see* **Note 12**).
17. Centrifuge samples at 3,000 × g for 5 min at 4 °C.
18. Remove the supernatant completely by pipetting.
19. Repeat washing with 1 mL of cold water and centrifugation (*see* **Note 13**).
20. Remove the supernatant completely.
21. Measure the fresh weight of the cells (*see* **Note 14**).

3.2 Extraction of Substrates from Yeast Cells

1. Add 0.2 g acid-washed glass beads and extraction buffer (2 μL/mg cell fresh weight).
2. Resuspend cells by vortexing (*see* **Note 15**).
3. Vortex the mixture vigorously for a total of 10 min, tapping the bottom of the tube every 2 min (*see* **Note 16**).
4. Centrifuge the mixture for 15 min at maximum speed (approximately 20,000 × g) at 4 °C.
5. Transfer the supernatant to a fresh microcentrifuge tube.
6. Repeat the centrifugation to completely remove any small precipitates.
7. Carefully transfer the supernatant to a fresh microcentrifuge (*see* **Notes 17** and **18**).
8. Analyze the substrate concentration with analytical techniques, such as high-performance liquid chromatography or gas chromatography (*see* **Note 19**).

4 Notes

1. Weigh glass beads and put them into a glass vessel (conical flask or bottle). Add an equal weight of concentrated nitric acid. Incubate at room temperature for 16 h with gentle shaking. Carefully remove the nitric acid and wash the glass beads with distilled water several times until the washed water has a neutral pH. Dry the glass beads and store them at room temperature.
2. In wild-type yeast cells, several endogenous transporters could potentially function in export or import of plant secondary metabolites. As a couple of yeast ABC transporters with broad specificity are known to confer multidrug resistance for substrates, including plant secondary metabolites, we use the AD12345678 strain, which lacks major ABC transporter-related genes, to decrease the effect of endogenous transporters. This strain was originally established by Dr. A. Goffeau (Université Catholique de Louvain, Belgium) [29].
3. A foreign gene is constitutively expressed by the yeast plasma membrane H^+-ATPase PMA1 promoter in pDR196, a kind gift of Dr. W. Frommer (Carnegie Institution, Stanford, CA, USA). Although growth retardation is observed in transporter-expressing yeast cells compared to vector control cells, the difference is not very large. When the cell growth is strongly inhibited by constitutive expression of a particular transporter molecule, it is possible to use an inducible promoter in another vector, such as pYES52 (Invitrogen, Carlsbad, CA, USA). For the induction method of the transporter gene, follow the manufacturer's instruction manual.
4. To confirm the plasma membrane localization of the transporter in the yeast cells, we recommend creating GFP-fusion proteins and observing the fluorescence by confocal microscopy (Fig. 1c). If the fusion protein is mainly localized to the tonoplast and/or endomembranes, the method is not applicable. In the case of tonoplast localization, a vesicular transport assay with transformed yeast cells is advised [21–24].
5. Always dilute nicotine because it is a labile compound. The dilution solvent depends on the substrate used; for instance, berberine was dissolved in water.
6. The extraction buffer should consist of solvents that efficiently dissolve the substrate; for instance 50 % (v/v) methanol (0.01 M HCl) was used as extraction buffer in the berberine transport assay.
7. If the OD_{600} value of an overnight culture is assumed to be 2.0 OD/mL, the volume of the overnight culture necessary to

inoculate a 40-mL culture at OD_{600} = 0.25 is 0.25 (OD/mL) × 40 (mL)/2.0 (OD/mL) = 5 mL.

8. Usually, this experiment is done with three replicates.
9. In general, an OD_{600} value of 0.7–0.9 is obtained. Adjust the OD_{600} difference among the transformants to less than 10 %.
10. We used 0.5, 1, or 3 mM nicotine as the final concentration. As some secondary metabolites are very toxic, we recommend checking the growth of yeast cells under several substrate concentrations after incubation with the substrates. The concentrations that inhibit cell growth should not be used. Therefore very toxic compounds are not applicable to this assay method. This point is a critical point and also a limiting factor of the cellular transport assay. The appropriate substrate concentration depends on the properties of the substrate and transporter. Careful examination of the relevant substrate concentrations and incubation time is advisable. For instance, berberine was added at 5 μM in the case of CjABCB2. Another critical point regarding this method is that compounds impermeable at the plasma membrane, due to, for example, high hydrophilicity, are not applicable to the export assay.
11. For time-course experiments to investigate NtJAT1-mediated nicotine transport, we harvested the yeast transformants at 2, 4, 6, and 8 h. A difference in cellular content between transformants and control cells was visible at all time points from 4 to 8 h (Fig. 1b).
12. The weight of each vacant microcentrifuge tube should be checked in advance.
13. After an incubation of approximately 6 or 8 h, the efficiency of the subsequent extraction may be decreased due to the large volume of cells in the extraction buffer in the small microcentrifuge tube. In such cases, reduce the cell suspension volume (e.g., to 60–80 %) per tube after complete mixture with cold water, to decrease the scale for the subsequent extraction step.
14. Cells can be stored at −20 °C prior to extraction.
15. The extraction buffer volume can be scaled up to extract the substrate efficiently.
16. The cells are sometimes difficult to suspend when some organic solvents are used, such as ethanol or methanol, instead of water. Special care should be taken for full suspension of the cells by tapping the tube bottom or by using a vortex.
17. In most cases, after a single extraction procedure, the supernatant can be used directly for quantification with high-performance liquid chromatography or gas chromatography. However, when the supernatant contains some small precipitates that inhibit the liquid chromatography analysis, remove these

substances completely from the supernatant by means of filters such as Cosmospin Filter H (Nacalai Tesque, Kyoto, Japan) to protect the analytical apparatus.

18. Supernatants can be stored at −20 °C prior to the next analysis.
19. To quantify the substrate specifically, liquid chromatography/mass spectrometry is suitable because the supernatants contain mixtures of various metabolites that are extracted from the yeast cells.

Acknowledgments

This work was supported in part by grants from the Japanese Society for the Promotion of Science Grant-in-Aid for Scientific Research (no. 19039019 to K.Y.) and Grant-in-Aid for Young Scientists (B) (no. 23780109 to N.S.), the Takeda Science Foundation, the Hyogo Science and Technology Association, the Noda Institute for Scientific Research, the Nippon Life Insurance Foundation, and the Research Institute for Sustainable Humanosphere (Kyoto University) for a grant for Exploratory Research on Sustainable Humanosphere Science.

References

1. Koo AJK, Howe GA (2009) The wound hormone jasmonate. Phytochemistry 70: 1571–1580
2. Gundlach H, Müller MJ, Kutchan TM, Zenk MH (1992) Jasmonic acid is a signal transducer in elicitor-induced plant cell cultures. Proc Natl Acad Sci USA 89:2389–2393
3. Yukimune Y, Tabata H, Higashi Y, Hara Y (1996) Methyl jasmonate-induced overproduction of paclitaxel and baccatin III in *Taxus* cell suspension cultures. Nat Biotechnol 14:1129–1132
4. Menke FLH, Champion A, Kijne JW, Memelink J (1999) A novel jasmonate- and elicitor-responsive element in the periwinkle secondary metabolite biosynthetic gene *Str* interacts with a jasmonate- and elicitor-inducible AP2-domain transcription factor, ORCA2. EMBO J 18:4455–4463
5. Baldwin IT (1998) Jasmonate-induced responses are costly but benefit plants under attack in native populations. Proc Natl Acad Sci USA 95:8113–8118
6. Croteau R, Kutchan TM, Lewis NG (2000) Natural products (secondary metabolites). In: Buchanan BB, Gruissem W, Jones RL (eds) Biochemistry and molecular biology of plants. American Society of Plant Physiologists, Rockville, MD, pp 1250–1318
7. Yazaki K (2004) Natural products and metabolites. In: Christou P, Klee H (eds) *Handbook of Plant Biotechnology*, Volume 2. Chichester, UK, John Wiley & Sons, pp 811–857
8. Shitan N, Bazin I, Dan K, Obata K, Kigawa K, Ueda K, Sato F, Forestier C, Yazaki K (2003) Involvement of CjMDR1, a plant multidrug-resistance-type ATP-binding cassette protein, in alkaloid transport in *Coptis japonica*. Proc Natl Acad Sci USA 100:751–756
9. Shitan N, Dalmas F, Dan K, Kato N, Ueda K, Sato F, Forestier C, Yazaki K (2012) Characterization of *Coptis japonica* CjABCB2, an ATP-binding cassette protein involved in alkaloid transport. Phytochemistry (doi: 10.1016/j.chtochem.2012.02.012)
10. Goodman CD, Casati P, Walbot V (2004) A multidrug resistance-associated protein involved in anthocyanin transport in *Zea mays*. Plant Cell 16:1812–1826
11. Terasaka K, Blakeslee JJ, Titapiwatanakun B et al (2005) PGP4, an ATP binding cassette P-glycoprotein, catalyzes auxin transport in *Arabidopsis thaliana* roots. Plant Cell 17: 2922–2939

12. Cho M, Lee SH, Cho HT (2007) P-glycoprotein4 displays auxin efflux transporter-like action in Arabidopsis root hair cells and tobacco cells. Plant Cell 19:3930–3943

13. Lee EK, Kwon M, Ko JH et al (2004) Binding of sulfonylurea by AtMRP5, an Arabidopsis multidrug resistance-related protein that functions in salt tolerance. Plant Physiol 134: 528–538

14. Nagy R, Grob H, Weder B et al (2009) The Arabidopsis ATP-binding cassette protein AtMRP5/AtABCC5 is a high affinity inositol hexakisphosphate transporter involved in guard cell signaling and phytate storage. J Biol Chem 284:33614–33622

15. Jasiński M, Stukkens Y, Degand H, Purnelle B, Marchand-Brynaert J, Boutry M (2001) A plant plasma membrane ATP binding cassette-type transporter is involved in antifungal terpenoid secretion. Plant Cell 13:1095–1107

16. Grec S, Vanham D, Christyn de Ribaucourt J, Purnelle B, Boutry M (2003) Identification of regulatory sequence elements within the transcription promoter region of *NpABC1*, a gene encoding a plant ABC transporter induced by diterpenes. Plant J 35:237–250

17. van den Brûle S, Müller A, Fleming AJ, Smart CC (2002) The ABC transporter SpTUR2 confers resistance to the antifungal diterpene sclareol. Plant J 30:649–662

18. Campbell EJ, Schenk PM, Kazan K, Penninckx IAMA, Anderson JP, Maclean DJ, Cammue BPA, Ebert PR, Manners JM (2003) Pathogen-responsive expression of a putative ATP-binding cassette transporter gene conferring resistance to the diterpenoid sclareol is regulated by multiple defense signaling pathways in Arabidopsis. Plant Physiol 133:1272–1284

19. Morita M, Shitan N, Sawada K, Van Montagu CE, Inzé D, Rischer H, Goossens A, Oksman-Caldentey K-M, Moriyama Y, Yazaki K (2009) Vacuolar transport of nicotine is mediated by a multidrug and toxic compound extrusion (MATE) transporter in *Nicotiana tabacum*. Proc Natl Acad Sci USA 106:2447–2452

20. Shoji T, Inai K, Yazaki Y, Sato Y, Takase H, Shitan N, Yazaki K, Goto Y, Toyooka K, Matsuoka K, Hashimoto T (2009) Multidrug and toxic compound extrusion-type transporters implicated in vacuolar sequestration of nicotine in tobacco roots. Plant Physiol 149:708–718

21. Gomez C, Terrier N, Torregrosa L, Vialet S, Fournier-Level A, Verriès C, Souquet J-M, Mazauric J-P, Klein M, Cheynier V, Ageorges A (2009) Grapevine MATE-type proteins act as vacuolar H^+-dependent acylated anthocyanin transporters. Plant Physiol 150:402–415

22. Marinova K, Pourcel L, Weder B, Schwarz M, Barron D, Routaboul J-M, Debeaujon I, Klein M (2007) The *Arabidopsis* MATE transporter TT12 acts as a vacuolar flavonoid/H^+-antiporter active in proanthocyanidin-accumulating cells of the seed coat. Plant Cell 19: 2023–2038

23. Zhao J, Dixon RA (2009) MATE transporters facilitate vacuolar uptake of epicatechin 3'-*O*-glucoside for proanthocyanidin biosynthesis in *Medicago truncatula* and *Arabidopsis*. Plant Cell 21:2323–2340

24. Zhao J, Huhman D, Shadle G, He X-Z, Sumner LW, Tang Y, Dixon RA (2011) MATE2 mediates vacuolar sequestration of flavonoid glycosides and glycoside malonates in *Medicago truncatula*. Plant Cell 23: 1536–1555

25. Hildreth SB, Gehman EA, Yang H, Lu R-H, Ritesh KC, Harich KC, Yu S, Lin J, Sandoe JL, Okumoto S, Murphy AS, Jelesko JG (2011) Tobacco nicotine uptake permease (NUP1) affects alkaloid metabolism. Proc Natl Acad Sci USA 108:18179–18184

26. Otani M, Shitan N, Sakai K, Martinoia E, Sato F, Yazaki K (2005) Characterization of vacuolar transport of the endogenous alkaloid berberine in *Coptis japonica*. Plant Physiol 138:1939–1946

27. Sugiyama A, Shitan N, Yazaki K (2007) Involvement of a soybean ATP-binding cassette-type transporter in the secretion of genistein, a signal flavonoid in legume-*Rhizobium* symbiosis. Plant Physiol 144:2000–2008

28. Goossens A, Häkkinen ST, Laakso I, Seppänen-Laakso T, Biondi S, De Sutter V, Lammertyn F, Nuutila AM, Söderlund H, Zabeau M, Inzé D, Oksman-Caldentey KM (2003) A functional genomics approach toward the understanding of secondary metabolism in plant cells. Proc Natl Acad Sci USA 100: 8595–8600

29. Decottignies A, Grant AM, Nichols JW, de Wet H, McIntosh DB, Goffeau A (1998) ATPase and multidrug transport activities of the overexpressed yeast ABC protein Yor1p. J Biol Chem 273:12612–12622

Chapter 20

Expression Analysis of Jasmonate-Responsive Lectins in Plants

Nausicaä Lannoo and Els J.M. Van Damme

Abstract

The *Nicotiana tabacum* lectin, also designated Nictaba, is a nucleocytoplasmic carbohydrate-binding protein produced in tobacco leaves after application of specific jasmonates and upon insect herbivory. Here, we describe different techniques by which lectin production can be induced through exogenous jasmonate application on tobacco plants. Furthermore, we elaborate on the assays to detect *Nictaba* expression at RNA and protein levels as well as on the agglutination assays to identify the lectin activity.

Key words Agglutination activity, Glycan binding, Nictaba, Plant lectin, Tobacco

1 Introduction

Lectins are a very diverse and heterogeneous group of nonenzymatic proteins that bind reversibly to specific free sugars or glycans present on glycoproteins and glycolipids without altering the glycan structure [1]. Lectins take their name from the Latin verb "legere" that means "to select." Since their discovery in plants in 1888, lectins have been found in all life kingdoms, ranging from viruses, fungi, and bacteria to animals and plants. For hundreds of plant lectins, the molecular structure, biochemical and physicochemical properties, glycan-binding specificities, and biological activities have been characterized [2–5]. Depending on their specificity and interaction with particular carbohydrates, several plant lectins have been developed into powerful tools to purify and analyze glycan structures and glycoproteins present in different cells, tissues, and organisms [6–8].

To organize the seemingly very heterogeneous group of plant lectins, a classification system has been elaborated in which plant lectins are subdivided into 12 distinct families of evolutionarily and structurally related carbohydrate-binding domains [5], each

Alain Goossens and Laurens Pauwels (eds.), *Jasmonate Signaling: Methods and Protocols*, Methods in Molecular Biology, vol. 1011, DOI 10.1007/978-1-62703-414-2_20, © Springer Science+Business Media, LLC 2013

typified by its own three-dimensional fold. Different carbohydrate-binding domains are able to recognize similar glycan structures with different affinities. The occurrence of these carbohydrate-binding domains is not restricted to certain plant species; most lectin domains are widespread throughout the plant kingdom.

For a long time, plant lectin research has been focused on the purification and characterization of lectins that are highly abundant in seeds and vegetative storage tissues, such as bulbs, bark, or fruits [2]. Most of these abundant lectins are synthesized with a signal peptide and targeted via the secretory pathway into the vacuolar or extracellular compartment, where they accumulate during a certain developmental stage independently from external stress conditions. Evidence has been gained that many of these so-called classical lectins play a role as storage and/or defense proteins in plants [3].

However, recently, it has become clear that plants also synthesize minute amounts of lectins in response to specific stresses and changing environmental conditions (e.g., plant hormone treatment, drought stress, heat stress, salt stress, insect herbivory, and pathogen infection) [9–11]. As these lectins cannot be detected in normal (untreated) tissues, but occur after certain stress treatments, these lectins are referred to as "inducible lectins." The class of inducible lectins can also be distinguished from the classical vacuole-located lectins by their localization in the plant cell nucleus and cytoplasm. Currently, these new lectins are generally believed to be involved in (nucleo) cytoplasmic protein–glycan interaction signaling pathways in the plant cell [12–15].

One of these inducible plant lectins is the tobacco (*Nicotiana tabacum*) agglutinin, further referred to as Nictaba, identified in 2002 [16]. Tobacco plants start to accumulate the Nictaba lectin in leaves only after treatment with jasmonic acid (JA) or its methyl ester (MeJA). When grown under normal environmental conditions, the tobacco plants do not show *Nictaba* expression or lectin activity. Treatment of tobacco plants with other plant hormones, such as salicylic acid, gibberellic acid, abscisic acid, ethylene, indole-3-acetic acid, and 6-benzylaminopurine, revealed that only the application of jasmonates was able to induce *Nictaba* expression. Herbivory of chewing insects, which provokes an increase in the endogenous JA levels, also results in Nictaba accumulation [10, 11].

Nictaba is a homodimeric protein built up of two identical nonglycosylated subunits of 19 kDa. The protein is synthesized on free ribosomes in the cytoplasm and does not undergo posttranslational modifications, besides acetylation of the N-terminal methionine [16]. Molecular cloning and analysis of the coding sequence demonstrated that the nuclear localization signal present in the polypeptide chain is necessary, but also sufficient, to transfer the cytoplasmic protein partly to the nucleus [17].

Hapten inhibition assays showed that Nictaba specifically recognizes oligomers of *N*-acetylglucosamine (GlcNAc) [16]. Therefore, Nictaba had been considered originally a chitin-binding lectin. Glycan array screening confirmed the interaction with $(GlcNAc)_n$, but also revealed the high specificity of Nictaba for high-mannose and complex *N*-glycans [17]. By pull-down assays and far-Western blot analyses with nuclear protein extracts, the in vitro interaction of Nictaba with some *O*-GlcNAcylated histone proteins could be demonstrated [18]. Recently, this interaction was confirmed in vivo by confocal microscopy and bimolecular fluorescence complementation assays (unpublished results). All these data suggest that Nictaba might fulfill a signaling role in response to stress by interacting with *O*-GlcNAcylated nuclear proteins. We hypothesize that lectin–histone interactions will alter the chromatin conformation and folding and, hence, the gene expression.

This chapter describes the different assays that have been developed to induce Nictaba expression in tobacco plants by jasmonate treatment and the corresponding protocols to detect it at the RNA and protein levels and to test the lectin activity.

2 Materials

2.1 Materials for Plant Growth

1. Seeds of *Nicotiana tabacum* cv Samsun NN.
2. Petri dishes (90 mm diameter).
3. Urgopore tape.
4. 15-mL Falcon tubes.
5. Sterile ddH_2O.
6. 70 % (v/v) ethanol (in sterile ddH_2O).
7. 6 % (v/v) NaOCl (bleach) (in sterile ddH_2O).
8. Flower pots, pot soil, fertilizer (Substral, Scotts Benelux, Sint-Niklaas, Belgium).
9. Rotary shaker.
10. Tissue culture hood.
11. Acclimatized plant growth chamber at 28 °C, with 70 % relative humidity and a 16-h photoperiod.

2.2 MeJA Treatments in Floating Experiments

1. 50 μM MeJA solution: 1.1 μL of 4.6 M MeJA (Duchefa, Haarlem, The Netherlands) dissolved in 1 mL of 100 % (v/v) EtOH and transferred into 99 mL H_2O.
2. ddH_2O containing 1 % (v/v) EtOH.
3. Petri dishes.

2.3 MeJA Treatments Through the Gas Phase

1. Make 10 % MeJA solution (Duchefa) in 100 % EtOH.
2. Whatman filter paper of 10 cm × 10 cm.
3. Transparent plastic bag of 50 L content.

2.4 Jasmonate Treatments in Lanolin Paste

1. Lanolin paste (Roth, Karlsruhe, Germany).
2. Jasmonates (for source and synthesis, *see* ref. 19): 12-oxo-phytodienoic acid (OPDA), jasmonate (JA), MeJA, hydroxylated jasmonate (12-OH-JA), sulfate of 12-OH-JA (12-HSO_4-JA), glucoside of 12-OH-JA (12-*O*-Gluc-JA).
3. Jasmonate solutions: Mix 7.5 mg of each compound into 1 mL of the lanolin paste, giving final concentrations of 26 mM OPDA, 36 mM JA, 33 mM MeJA, 33 mM 12-OH-JA, 25 mM 12-HSO_4-JA, and 19 mM 12-*O*-Gluc-JA.

2.5 Reverse-Transcriptase Polymerase Chain Reaction

1. Liquid nitrogen.
2. Mortar and pestle.
3. Trizol® Reagent (Invitrogen, Carlsbad, CA, USA).
4. DNaseI (1 U/μL) (Thermo Scientific, Waltham, MA, USA).
5. 1.5 % agarose (Invitrogen) gel prepared in 0.5× tris(hydroxymethyl)aminomethane (Tris)-acetate-ethylenediaminetetraacetic acid (EDTA) (TAE) buffer.
6. Ethidium bromide (Thermo Scientific) (1/1,000 diluted in 0.5× TAE).
7. M-MLV reverse transcriptase (200 U/μL) (Invitrogen).
8. 10 mM dNTP mix, 10× Extra buffer, and Taq DNA polymerase (all from VWR, Leuven, Belgium).
9. Primers (5 μM) (Invitrogen).
   (a) Specific primers to amplify the full-length cDNA sequence encoding Nictaba (Genbank accession No AF389848): Forward primer 5'-GATAGCATCATATCATATA-3' and reverse primer 5'-AGAAAATCATAAAGACAAAC-3'.
   (b) Specific primers to amplify part of the cDNA sequence of the RIBOSOMAL PROTEIN L25 (RL25; Genbank accession No L18908): Forward primer 5'-TGCAATGAAGAAGATTGAGGACAACA-3' and reverse primer 5'-CCATTCAAGTGTATCTAGTAACTCAAATCCAAG-3'.
10. Spectrophotometer.
11. Electrophoresis equipment.
12. PCR apparatus.

2.6 Western Blot

1. Mortar and pestle.
2. 20 mM 1,3-propane diamine (VWR).
3. Coomassie Protein Assay kit (Thermo Fisher Scientific).

4. 15 % acrylamide gel.
5. Coomassie Brilliant blue R250 (VWR).
6. 0.45-μm polyvinylidene fluoride (PVDF) transfer membranes (Pall Immobilon, Port Washington, NY, USA).
7. Tris-buffered saline (TBS): 10 mM Tris–HCl, 150 mM NaCl, 0.1 % (v/v) Triton X-100 (pH 7.6).
8. Blocking buffer: 5 % (w/v) nonfat milk powder (Applichem, Darmstadt, Germany) in TBS.
9. Primary antibody solution: TBS containing 1:80 diluted polyclonal antibody directed against Nictaba, raised in rabbit [20].
10. Secondary antibody solution: TBS containing 1:300 diluted horseradish peroxidase (HRP)-coupled goat anti-rabbit IgG (Sigma-Aldrich, St. Louis, MO, USA).
11. Peroxidase-antiperoxidase (PAP) solution: TBS containing 1:400 diluted PAP complex (Sigma-Aldrich).
12. Detection buffer: 0.1 M Tris–HCl (pH 7.6) containing 700 μM diaminobenzidine (Acros-Organics, Geel, Belgium) and 0.03 % (v/v) H_2O_2.

2.7 Agglutination Assay

1. Mortar and pestle.
2. 20 mM 1,3-propane diamine.
3. Phosphate-buffered saline (PBS): 135 mM NaCl, 3 mM KCl, 1.5 mM KH_2PO_4, 8 mM Na_2HPO_4, pH 7.5.
4. Small U-shape bottom glass tubes (0.5 cm diameter).
5. Polystyrene 96 U-welled microtiter plates.
6. Trypsin-treated rabbit erythrocytes (BioMérieux, Marcy l'Etoile, France): Add 10 mg trypsin to 200 μL erythrocytes suspended in 1 mL of PBS and incubate at 37 °C for 1 h; wash three times in PBS; centrifuge at 1,000 ×*g* between washing steps and remove the supernatant; suspend the erythrocytes in 1 mL of PBS by gentle mixing.
7. 1 M ammonium sulfate.

2.8 Immunosorbent Assay (ELISA)

1. Mortar and pestle.
2. 20 mM 1,3-propane diamine.
3. Maxisorp F96 Nunc microtiter plates (VWR).
4. Automatic 96-well microtiter plate reader (Powerwave X340, Bio-Tek Instruments, Winooski, VT, USA).
5. Coating buffer: 15 mM sodium carbonate, 35 mM sodium bicarbonate, 3 mM sodium azide, pH 9.6.
6. PBS containing 0.1 % (v/v) Tween-20 (PBST).
7. Blocking buffer: 5 % (w/v) nonfat milk powder in PBS.

8. Primary antibody solution: PBS containing 1:500 diluted affinity purified polyclonal antibody directed against Nictaba, raised in rabbit [20].
9. Secondary antibody solution: PBS containing 1:10,000 diluted HRP-coupled goat antirabbit immunoglobulin (Sigma-Aldrich).
10. Substrate buffer: 0.4 mg/mL *O*-phenylenediamine hydrochloride (Sigma-Aldrich) dissolved in 0.15 M citrate buffer (pH 5) containing 0.04 % (v/v) H_2O_2.

3 Methods

Carry out all procedures at room temperature unless otherwise specified.

3.1 Surface Sterilization of Tobacco Seeds

The sterilization protocol should be performed in a tissue culture hood.

1. Place tobacco seeds in a 15-mL Falcon tube.
2. Add 10 mL of 70 % (v/v) ethanol (in sterile ddH_2O).
3. Shake vigorously to mix the seeds in the ethanol solution.
4. Incubate the 15-mL Falcon tube for 2 min on a rotary shaker.
5. Let the seeds settle to the bottom of the tube, and then remove the ethanol solution.
6. Add 10 mL of 6 % (v/v) NaOCl (bleach) (in sterile ddH_2O).
7. Shake vigorously to mix the seeds in the bleach solution.
8. Incubate the 15-mL Falcon tube for 10 min on a rotary shaker.
9. Let the seeds settle, and then remove the bleach solution.
10. Add 10 mL of sterile ddH_2O.
11. Shake vigorously to mix the seeds in ddH_2O.
12. Incubate the 15-mL Falcon tube for 1 min on a rotary shaker.
13. Let the seeds settle, and then remove the ddH_2O.
14. Repeat **steps 10–13** four times.
15. Let the seeds completely dry in an open sterile Petri dish before using them.
16. Store sterilized seeds at room temperature in a sterile container (e.g., 15-mL Falcon tube or Petri dish).

3.2 Growth of Tobacco Plants in Soil

1. Sow ± 50 surface-sterilized seeds of *Nicotiana tabacum* cv Samsun NN in Petri dishes filled with pot soil and incubate the plates for 2 weeks in an acclimatized plant growth chamber at 28 °C, 70 % relative humidity, and a 16-h light/8-h dark photoperiod (*see* **Note 1**).

2. Close the Petri dishes with Urgopore tape to permit airflow to the seeds.
3. Water the seeds regularly (twice a week) with 2 mL tap water (*see* **Note 2**).
4. After appearance of the cotyledons and first leaves, transfer the plantlets to open flower pots filled with soil (maximum 2–3 plants per pot).
5. Incubate the pots for several weeks in the acclimatized plant growth chamber.
6. Water the plants twice a week with 50 mL tap water (*see* **Note 2**).
7. Add fertilizer (dissolved in tap water) once a month.

3.3 Treatment of Excised Tobacco Leaves with MeJA in Floating Experiments

1. Cut green healthy leaves from 6- to 8-week-old tobacco plants (*see* **Note 3**).
2. Transfer each leaf to a Petri dish filled with 15 mL of 50 μM MeJA solution and allow floating on the MeJA solution.
3. Transfer control leaves to Petri dishes filled with 15 mL of ddH_2O containing 1 % (v/v) EtOH.
4. Close all Petri dishes (no tape) and incubate them for 24 h (in case of RT-PCR experiments) or 72 h (in case of lectin assays) at room temperature (*see* **Note 4**).
5. After incubation, rinse the MeJA-treated leaves in ddH_2O and blot them dry.
6. Determine the total weight of each leaf before further use.
7. Use immediately for total RNA and protein extraction or freeze at −80 °C for later use.

3.4 Treatment of Tobacco Plants with MeJA Through the Gas Phase

1. Use 4- to 16-week-old tobacco plants.
2. Transfer healthy plants into a 50-L transparent plastic bag containing a piece of filter paper (10×10 cm) on which 100 μL of a 10 % (v/v) solution of MeJA (dissolved in ethanol) is spotted.
3. Replace with new filter paper on which MeJA is spotted for 3 or 4 consecutive days every 24 h (*see* **Note 5**).
4. After incubation, collect all the leaves from the tobacco plants and determine their fresh weight before use (*see* **Notes 3** and **6**).
5. Use leaves immediately for total RNA and protein extraction or freeze at −80 °C for later use.

3.5 Treatment of Tobacco Leaves with Jasmonates in Lanolin Paste

1. Use 6- to 8-week-old healthy tobacco plants.
2. Take only undamaged leaves, attached to the middle part of the tobacco plant.
3. Perform each treatment on separate plants with three leaves per plant as biological replicates.

4. Apply two droplets of 10 μL of a jasmonate mixture (in lanolin paste) to each side of the main vein on the upper side of the selected leaves.
5. Spread the mixture equally over the leaf surface by gentle rubbing (*see* **Note 7**).
6. Similarly, apply two droplets of 10 μL of pure lanolin paste on control leaves present on a separate plant (*see* **Note 8**).
7. Collect the treated leaves after 3 days of treatment and determine their fresh weight before use.
8. Use leaves immediately for total RNA and protein extraction or freeze at −80 °C for later use.

3.6 Detection of Lectin Expression with Classical RT-PCR

1. Homogenize 200 mg of leaf material in liquid nitrogen with mortar and pestle.
2. Extract total RNA with Trizol® reagent according to the manufacturer's instructions.
3. Remove residual DNA by treatment with 2 U of DNaseI for 30 min at 37 °C.
4. To check the RNA quality, load equal amounts of RNA onto 1.5 % TAE-agarose gel.
5. Perform electrophoresis and visualize RNA after ethidium bromide staining.
6. Determine the RNA content of the samples with a spectrophotometer.
7. Synthesize single-stranded cDNA from 1 μg of total RNA with M-MLV reverse transcriptase as described by the manufacturer.
8. Set up RT-PCR on 2 μL of synthesized single-stranded cDNA with Taq DNA polymerase.
9. Use the *Nictaba*-specific forward and reverse primers to amplify the 608-bp *Nictaba* sequence.
10. Use the *RL25*-specific forward and reverse primers to amplify 287 bp of the coding sequence of the *RL25*.
11. Perform RT-PCR according to the program: 2 min at 94 °C, followed by 25 cycles of 15 s at 94 °C, 30 s at 50 °C, and 1 min at 72 °C, ending with 5 min at 72 °C.

3.7 Detection of Lectin Expression with Western Blot

1. Homogenize collected leaves in 20 mM 1,3-propane diamine with mortar and pestle.
2. Add 5 mL of extraction buffer per gram fresh weight of leaf material.
3. Transfer the homogenates to centrifuge tubes and centrifuge for 10 min at 3,000 ×*g*.

4. Collect the supernatant (=protein extract) and keep at 4 °C until use.
5. Determine the total protein concentration with the Coomassie Protein Assay kit.
6. Load 50 μg of total protein on 15 % acrylamide gel.
7. Run electrophoresis for 1 h at 150 V as described [21].
8. After electrophoresis, visualize the proteins by staining gels in Coomassie Brilliant blue R250 or blot the proteins on 0.45-μm PVDF transfer membranes for Western blotting.
9. Perform immunoblotting by blocking the membrane first with TBS containing 5 % (w/v) nonfat milk powder, followed by consecutive incubation in TBS supplemented with a primary rabbit antibody directed against Nictaba (1:80 diluted), an HRP-coupled goat antirabbit immonuglobulin (1:300 diluted), and the PAP complex (1:400 diluted).
10. Wash the membrane between every incubation step three times for 5 min in TBS.
11. Perform immunodetection with the detection buffer, but not for more than 10 min.
12. Stop the coloring reaction by washing the membrane with water.

3.8 Detection of Lectin Activity with Agglutination Assays

1. Homogenize collected leaves in 20 mM 1,3-propane diamine with mortar and pestle.
2. Use 5 mL of extraction buffer per gram fresh weight of leaf material.
3. Transfer the homogenates to centrifuge tubes and centrifuge them for 10 min at 3,000 ×*g*.
4. Collect the supernatant (=protein extract) in fresh tubes and keep at 4 °C until use.
5. For agglutination assays, mix 10 μL of crude protein extract with 10 μL of 1 M ammonium sulfate and 30 μL of a 2 % (v/v) solution of trypsin-treated rabbit erythrocytes (*see* **Note 9**).
6. To obtain a negative control, mix 20 μL of 1 M ammonium sulfate and 30 μL of a 2 % solution of trypsin-treated rabbit erythrocytes.
7. To obtain a positive control, mix 10 μL of a purified lectin solution (1 μg/μL) with 10 μL of 1 M ammonium sulfate and 30 μL of a 2 % solution of trypsin-treated rabbit erythrocytes.
8. Incubate the samples in a small U-shaped bottom glass tube (0.5 cm diameter) at room temperature (*see* **Note 10**).
9. For a semiquantitative estimate of the lectin content, make serial dilutions of the lectin samples in 1 M ammonium sulfate with twofold increments.

10. Transfer 10-μL aliquots of the diluted extracts to glass tubes or polystyrene 96 U-welled microtiter plates and supplement with 40 μL of a 2 % solution of trypsin-treated rabbit erythrocytes.
11. Include also a dilution series of a purified Nictaba solution [20] at a known concentration to determine the absolute lectin content of the extracts.
12. Incubate the samples for 1 h at room temperature before assessment of agglutination (*see* **Note 11**).

3.9 Quantification of Lectin Expression with ELISA

1. Coat the microtiter plates overnight at 4 °C with 25 μg of crude protein extracts or purified Nictaba serially diluted from 300 to 2 ng in coating buffer.
2. After coating, wash the plates twice with PBST.
3. Block for 3 h at 37 °C with 5 % (w/v) nonfat milk powder in PBS.
4. Wash the wells three times with PBST.
5. Incubate the wells with an affinity-purified rabbit antibody directed against Nictaba (1:500 diluted in PBS) for 1 h at 37 °C.
6. Wash the wells five times with PBST before the next incubation with an HRP-coupled goat antirabbit immunoglobulin (1:10,000 diluted in PBS) for 1 h at 37 °C.
7. Wash the wells five times with PBST before addition of the substrate buffer.
8. After 30 min of incubation in the dark, measure the absorbance of all samples at 450 nm with a microtiter plate reader.
9. For each sample, use the average absorbance measured for three replicate samples to determine its Nictaba content.
10. Use a reference curve to calculate the lectin content in the sample (*see* **Note 12**).

4 Notes

1. After 2 weeks of incubation, the tobacco seeds should have germinated and cotyledons and first leaves should appear on the plantlets. If not, plates can be incubated longer.
2. Do not water the plants too much to avoid fungal and bacterial growth in the pot soil.
3. Use only fresh-looking green leaves that have developed in the middle part of healthy tobacco plants. Avoid using old yellowish-looking leaves because these samples will show only a limited response to jasmonate treatment and the protein will be more difficult to extract.

4. Jasmonates are volatile compounds. Therefore, keep plates with jasmonate-treated leaves separated from the control leaves to prevent jasmonate-induced gene expression in the control leaves. Do not prepare the jasmonate solutions or perform the treatments in the same room where tobacco plants are cultivated because this handling might already affect the gene expression in neighboring plants. Also keep control (noninduced) and jasmonate-treated plants in separate rooms at all times.
5. Before the start of the MeJA treatment, add tap water to the pot soil of the selected plants so that plant roots are incubated in wet soil. This handling enhances water transport and transpiration throughout the plants and ameliorates the plant's uptake of volatile MeJA through the stomata.
6. MeJA acts systemically on intact plants, but very locally in excised leaves. Kinetic analyses of adult tobacco plants indicated that the lectin is synthesized within 12-h exposure time to MeJA, reaching a maximum after 60 h. After removal of MeJA, the lectin progressively disappears from the leaf tissue [10].
7. Lanolin is a waxy substance without any adverse effect on plant development. It can be mixed with growth hormones and allows easy application of the compound to be tested.
8. A control group of plants that receive an application of pure lanolin should always be included to account for any irregularities in plant growth caused by the lanolin.
9. Red blood cells tend to clump together in the presence of lectins that show specificity toward the glycan structures present on the surface of these cells. In glass tubes, hemagglutination is visible as clumps of red blood cells at the bottom of tube. In the wells of polystyrene microtiter plates, the network made of red blood cells and lectins will also attach and, hence, will form a homogeneous suspension. In contrast, when no agglutination takes place (in the absence of lectin or at very low lectin concentrations), the erythrocytes settle at the bottom of the wells of the microtiter plate. Agglutination assays should be performed with a fresh solution of rabbit erythrocytes. Trypsin treatment of red blood cells can help to increase the sensitivity of the assay, but should be done just before the agglutination assay. Usually, agglutination of red blood cells is visible within a few minutes (depending on the amount of lectin).
10. The lectins in the samples will bind to the carbohydrates present at the surface of the red blood cells. As such, agglutination should be visible as clear clumps of red blood cells settled down at the bottom of the glass tube (*see* **Note 9**).
11. This semiquantitative method allows detection of lectin concentrations as low as 0.6 μg/mL with an error range < 12.5 %.

12. The amount of *Nictaba* that is expressed in tobacco leaves after MeJA treatment varies from 2 to 10 mg lectin/g fresh weight of leaf material. The lectin concentration will depend on the age of the tobacco plants (6- to 16-week-old plants) and the position of the leaf on the plant [10, 11].

Acknowledgments

This work was supported by grants from Ghent University and the Fund for Scientific Research—Flanders.

References

1. Peumans WJ, Van Damme EJM (1995) Lectins as plant defense proteins. Plant Physiol 109: 347–352
2. Van Damme EJM, Peumans WJ, Barre A, Rougé P (1998) Plant lectins: a composite of several distinct families of structurally and evolutionary related proteins with diverse biological roles. Crit Rev Plant Sci 17:575–692
3. Van Damme EJM, Peumans WJ, Pusztai A, Bardocz S (1998) Handbook of Plant Lectins: Properties and Biomedical Applications. John Wiley & Sons, Chichester, p 452
4. Van Damme EJM, Rougé P, Peumans WJ (2007) Plant lectins. In: Kamerling JP, Boons G-J, Lee Y, Suzuki A, Taniguchi N, Voragen AJG (eds) *Comprehensive Glycoscience: From Chemistry to Systems Biology*, Volume 3: Biochemistry of Glycoconjugate Glycans: Carbohydrate-Mediated Interactions. Elsevier, Amsterdam, pp 563–599
5. Van Damme EJM, Lannoo N, Peumans WJ (2008) Plant lectins. Adv Bot Res 48: 107–209
6. Hirabayashi J (2008) Concept, strategy and realization of lectin-based glycan profiling. J Biochem 144:139–147
7. Van Damme EJM (2011) Lectins as tools to select for glycosylated proteins. In: Gevaert K, Vandekerckhove J (eds) Gel-free proteomics: methods and protocols, vol 753, Methods in molecular biology. Humana Press, New York, pp 289–297
8. Wu AM, Lisowska E, Duk M, Yang Z (2009) Lectins as tools in glycoconjugate research. Glycoconjugate J 26:899–913
9. Fouquaert E, Peumans WJ, Smith DF, Proost P, Savvides SN, Van Damme EJM (2008) The "old" *Euonymus europaeus* agglutinin represents a novel family of ubiquitous plant proteins. Plant Physiol 147:1316–1324
10. Lannoo N, Vandenborre G, Miersch O, Smagghe G, Wasternack C, Peumans WJ, Van Damme EJM (2007) The jasmonate-induced expression of the *Nicotiana tabacum* leaf lectin. Plant Cell Physiol 48:1207–1218
11. Vandenborre G, Miersch O, Hause B, Smagghe G, Wasternack C, Van Damme EJM (2009) *Spodoptera littoralis*-induced lectin expression in tobacco. Plant Cell Physiol 50:1142–1155
12. Lannoo N, Van Damme EJM (2010) Nucleocytoplasmic plant lectins. Biochim Biophys Acta 1800:190–201
13. Van Damme EJM, Barre A, Rougé P, Peumans WJ (2004) Cytoplasmic/nuclear plant lectins: a new story. Trends Plant Sci 9:484–489
14. Van Damme EJM, Lannoo N, Fouquaert E, Peumans WJ (2004) The identification of inducible cytoplasmic/nuclear carbohydrate-binding proteins urges to develop novel concepts about the role of plant lectins. Glycoconjugate J 20:449–460
15. Van Damme EJM, Fouquaert E, Lannoo N, Vandenborre G, Schouppe D, Peumans WJ (2011) Novel concepts about the role of lectins in the plant cell. In: Wu AM (ed) The molecular immunology of complex carbohydrates-3, vol 705, Advances in experimental medicine and biology. Springer, New York, pp 271–294
16. Chen Y, Peumans WJ, Hause B, Bras J, Kumar M, Proost P, Barre A, Rougé P, Van Damme EJM (2002) Jasmonic acid methyl ester induces the synthesis of a cytoplasmic/nuclear chito-oligosaccharide-binding lectin in tobacco leaves. FASEB J 16:905–907
17. Lannoo N, Peumans WJ, Van Pamel E, Alvarez R, Xiong T-C, Hause G, Mazars C, Van Damme EJM (2006) Localization and in vitro binding studies suggest that the cytoplasmic/nuclear tobacco lectin can interact in situ with high-mannose and complex *N*-glycans. FEBS Lett 580:6329–6337

18. Schouppe D, Ghesquière B, Menschaert G, De Vos WH, Bourque S, Trooskens G, Proost P, Gevaert K, Van Damme EJM (2011) Interaction of the tobacco lectin with histone proteins. Plant Physiol 155:1091–1102
19. Miersch O, Neumerkel J, Dippe M, Stenzel I, Wasternack C (2008) Hydroxylated jasmonates are commonly occurring metabolites of jasmonic acid and contribute to a partial switch-off in jasmonate signaling. New Phytol 177: 114–127
20. Vandenborre G, Van Damme EJM, Smagghe G (2009) *Nicotiana tabacum* agglutinin expression in response to different biotic challengers. Arthropod Plant Interact 3:193–202
21. Laemmli UK (1970) Cleavage of structural proteins during the assembly of the head of bacteriophage T4. Nature 227:680–685

Part IV

Omics

Chapter 21

Profiling the Jasmonic Acid Responses by Nuclear Magnetic Resonance-Based Metabolomics

Hye Kyong Kim, Young Hae Choi, and Robert Verpoorte

Abstract

Metabolomics based on nuclear magnetic resonance (NMR) can be used to monitor the metabolite response of plants to jasmonic acid. Metabolomics experiments consist of three important steps: sample preparation, NMR analysis, and data mining. In sample preparation, a very critical factor is the selection of a proper solvent for plant material extraction because the presence of metabolites will vary according to the solvent used. For NMR analysis, an intermediate-to-polar solvent, such as a mixture of methanol and water, is a good choice. In general, ^{1}H-NMR spectroscopy is the standard method for metabolite profiling, whereas two-dimensional spectroscopy provides more detailed information on the metabolites. Finally, various chemometric methods can be used for data mining. Here, we describe all three steps of metabolomic analysis by means of NMR spectroscopy.

Key words Metabolomics, NMR spectroscopy, Jasmonic acid, Extraction, Metabolite identification, Data mining

1 Introduction

Jasmonic acid (JA) is a well-known plant signaling molecule involved in various physiological functions of plants, particularly in plant defense responses [1]. The signaling of jasmonate has been thoroughly investigated in the past. JA has been shown to act as an elicitor to induce plant secondary metabolism via extensive transcriptional reprogramming [2, 3], activating entire metabolic pathways. Hence, metabolic responses of plants to JA are very complex and diverse.

Recent advances in analytical technology allow us to analyze organisms comprehensively. One such method is metabolomics, a tool that can be used to draw a whole picture of the organismal behavior under certain conditions [4]. Thus, it can be applied to examine the JA response in plants. Mass spectroscopy and nuclear magnetic resonance (NMR) have emerged as the most suitable

Alain Goossens and Laurens Pauwels (eds.), *Jasmonate Signaling: Methods and Protocols*, Methods in Molecular Biology, vol. 1011, DOI 10.1007/978-1-62703-414-2_21, © Springer Science+Business Media, LLC 2013

platforms among many analytical techniques. The advantage of mass spectroscopy is its high sensitivity, but its drawback is the limited range of metabolites that can be detected [5]. In contrast, NMR can detect diverse groups of metabolites, although its low sensitivity still remains a major shortcoming. Considering its high reproducibility, simple sample preparation, and short analysis time, NMR is most suited for high-throughput processes. For example, in case of time courses, in which numbers of samples have to be analyzed to follow the dynamics of the plant's responses to JA treatments over time, NMR is the method of choice [6].

In this chapter, we describe a method for metabolite profiling both in intact plants and cell cultures. In general, experiments on metabolite profiling consist of three steps: (1) sample preparation, (2) sample analysis, and (3) data mining. The experiments start with the JA treatment to plants, subsequent harvest, and extraction prior to the NMR measurement. The NMR analysis itself is quite straightforward, because, whereas ^{1}H NMR is the method of choice for standard profiling, other two-dimensional NMR techniques are utilized to identify the metabolites [7, 8]. Diverse chemometric methods can be used for data mining [9, 10], extracting all information from the NMR data set, such as increasing or decreasing levels of metabolites after JA treatment.

2 Materials

2.1 Equipment

1. Glass rods.
2. Growth chamber.
3. Culture flasks.
4. Pestle and mortar.
5. Plastic tubes.
6. Spatula.
7. Freeze drying machine.
8. Vacuum pump.
9. 1.5-mL Eppendorf tubes.
10. Vortex machine.
11. Ultrasonicator.
12. Microcentrifuge.
13. 5-mm NMR tubes.
14. Spectrometer.
15. NMR spectrometer (Bruker BioSpin GmbH, Rheinstetten, Germany), including software, such as AMIX (Bruker BioSpin) and SIMCA-P (Umetrics AB, Umeå, Sweden).

2.2 Solvents and Solutions

2.2.1 NMR

1. Deuterium oxide (D_2O).
2. 1.0 M NaOD: 1 mL of NaOD (40 % (w/v) in D_2O, corresponding to 10 M) added to 9 mL of D_2O.
3. 90 mM phosphate buffer (pH 6.0): 1.232 g of KH_2PO_4 and 10 mg (0.01 % (w/v) 3-trimethylsilyl)propionic-2,2,3,3-d_4 acid sodium salt (TPS) added to 100 mL D_2O; mix them until completely dissolved and adjust pH with 1.0 M NaOD.
4. CH_3OH-*d4*.

2.2.2 JA

1. Mock solution: 20 % (v/v) EtOH.
2. For plant cell cultures, 100 μM MeJA stock solution: 22.4 mg of MeJA (=21.6 μL) dissolved in 1 mL EtOH; 50 μL of this stock solution to be added to 50 mL of culture-containing flasks [11].
3. For intact plants, 5 mM MeJA solution: 11.2 mg (=10.8 μL) of MeJA, dissolved in 20 % EtOH to obtain 10 mL [12].

3 Methods

Carry out all procedures at room temperature, unless otherwise specified.

3.1 JA Treatment

3.1.1 Intact Plants

The MeJA solution should be applied topically on the surface of individual leaves.

1. Spread the MeJA solution evenly on the surface with a glass rod.
2. Spray mock solution (20 % EtOH) on the control plants evenly with a glass rod.
3. Place the plant in the growth room (25 °C, 50 % humidity) (*see* **Note 1**).

3.1.2 Plant Cell Cultures

1. Add 250 μL of the MeJA solution to the culture flasks.
2. Add an equal amount of the mock solution (ethanol) to the controls.
3. Place the flasks in the growth chamber separately.

3.2 Sample Preparation

3.2.1 Intact Plants

1. After treatment, harvest plants at fixed time points (e.g., at 0 h, 12 h, 1 day, 2 days, 3 days, 7 days, and 14 days) and freeze them immediately in liquid nitrogen (*see* **Note 2**).
2. Precool a pestle and mortar with liquid nitrogen.
3. Grind the frozen leaves with the precooled pestle and mortar under liquid nitrogen (*see* **Note 3**).
4. Transfer ground samples to a plastic tube with a spatula.
5. Keep in the deep-freezer before drying.
6. Place frozen samples in the freeze dryer for 1–2 days (*see* **Note 4**).

3.2.2 Plant Cell Cultures

1. Harvest cultures by vacuum filtration.
2. Collect cells with a spatula and transfer to the tubes.
3. Transfer tubes to the freezer at least for 1 day (*see* **Note 2**).
4. Place frozen samples in the freeze dryer for 1–2 days (*see* **Note 4**).

3.3 Sample Extraction

1. Weigh the freeze-dried sample in an Eppendorf tube.
2. Add 0.75 ml of CH_3OH-*d4* and 0.75 ml of KH_2PO_4 buffer in D_2O (pH 6.0) containing 0.1 % (w/w) TSP (*see* **Note 5**).
3. Vortex for 1 min.
4. Ultrasonicate for 10–20 min.
5. Centrifuge at room temperature for 5–10 min with a microtube centrifuge (17,000 × *g*) (*see* **Note 6**).
6. Transfer the supernatant (more than 1 mL) to a 1.5-mL Eppendorf tube.
7. If the supernatant is not clear, repeat centrifugation (17,000 × *g* for 1 min at room temperature).
8. Transfer 800 μL of the supernatant to a 5-mm NMR tube.

3.4 NMR Measurement

1. Load the NMR tube into the spectrometer.
2. Set the sample temperature at 298 K (25 °C) and leave a few minutes for thermal equilibration.
3. Tune and match the NMR tube.
4. Lock the spectrometer frequency to the deuterium resonance arising from the NMR solvents (either MeOD or D_2O; preferably MeOD).
5. Shim the sample either manually or with an automation system (*see* **Note 7**).
6. Start acquisition for 64 scans (*see* **Note 8**).
7. Set the parameters for standard NMR (*see* **Note 9**).
8. After acquisition, perform Fourier transformation.
9. Correct phase and baseline and calibrate the spectrum by setting the TSP peak at 0.00 ppm or MeOD at 3.3 ppm (Fig. 1a).

3.5 NMR Data Analysis

1. Convert the NMR spectra to a suitable form for further multivariate analysis (Fig. 1b), for instance with the commonly used software AMIX for the conversion to an ASCII file (*see* **Note 10**) (Fig. 1c).
2. Perform a principal component analysis with the SIMCA-P software or an equivalent software (*see* **Note 11**) (Fig. 2).

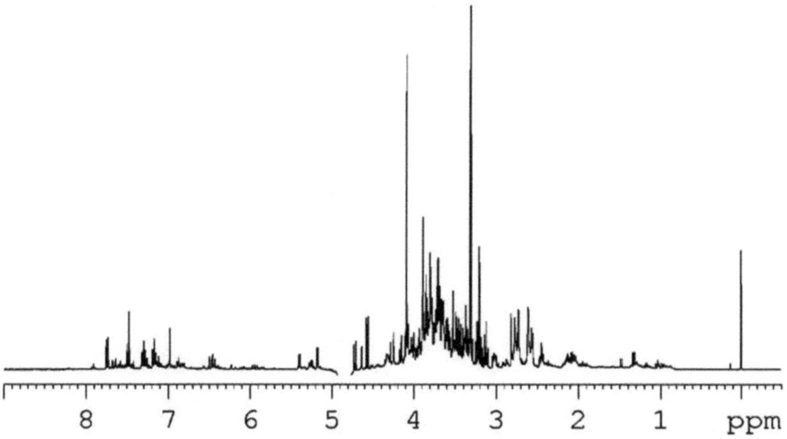

b

Primary ID	Obs. Sec. ID	10.00	9.96	9.92	9.88	9.84	9.80	9.76	9.72
C0-1	C0	0.00085	0	0	0	0	0	0	0.00107
C0-2	C0	0	0	0.01792	0.00709	0.02182	0.00031	0.00099	0.00183
C0-3	C0	0	0	0.00308	0.00283	0.00032	0.02552	0	0.00508
M0-1	M0	0	0	0	0	0	0	0	0
M0-2	M0	0.00778	0	0	0.00195	0.01385	0	0.00721	0.02529
M0-3	M0	0.02844	0.01783	0.04343	0.01691	0.03784	0.0297	0.01503	0.02375
M0-4	M0	0.00092	0.01728	0.01412	0.00514	0.05688	0.00881	0.0345	0.03938
C12-1	C0.5	0	0	0	0	0	0	0	0
C12-2	C0.5	0	0	0	0	0	0	0	0
C12-3	C0.5	0	0	0	0	0	0	0	0
M12-1	M0.5	0.00104	0	0	0	0	0.00064	0	0
M12-2	M0.5	0.00268	0	0	0	0.01143	0.00347	0.00148	0.00762
M12-3	M0.5	0	0	0	0	0	0	0	0
M12-4	M0.5	0.08161	0.1158	0.12959	0.10618	0.0473	0.03822	0.15212	0.24314
C1D-1	C1	0	0	0	0	0	0	0	0

c

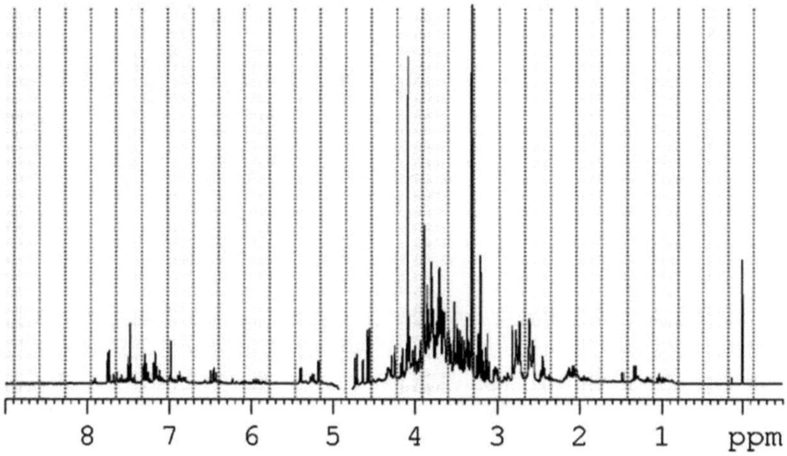

Fig. 1 ^{1}H-NMR spectra and bucketing with small bins (0.04 ppm). (**a**) ^{1}H-NMR spectrum of MeJA-treated *Brassica rapa* (turnip) (adapted from ref. 13 with permission © Elsevier B.V.; http://www.journals.elsevier.com/phytochemistry/). (**b**) Bucketed table with the peak intensity of each bin created with the AMIX software. (**c**) Illustration of bucketing

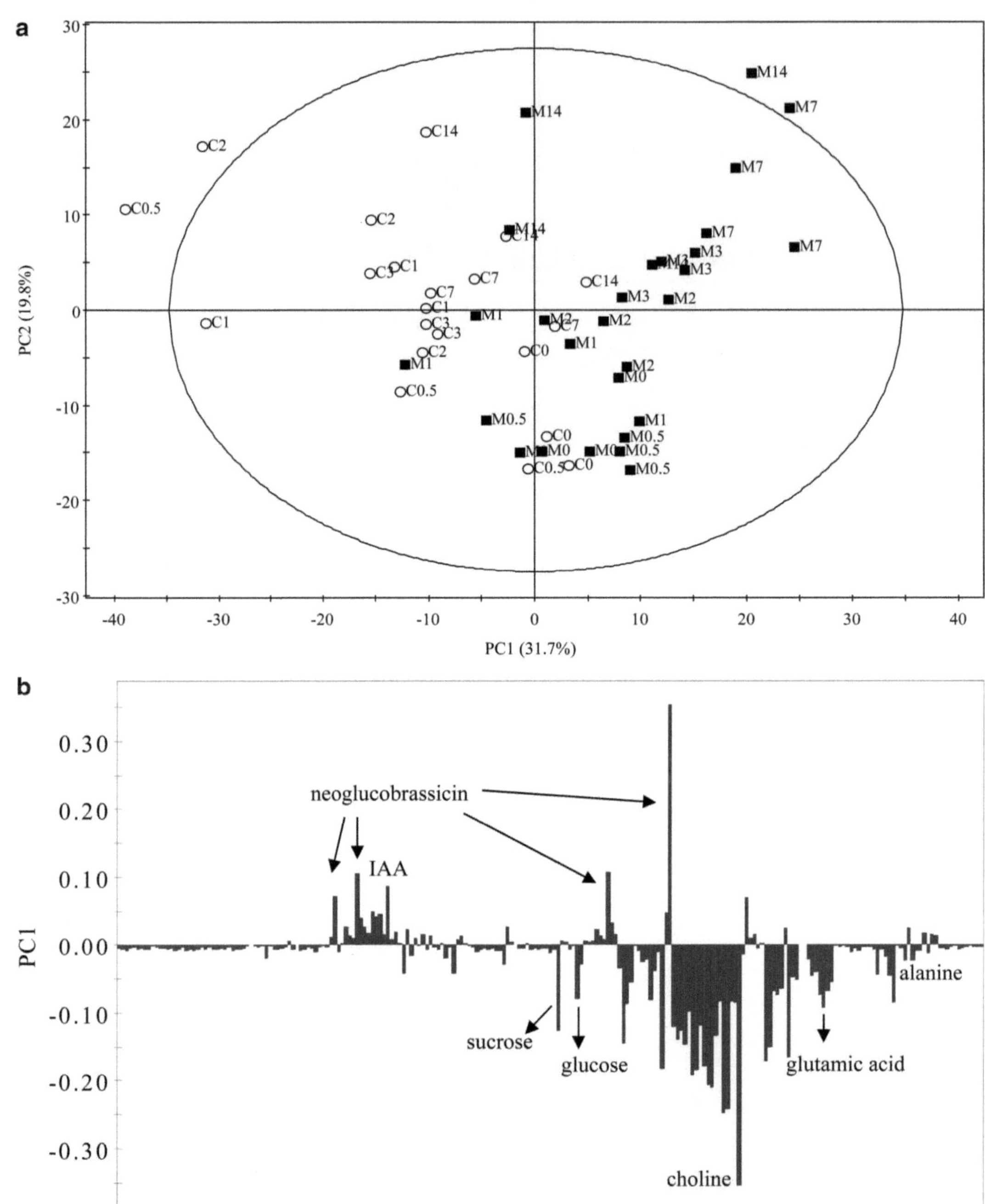

Fig. 2 NMR data from MeJA-treated *Brassica rapa* and corresponding control plants. (**a**) Score plot. (**b**) Loading plot. Several metabolites are indicated that increased with MeJA treatment in *Brassica rapa* (adapted from ref. 13 with permission © Elsevier B.V.; http://www.journals.elsevier.com/phytochemistry/)

4 Notes

1. To avoid the effect of volatile MeJA on control plants, it is recommended to place MeJA-treated plants in a separate growth chamber under conditions identical to those of the control chamber. To ensure reproducible experiments, please keep in mind the various biological variables. Metabolic changes can happen during the day, so the elicitation and harvest times, for instance, should be fixed. Other factors, such as the age of the leaves to be used or the different developmental stages of the plant, will also be reflected in the metabolome. Therefore, they should all be standardized in each experiment.
2. Harvested plant material can be stored at −80 °C for several weeks before extraction.
3. Alternatively, a grinder (ball mill) can be used. However, it is important to always use the same method, because differences can occur depending on the grinding methods.
4. The time required for drying with a freeze dryer can vary according to the machine and environment.
5. The choice of the extraction solvent is very critical, because the type of metabolites extracted depends on it. Intermediate-polar and polar metabolites are well detected by NMR, whereas nonpolar metabolites (fatty acids and terpenoids) are difficult to analyze individually. Although some solvents allow a good differentiation of the metabolites, some others fail to do so, probably not only because of the characteristics of the solvent itself but also because of the possible saturation of metabolites in the extraction solvent. Therefore, we recommend to test various solvents to determine the most suitable conditions for the proposed study. In our protocol, a MeOH–water mixture proved to be the most suitable in terms of the range of covered metabolites [14, 15]. If the full metabolome is to be extracted, a comprehensive extraction method can be considered [16].
6. Variable speeds (14,000–17,000 × *g*) can be used to obtain a clear supernatant.
7. All these steps can be set up in the automation system (for instance, ICON NMR; Bruker BioSpin).
8. The number of scans can vary depending on the concentration of samples. Although 64 scans are the most commonly used, they can be increased to 128 scans. It takes approximately 5–6 min to obtain 64 scans and 10–12 min for 128 scans.
9. Parameters for standard ^{1}H NMR spectroscopy are as follows: Set up pulse sequence comprising relaxation delay-60°-acquire, in which the pulse power is set to achieve a 60° flip angle, 10 kHz spectral width, and water presaturation applied with

1.5-s relaxation delay. Processing parameters are as follows: Zero-fill to 64 k data points with exponential line broadening of 0.3 Hz.

10. Other software, such as MestReNova (Mestrelab Research, Santiago de Compostela, Spain; www.mestrelab.com) and Chenomx (Edmonton, Alberta, Canada; www.chenomx.com), can be used for the file conversion. In this step, the peak is integrated into a small bin (bucket), the size of which is defined by the user. The size is preferably 0.04 ppm to avoid the effect of signal fluctuation due to pH or concentration. The signals of remaining solvents have to be removed for the statistical analysis.
11. For instance, equivalent software can be used as well, such as the Statistical Package for the Social Sciences (SPSS; IBM Corporation, Armonk, NY, USA: http://www-01.ibm.com/software/analytics/spss/), The UnScrambler® (Camo Software AS, Olso, Norway; www.camo.com), and MATLAB (Mathworks, Natick, MA, USA; www.mathworks.com) with statistics toolbox.

Acknowledgments

The authors wish to thank Dr. E. Wilson for reviewing the manuscript and providing helpful comments. This research was supported by the European Community's Seventh Framework Programme (FP7/2007-2013-217895 and FP7 project 222716 SmartCell) and Stichting voor de Technische Wetenschappen (STW Genbiotics project No. 10467 GENEXPAND to H.K.K.).

References

1. Farmer EE, Alméras E, Krishnamurthy V (2003) Jasmonates and related oxylipins in plant responses to pathogenesis and herbivory. Curr Opin Plant Biol 6:372–378
2. De Geyter N, Gholami A, Goormachtig S, Goossens A (2012) Transcriptional machineries in jasmonate-elicited plant secondary metabolism. Trends Plant Sci 17:349–359
3. Pauwels L, Inzé D, Goossens A (2009) Jasmonate-inducible gene: what does it mean? Trends Plant Sci 14:87–91
4. Sumner LW, Mendes P, Dixon RA (2003) Plant metabolomics: large-scale phytochemistry in the functional genomics era. Phytochemistry 62:817–836
5. Verpoorte R, Choi YH, Mustafa NR, Kim HK (2008) Metabolomics: back to basics. Phytochem Rev 7:525–537
6. Seger C, Sturm S (2007) Analytical aspects of plant metabolite profiling platforms: current standings and future aims. J Proteome Res 6:480–497
7. Colquhoun IJ (2007) Use of NMR for metabolic profiling in plant systems. J Pestic Sci 32:200–212
8. Fan TW-M (1996) Metabolite profiling by one- and two-dimensional NMR analysis of complex mixtures. Prog Nucl Magn Reson Spectrosc 28:161–219
9. Trygg J, Lundstedt T (2007) Chemometrics techniques for metabonomics. In: Lindon JC, Nicholson JK, Holmes E (eds) The Handbook of Metabonomics and Metabolomics. Elsevier, Amsterdam, pp 171–199
10. van den Berg RA, Hoefsloot HCJ, Westerhuis JA, Smilde AK, van der Werf MJ (2006) Centering, scaling, and transformations:

improving the biological information content of metabolomics data. BMC Genomics 7:142

11. Sánchez-Sampedro A, Kim HK, Choi YH, Verpoorte R, Corchete P (2007) Metabolomic alterations in elicitor treated *Silybum marianum* suspension cultures monitored by nuclear magnetic resonance spectroscopy. J Biotechnol 130:133–142
12. Hendrawati O, Yao Q, Kim HK, Linthorst HJM, Erkelens C, Lefeber AWM, Choi YH, Verpoorte R (2006) Metabolic differentiation of *Arabidopsis* treated with methyl jasmonate using nuclear magnetic resonance spectroscopy. Plant Sci 170:1118–1124
13. Liang Y-S, Choi YH, Kim HK, Linthorst HJM, Verpoorte R (2006) Metabolomic analysis of methyl jasmonate treated *Brassica rapa* leaves by 2-dimensional NMR spectroscopy. Phytochemistry 67:2503–2511
14. Kim HK, Verpoorte R (2010) Sample preparation for plant metabolomics. Phytochem Anal 21:4–13
15. Kim HK, Choi YH, Verpoorte R (2010) NMR-based metabolomic analysis of plants. Nat Protoc 5:536–549
16. Yuliana ND, Khatib A, Verpoorte R, Choi YH (2011) Comprehensive extraction method integrated with NMR metabolomics: a new bioactivity screening method for plants, adenosine A1 receptor binding compounds in *Orthosiphon stamineus* Benth. Anal Chem 83:6902–6906

Chapter 22

Metabolite Profiling of Plant Tissues by Liquid Chromatography Fourier Transform Ion Cyclotron Resonance Mass Spectrometry

Jacob Pollier and Alain Goossens

Abstract

Plants accumulate an overwhelming variety of secondary metabolites that play important roles in defense and interaction of the plant with its environment. To investigate the dynamics of plant secondary metabolism, large-scale untargeted metabolite profiling (metabolomics) is mandatory. Here, we describe a detailed protocol for untargeted metabolite profiling in which methanol extracts of jasmonate-treated plant tissues are analyzed by reversed-phase liquid chromatography coupled to negative-ion electrospray ionization Fourier transform ion cyclotron resonance mass spectrometry (MS). By means of dedicated integration and alignment software, the relative abundance of thousands of mass peaks, corresponding to hundreds of compounds, is calculated, and mass peaks of which the area is significantly changed by jasmonate treatment are identified. Subsequently, the metabolites corresponding to the significantly changed peaks are tentatively annotated using the accurate mass prediction of the Fourier transform-MS and the generated MS/MS data. Via this method, compounds of medium polarity, such as glucosinolates, alkaloids, phenylpropanoids, flavonoids, polyamines, and saponins, can be analyzed.

Key words Metabolite profiling, Metabolomics, LC-MS, FT-MS, XCMS

1 Introduction

Plants defend themselves against herbivores or pathogen attacks by activating specific defense programs that include the production of bioactive secondary metabolites to eliminate or deter the attackers, a process regulated by a signaling cascade that involves the jasmonate hormones. The application of jasmonates on plants or plant cell cultures triggers this signaling cascade and leads to drastic changes in the metabolic profile of the cells. For instance, the treatment of *Medicago truncatula* cell cultures with methyl jasmonate (MeJA) induced the accumulation of triterpene saponins, compounds that are nearly absent in untreated cells [1]. Similar effects were observed for other classes of secondary metabolites [2, 3], such as the increased accumulation of terpenoid indole alkaloids

Alain Goossens and Laurens Pauwels (eds.), *Jasmonate Signaling: Methods and Protocols*, Methods in Molecular Biology, vol. 1011, DOI 10.1007/978-1-62703-414-2_22, © Springer Science+Business Media, LLC 2013

and monolignols in *Catharanthus roseus* (Madagascar periwinkle) and *Arabidopsis thaliana* cell cultures, respectively, after MeJA treatment [4, 5].

To analyze the dynamics of the metabolome after perturbations, such as MeJA treatment, large-scale untargeted metabolite profiling (metabolomics) is mandatory [6]. However, owing to the chemical complexity of the metabolome and to the large variations in the relative metabolite concentrations, profiling all the metabolites at once is impossible [6–8]. For profiling of very polar or very apolar compounds, gas-chromatography-mass spectrometry (GC-MS) is the most suitable method, whereas liquid chromatography (LC)-MS is applied mainly for profiling compounds of medium polarity [9–11]. As many key classes of plant secondary metabolites, such as alkaloids, saponins, flavonoids, glucosinolates, phenylpropanoids, and polyamines, belong to the latter category, plant secondary metabolism can be best explored with LC-MS [8, 12]. The use of LC coupled to a soft ionization technique, such as atmospheric pressure chemical ionization (APCI) or electrospray ionization (ESI), prior to MS allows the detection of molecular ions and characteristic fragment ions. To aid the identification of the detected compounds, LC-MS-based metabolomics benefit from the use of highly accurate mass spectrometers, such as Fourier transform ion cyclotron resonance MS (FT-ICR-MS) instruments, that enable reliable prediction of the molecular formula of the detected ions [8, 13].

By means of LC-MS, several hundreds of compounds from a single crude plant extract can be profiled in one run. However, no standards are available for most secondary metabolites; hence, metabolite profiling based on the analysis of compounds for which standards are available, as is common for primary metabolites, would not exploit the full potential of LC-MS in plants. Therefore, software for unbiased peak alignment and integration is required to obtain information on as many metabolites as possible [8]. Processing software for peak alignment and integration of LC-MS data, such as XCMS [14] or MetAlign [15], transform the peaks into three-dimensional coordinates based on the mass, retention time, and signal amplitude. Subsequently, the coordinates are aligned across all the samples and relative peak intensities are generated as output [14]. Using statistics, differential peaks that correlate with a trait or a perturbation, such as MeJA treatment, can be identified. Finally, the metabolites corresponding to the differential peaks can be tentatively annotated based on the accurate mass and the MS/MS fragmentation. When available, standards can confirm the identity of the compounds [8].

Here, we describe a protocol for large-scale untargeted metabolite profiling of plant tissues by reversed-phase LC coupled to negative-ion ESI FT-ICR-MS. To start the procedure, plant material is harvested by freezing in liquid nitrogen and homogenized. Subsequently, the metabolites are extracted with methanol, injected on a C18 column, and separated via high-performance LC with an acetonitrile gradient. After ESI of the separated metabolites, the compound mass is determined with the FT-ICR-MS. The resulting LC-MS chromatograms are processed with the XCMS package [14] that calculates the relative abundance of thousands of mass peaks and performs statistics to indicate the differences between sample groups. The identified differential peaks are tentatively annotated based on the accurate mass and the MS/MS fragmentation.

2 Materials

2.1 Equipment

1. Mortar and pestle.
2. LC vials: 300-μL screw-top autosampler vials with cap and bonded polytetrafluoroethylene/silicone septa (Waters, Milford, MA, USA).
3. Vacuum drier.
4. Ultra-high-performance Accela LC system (Thermo Electron Corporation, Waltham, MA, USA) consisting of an Accela pump and an Accela autosampler that is connected to an LTQ FT Ultra via an ESI source.
5. Acquity UPLC BEH C18 column (150 × 2.1 mm, 1.7 μm; Waters).
6. 2-mL Eppendorf tubes.
7. Long, thin pipette tips.

2.2 Solutions, Buffers, and Media

1. Liquid nitrogen.
2. Methanol, ULC/MS grade (Biosolve, Valkenswaard, The Netherlands).
3. Ultrapure water, ULC/MS grade (Biosolve).
4. Cyclohexane, HPLC grade (Sigma-Aldrich, St. Louis, MO, USA).
5. Acetonitrile, ULC/MS grade (Biosolve).
6. Acetic acid (glacial), ULC/MS grade (Biosolve).
7. HPLC buffer A: Add 10 mL of acetonitrile to 990 mL of ultrapure water, supplemented with 1 mL of (glacial) acetic acid and mix well.
8. HPLC buffer B: Add 10 mL of ultrapure water to 990 mL of acetonitrile, supplemented with 1 mL of (glacial) acetic acid and mix well.

2.3 Programs and Software

1. RecalOffline software package (Thermo Electron Corporation) that is part of the FT-Programs software tools (freely available at http://sjsupport.thermofinnigan.com/public/detail.asp?id=773).
2. Xcalibur software package (Thermo Electron Corporation).
3. XCMS [14] software package installed in R, a freely available, open-source programming language and environment for statistical computing and graphics (http://cran.r-project.org/).

3 Methods

3.1 Sampling

1. Harvest the plant material by flash-freezing in liquid nitrogen (*see* **Note 1**).
2. Precool a mortar and pestle with liquid nitrogen.
3. Add the frozen plant material in the precooled mortar and grind to a fine powder (*see* **Note 2**).
4. Weigh the ground plant material in a precooled 2-mL Eppendorf tube (*see* **Note 3**).

3.2 Metabolite Extraction

1. Add 1 mL of methanol to the frozen plant material and vortex immediately (*see* **Note 4**).
2. Incubate the samples at room temperature for 10 min (*see* **Note 5**).
3. Centrifuge at ≥12,000 × *g* for 10 min at room temperature.
4. Transfer 750 μL of the supernatant to a fresh 2-mL Eppendorf tube.
5. Vacuum dry until all methanol is evaporated.
6. Add 400 μL of water and 400 μL of cyclohexane to the dry pellet (*see* **Note 6**).
7. Vortex the samples vigorously until all metabolites are redissolved.
8. Centrifuge the samples at ≥12,000 × *g* for 10 min at room temperature (*see* **Note 7**).
9. Transfer 200 μL of the lower aqueous phase to an LC vial (*see* **Note 8**).

3.3 LC ESI FT-ICR-MS Analysis

1. Install the analytical column and place it in the column oven conditioned at 80 °C.
2. Purge the lines with the HPLC buffers to avoid air bubbles in the tubes.
3. To condition the column, pump 95 % HPLC buffer B through the column for 5–10 min, until the pressure is constant.

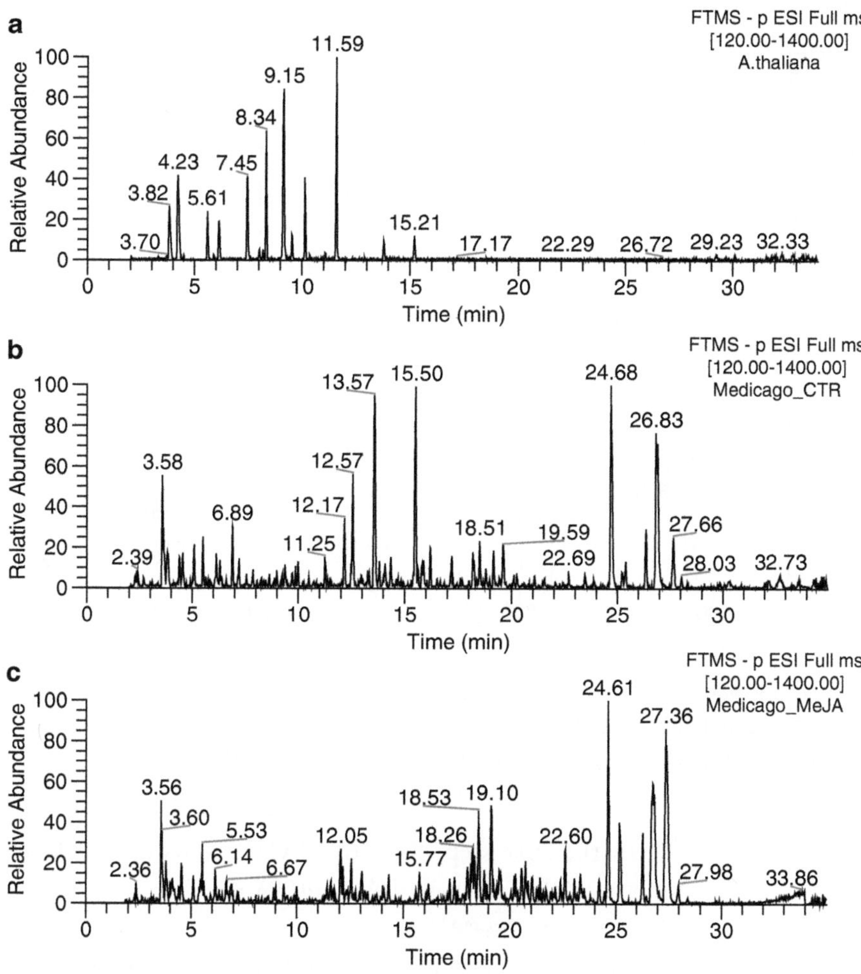

Fig. 1 LC ESI FT-ICR-MS chromatograms of metabolite extracts of (**a**) *Arabidopsis thaliana* leaves, (**b**) *Medicago truncatula* hairy roots, and (**c**) *M. truncatula* hairy roots treated with MeJA

4. Subsequently, increase the percentage of HPLC buffer A to the initial gradient conditions.
5. Place the LC vials in the autosampler (*see* **Note 9**).
6. Start the sample series (*see* **Note 10**).
7. Run at least one blank sample (water) to stabilize the LC ESI FT-ICR-MS system (*see* **Note 11** and Fig. 1).

3.4 Processing of the LC-MS Chromatograms

1. Slice the full FT-ICR-MS scans from each chromatogram with RecalOffline (*see* **Note 12**).
2. Convert the sliced chromatograms to netCDF format with Xcalibur File Convertor (*see* **Note 13**).
3. To integrate and align the chromatograms, use the XCMS [14] software package in R (*see* **Note 14**).

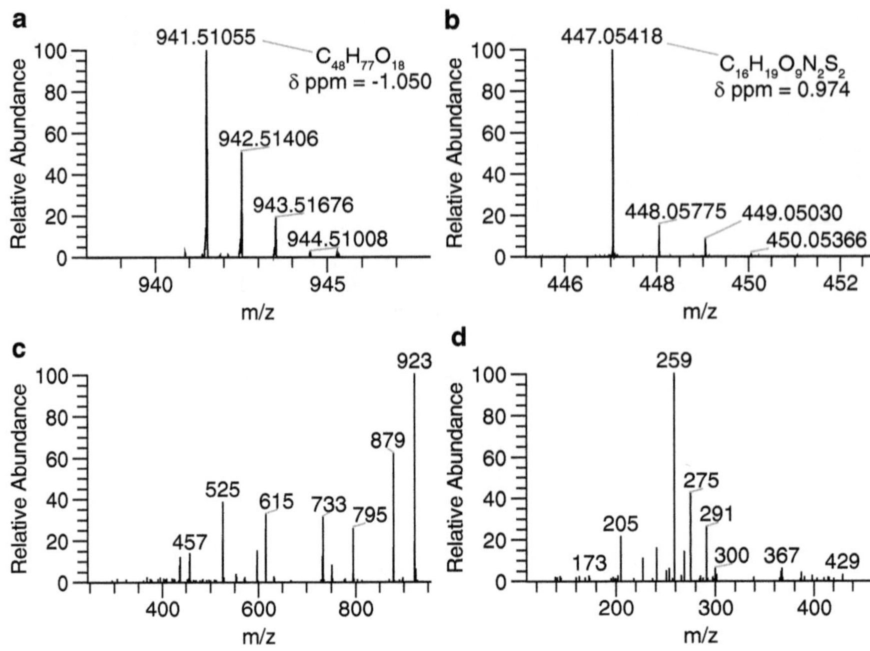

Fig. 2 Compound identification. Detected accurate masses of [M–H]$^-$ ions, additional isotope ions, and predicted molecular formula of soyasaponin I from *M. truncatula* (**a**) and of glucobrassicin from *A. thaliana* (**b**) MS/MS fragmentation spectra of soyasaponin I (**c**) and glucobrassicin (**d**)

4. In the first step, load the XCMS software package with the commands indicated with prefix >:

```
>library(xcms)
```

5. Run the peak detection algorithm (*see* **Note 15**):

```
>xset<-xcmsSet(fwhm=6,max=300,snthresh=2,mzdif
f=0.01)
```

6. Run the peak matching algorithm (*see* **Note 16**):

```
>xset<-group(xset,bw=8,max=300,mzwid=0.01)
```

7. Run the retention time alignment algorithm (*see* **Note 17**):

```
>xset2<-retcor(xset,method="loess",span=.2,family="
symmetric",plottype="mdevden")
```

8. Perform a second peak matching on the retention time-corrected samples:

```
>xset2<-group(xset2,bw=8,max=300,mzwid=0.01)
```

9. Generate the report file (*see* **Note 18**):

```
reporttab<-diffreport(xset2,"ctr","tmt","report")
```

3.5 Compound Identification

1. Manually search for the accurate mass of a compound of interest in the raw data file.
2. Predict the molecular formula of the compound of interest from the accurate mass and the isotope abundance (*see* **Note 19** and Fig. 2a, b).

3. Determine the structural information of the compound based on the MS/MS fragmentation pattern (Fig. 2c, d).

4 Notes

1. Changes in the cellular environment are almost immediately reflected by changes in the intracellular metabolite concentrations. To avoid enzymatic modification of metabolites upon manipulation while harvesting the plant material, quench all biochemical processes [16, 17] by rapidly freezing the plant material in liquid nitrogen. Harvested plant material can be stored for up to 1 year at −80 °C.
2. Thawing of the plant material during grinding of the samples should be avoided. If required, add some extra liquid nitrogen in the mortar and continue grinding after evaporation of the nitrogen. Alternatively, plant material, such as *Arabidopsis* seedlings, can be homogenized by ball-milling (Retsch, Haan, Germany) for 20 s at 25 Hz.
3. The amount of plant material for metabolite profiling depends on the species and type of material. For instance, 100 mg of plant material is used for *Arabidopsis.*
4. Allow the plant material to thaw in the supplemented methanol that will inactivate all enzymes and, thus, protect the metabolites from enzymatic degradation.
5. Mix the samples by vortexing briefly a few times during the metabolite extractions. In case green tissue is extracted, the cell debris should become white.
6. Cyclohexane treatment extracts apolar compounds that would otherwise block the HPLC column.
7. After centrifugation, the mixture is separated into three phases: the top organic phase containing apolar compounds, such as chlorophyll and fats; an interphase; and the bottom aqueous phase containing the secondary metabolites.
8. Use long, thin pipette tips to reach the lower aqueous phase without disturbing the upper layers.
9. For maximum reproducibility and stability of the samples, keep the autosampler at 5 °C.
10. Run the following gradient for all the samples: Time 0 min, 5 % buffer B; 30 min, 55 % buffer B; and 35 min, 100 % buffer B. Set the loop size, flow, and column temperature at 25 μL, 300 μL/min, and 80 °C, respectively. Apply full loop injection and set the following negative ionization parameter values:

Capillary temperature 150 °C, sheath gas 25 (arbitrary units), auxiliary gas 3 (arbitrary units), and spray voltage 4.5 kV. Record full FT-MS spectra between *m/z* 120 and 1,400 at a resolution of 100,000. For identification, interchange full MS spectra with a dependent MS^2 scan event in which the most abundant ion in the previous full MS scan is fragmented and two dependent MS^3 scan events of the two most abundant daughter ions. Set the collision energy at 35 %.

11. The resulting chromatograms (for examples, *see* Fig. 1) will be stored as .raw files.
12. To open the program, click Recalibrate Offline in the Tools section on the Xcalibur Roadmap window. To slice, open the chromatogram (.raw), select the FTMS–p ESI Full ms [120.00–1,400.00] scans, and save as a new .raw file.
13. To open the File Convertor, navigate to the Tools section on the Thermo Xcalibur Roadmap window. Select the .raw file as source data and .cdf as destination data type.
14. The working directory should contain the chromatograms in netCDF format in two folders, one folder containing the chromatograms of the control samples ("ctr") and the other the chromatograms of the MeJA-treated samples ("tmt").
15. The peak detection algorithm first cuts the LC-MS data into slices of 0.01 *m/z*. When a low-resolution mass spectrometer is used instead of the FT-MS, the peak width may be larger than 0.01 *m/z*, leading to bleeding of the peak signal across multiple slices. Hence, the mzdiff argument can be increased (for instance to 0.1 *m/z*). After the chromatograms are cut, the slices are filtered with matched filtration by means of a second derivative Gaussian as model peak shape, of which the characteristics should be similar to the peak shapes of the samples. For LC-MS data, a standard full width at half-maximum (fwhm) of 30 s is used for this model peak. When the peak width of the sample peaks is smaller, the fwhm argument can be decreased (for instance to 6–8 s for UPLC-MS data). Finally, after filtration of the slices, the peaks are selected with a signal-to-noise ratio cutoff set to 2 (snthresh = 2), but could be higher depending on the quality of the peak shape (for example, the XCMS default snthresh is 10).
16. With this command, the peaks identified in the individual samples are matched across all samples by grouping the peaks in bins of 0.25 *m/z* width and subsequent resolving peak groups with different retention times in each bin. The band width (bw) argument of peak groups should be set according to the retention time drifts of the samples. The lower the bw, the more chance a peak will split into two separate peaks, whereas too high a bw may merge two separate peaks.

17. With this command, the retention time deviation across the samples will be corrected in a single step. For the retention time correction, "well-behaved" peak groups are identified and used as temporary standards. To obtain a more precise alignment, the peak identification and retention time correction steps can be repeated iteratively.
18. In this procedure, XCMS will generate an output file containing mass and retention time information of the detected peaks and the corresponding peak intensities. In addition, XCMS performs a Student's *t*-test with Welch correction to indicate peaks that are significantly different between the control and the treatment groups.
19. The predicted molecular formula can be used to search the Internet or compound databases for possible candidates. However, due to the lack of reference compounds and the occurrence of many unknown compounds in plant metabolite extracts, spectroscopic methods are required to confirm the tentative identification of the compounds.

References

1. Suzuki H, Achnine L, Xu R, Matsuda SPT, Dixon RA (2002) A genomics approach to the early stages of triterpene saponin biosynthesis in *Medicago truncatula*. Plant J 32:1033–1048
2. De Geyter N, Gholami A, Goormachtig S, Goossens A (2012) Transcriptional machineries in jasmonate-elicited plant secondary metabolism. Trends Plant Sci 17:349–359
3. Pauwels L, Inzé D, Goossens A (2009) Jasmonate-inducible gene: what does it mean? Trends Plant Sci 14:87–91
4. Rischer H, Orešič M, Seppänen-Laakso T, Katajamaa M, Lammertyn F, Ardiles-Diaz W, Van Montagu MCE, Inzé D, Oksman-Caldentey K-M, Goossens A (2006) Gene-to-metabolite networks for terpenoid indole alkaloid biosynthesis in *Catharanthus roseus* cells. Proc Natl Acad Sci USA 103:5614–5619
5. Pauwels L, Morreel K, De Witte E, Lammertyn F, Van Montagu M, Boerjan W, Inzé D, Goossens A (2008) Mapping methyl jasmonate-mediated transcriptional reprogramming of metabolism and cell cycle progression in cultured *Arabidopsis* cells. Proc Natl Acad Sci USA 105:1380–1385
6. Bino RJ, Hall RD, Fiehn O, Kopka J, Saito K, Draper J, Nikolau BJ, Mendes P, Roessner-Tunali U, Beale MH, Trethewey RN, Lange BM, Syrkin Wurtele E, Sumner LW (2004) Potential of metabolomics as a functional genomics tool. Trends Plant Sci 9:418–425
7. Sumner LW, Mendes P, Dixon RA (2003) Plant metabolomics: large-scale phytochemistry in the functional genomics era. Phytochemistry 62:817–836
8. De Vos RCH, Moco S, Lommen A, Keurentjes JJB, Bino RJ, Hall RD (2007) Untargeted large-scale plant metabolomics using liquid chromatography coupled to mass spectrometry. Nat Protoc 2:778–791
9. von Roepenack-Lahaye E, Degenkolb T, Zerjeski M, Franz M, Roth U, Wessjohann L, Schmidt J, Scheel D, Clemens S (2004) Profiling of Arabidopsis secondary metabolites by capillary liquid chromatography coupled to electrospray ionization quadrupole time-of-flight mass spectrometry. Plant Physiol 134:548–559
10. Fiehn O, Weckwerth W (2003) Deciphering metabolic networks. Eur J Biochem 270:579–588
11. Halket JM, Waterman D, Przyborowska AM, Patel RKP, Fraser PD, Bramley PM (2005) Chemical derivatization and mass spectral libraries in metabolic profiling by GC/MS and LC/MS/MS. J Exp Bot 56:219–243
12. Hall RD (2006) Plant metabolomics: from holistic hope, to hype, to hot topic. New Phytol 169:453–468
13. Pollier J, Morreel K, Geelen D, Goossens A (2011) Metabolite profiling of triterpene saponins in *Medicago truncatula* hairy roots by liquid chromatography Fourier transform ion

cyclotron resonance mass spectrometry. J Nat Prod 74:1462–1476

14. Smith CA, Want EJ, O'Maille G, Abagyan R, Siuzdak G (2006) XCMS: processing mass spectrometry data for metabolite profiling using nonlinear peak alignment, matching, and identification. Anal Chem 78:779–787
15. Lommen A (2009) MetAlign: interface-driven, versatile metabolomics tool for hyphenated full-scan mass spectrometry data preprocessing. Anal Chem 81:3079–3086
16. Villas-Bôas SG, Mas S, Åkesson M, Smedsgaard J, Nielsen J (2005) Mass spectrometry in metabolome analysis. Mass Spectrom Rev 24: 613–646
17. Fiehn O (2002) Metabolomics—the link between genotypes and phenotypes. Plant Mol Biol 48:155–171

Chapter 23

cDNA-AFLP-Based Transcript Profiling for Genome-Wide Expression Analysis of Jasmonate-Treated Plants and Plant Cultures

Janine Colling, Jacob Pollier, Nokwanda P. Makunga, and Alain Goossens

Abstract

cDNA-AFLP is a commonly used, robust, and reproducible tool for genome-wide expression analysis in any species, without requirement of prior sequence knowledge. Quantitative expression data are generated by gel-based visualization of cDNA-AFLP fingerprints obtained by selective PCR amplification of subsets of restriction fragments from a double-stranded cDNA template. Differences in gene expression levels across the samples are reflected in different band intensities on the high-resolution polyacrylamide gels. The differentially expressed genes can be identified by direct sequencing of re-amplified cDNA-AFLP tags purified from the gels. The cDNA-AFLP technique is especially useful for profiling of transcriptional responses of jasmonate-treated plants or plant (tissue) cultures and the discovery of jasmonate-responsive genes.

Key words cDNA-AFLP, Transcript profiling, Jasmonate, Gene expression, Transcriptome, Elicitation

1 Introduction

Amplified fragment length polymorphism (AFLP) is a robust and reliable DNA-fingerprinting technique based on the polymerase chain reaction (PCR) amplification of restriction fragments of genomic DNA [1, 2]. Practically, the AFLP protocol can be divided into three distinct steps: (1) digestion of the DNA template with two restriction enzymes, followed by ligation of specific oligonucleotide adapters to the sticky ends of the digested DNA; (2) selective PCR amplification of a subset of the restriction fragments by means of AFLP primers with a few extra selective nucleotides besides the adapter and restriction site-specific sequences; and (3) visualization of the amplified DNA fragments on high-resolution polyacrylamide gels [1].

The cDNA-AFLP technique is derived from the AFLP protocol and has become a widely used, robust, and reproducible tool

Alain Goossens and Laurens Pauwels (eds.), *Jasmonate Signaling: Methods and Protocols*, Methods in Molecular Biology, vol. 1011, DOI 10.1007/978-1-62703-414-2_23,

for genome-wide expression analysis in any organism, without the need for prior sequence knowledge [3–5]. Similar to AFLP, the original cDNA-AFLP method [6] starts with a DNA template (double-stranded cDNA) that is digested with two restriction enzymes. After ligation of adapters to the restriction fragments, a subset of the restriction fragments is amplified by selective PCR. Finally, the amplified fragments are visualized on high-resolution polyacrylamide gels, with fragment intensities reflecting the relative abundance (copy number) of the corresponding genes across the samples [6]. To identify the differentially expressed genes, the corresponding cDNA-AFLP tags are purified from the polyacrylamide gel, re-amplified, and sequenced.

Since its development, several modifications of the original cDNA-AFLP protocol have been published [3, 5, 7, 8]. Here, we focus on the one-gene–one-tag variant of the original cDNA-AFLP method [3, 5, 8]. In contrast to the original method, in which multiple sequence tags can be obtained for a single gene (one-gene–multiple-tag), the one-gene–one-tag method includes the selection of the 3′ end restriction fragments of the transcripts prior to the selective PCR amplification, leading to a single diagnostic sequence tag per transcript [5]. This significantly reduces the total number of tags to be screened and, hence, the workload, but might lead to reduced transcriptome coverage or in sequence tags that do not cover the coding sequence, thereby hindering the functional annotation of the fragments [3, 5]. In addition, the one-gene–one-tag variant makes use of the *BstYI/MseI* restriction enzyme combination, instead of a combination of two tetracutters in the original cDNA-AFLP protocol. This results in a higher average fragment length that facilitates the functional annotation of the transcripts and the full-length cDNA cloning [5] (*see* Fig. 1). This method has proven successful for transcriptome analysis of several jasmonate-elicited medicinal plant species [4, 9, 10]. It is also useful to carry out pilot studies in model species, such as *Arabidopsis thaliana*, to screen large sample sets (e.g., time series) for the most relevant samples for full-transcriptome analysis by other methods, such as microarray analysis or RNA-sequencing [11].

2 Materials

2.1 Equipment

1. Magnetic stirrer.
2. Microwave oven.
3. Autoclave.
4. Dynabeads M-2800 Streptavidin (Invitrogen, Carlsbad, CA, USA).

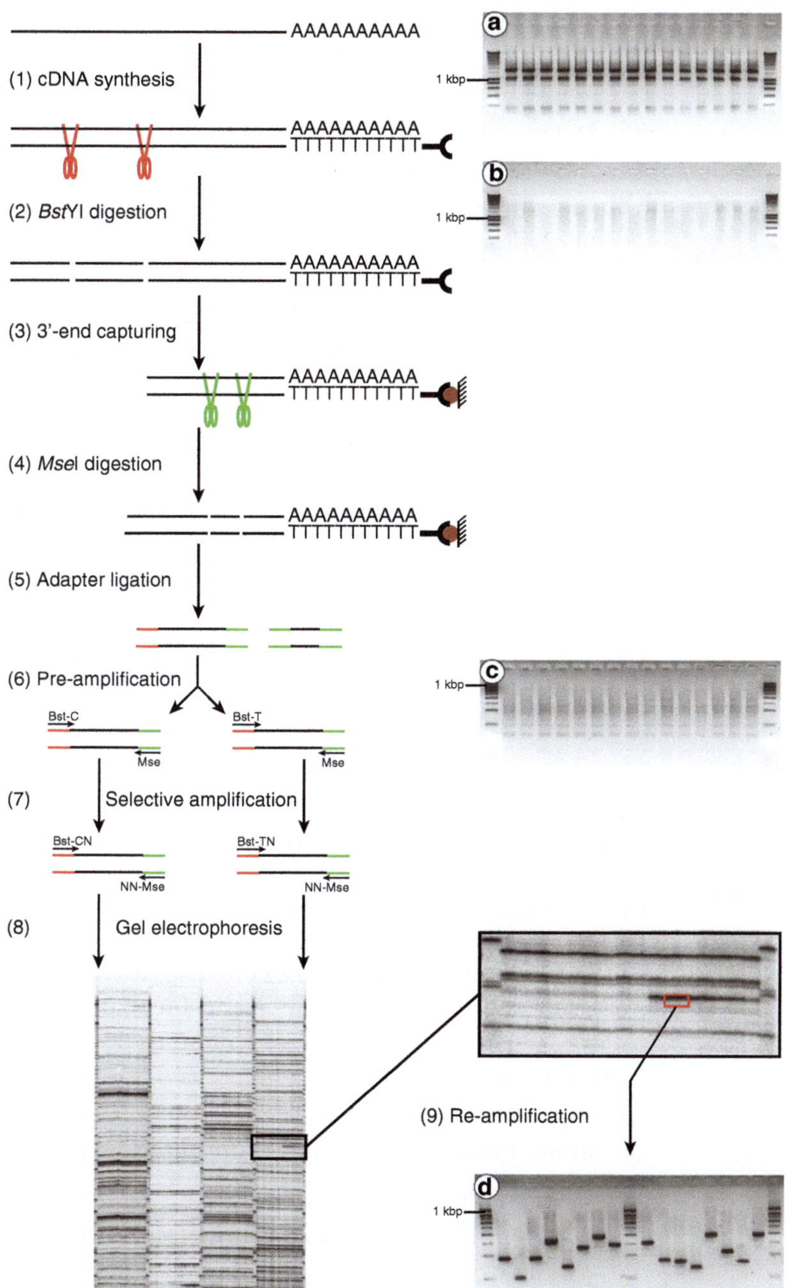

Fig. 1 Overview of the cDNA-AFLP procedure. The *left panel* shows the different steps in the procedure that include (1) synthesis of double-stranded cDNA from an RNA template with a biotinylated oligo-dT primer; (2) digestion of the cDNA with the restriction enzyme *Bst*YI; (3) 3′ end capturing of the digested cDNA by binding of the biotin to streptavidin-coated beads to isolate a single-sequence tag per transcript; (4) digestion of the captured cDNA fragments with the restriction enzyme *Mse*I; (5) ligation of specific *Bst*YI and *Mse*I adapters to the sticky ends of the digested DNA; (6) preamplification with the *Bst*YI + C or *Bst*YI + T primer in combination with the *Mse*I primer to reduce the complexity of the template mixture; (7) selective amplification of a subset of the transcript fragments by using *Bst*YI and *Mse*I primers with a few extra selective nucleotides; (8) visualization of the amplified DNA fragments on high-resolution polyacrylamide gels; and (9) purification, re-amplification, and sequencing of DNA fragments for identification of the differentially expressed genes. The *right panel* gives examples of good-quality RNA (**a**), cDNA (**b**), preamplifications (**c**), and re-amplification of 16 cDNA-AFLP tags (**d**)

5. Magnetic stands for isolation of Dynabeads (Invitrogen).
6. Power supply (PowerPac 3000; Bio-Rad Laboratories, Hercules, CA, USA).
7. Vacuum gel drier GD-1 (Heto-Holten Lab Equipments, Allerød, Denmark).
8. PhosphoImager scanning instrument and imaging plates (GE-Healthcare, Little Chalfont, UK).
9. Kodak BioMax MR film, 35 × 43 cm (Sigma-Aldrich, St. Louis, MO, USA).
10. Water repellent (Rain-X; Shell Car Care International Ltd., Manchester, UK).
11. Sequi-Gen GT electrophoresis system (38 × 50 cm) (Bio-Rad Laboratories).
12. Whatman pure cellulose blotting paper (3MM Chr, 35 × 43 cm) (GE-Healthcare).
13. Adhesive PCR foil seals (ABgene Ltd., Epsom, UK).
14. Nanodrop.

2.2 Buffers, Media, Solutions, and Reagents

Use ultrapure water (resistivity of 18 MΩ cm at 25 °C) and analytical grade reagents to prepare all solutions. Prepare and store all reagents at room temperature, unless stated otherwise. Products and buffers may be used for multiple steps in the protocol, but will only be described the first time they are needed.

2.2.1 Double-Stranded cDNA Synthesis

1. Diethylpyrocarbonate (DEPC)-treated water: Add 100 μL of DEPC to 100 mL of water and incubate for at least 1 h at 37 °C. Autoclave for at least 15 min to decompose the remaining traces of DEPC.
2. SuperScript™ II Reverse Transcriptase, supplied with 5× first-strand buffer and 100 mM dithiothreitol (DTT) (Invitrogen).
3. 10 mM dNTP mix.
4. Biotin-labeled oligo-dT25 primer (*see* **Note 1**).
5. *Escherichia coli* DNA ligase, supplied with 10× *E. coli* DNA ligase reaction buffer.
6. DNA polymerase I.
7. Ribonuclease H (RNase H).
8. cDNA purification kit.
9. 0.5 M ethylenediaminetetraacetic acid (EDTA) (pH 8.0): Add 18.61 g of EDTA to 80 mL of water and mix with a magnetic stirrer. Add NaOH pellets to adjust the pH to 8.0. Make up with water to 100 mL and autoclave (*see* **Note 2**).

10. 10× TAE buffer: Dissolve 48.4 g of tris(hydroxymethyl)aminomethane (Tris) in 500 mL of water. Add 11.44 mL of acetic acid (glacial) and 20 mL of 0.5 M EDTA (pH 8.0). Make up to 1 L with water.
11. 1.2 % (w/v) agarose gel: Add 3.6 g of agarose to 300 mL of 0.5× TAE buffer (*see* **Note 3**). Heat the solution to boiling in a microwave to dissolve the agarose. Cool the solution to approximately 60 °C and add a nucleic acid stain to allow visualization of the DNA after electrophoresis. Pour the gel solution in a casting tray containing a sample comb and allow the gel to harden at room temperature. The remaining gel solution can be stored for up to 1 month at 60 °C.

2.2.2 PCR Template Preparation

1. 1 M Tris-acetic acid (Tris–HAc) (pH 7.5): Dissolve 12.1 g of Tris in approximately 80 mL of water. Mix well and adjust the pH to 7.5 with acetic acid. Make up to 100 mL with water and autoclave.
2. 1 M magnesium acetate (MgAc): Dissolve 2.145 g of MgAc tetrahydrate in water. Make up to 10 mL with water and filter sterilize.
3. 4 M potassium acetate (KAc): Dissolve 3.926 g of KAc in water. Make up to 10 mL with water and filter sterilize. Store at −20 °C.
4. 1 M Tris–HCl (pH 8.0): Dissolve 12.1 g of Tris in approximately 80 mL of water. Mix well and adjust the pH to 8.0 with hydrochloric acid. Make up to 100 mL with water and autoclave.
5. 5 M sodium chloride (NaCl): Dissolve 29.22 g of NaCl in approximately 80 mL of water. Make up to 100 mL with water and autoclave.
6. 10× RL buffer: Mix 1 mL of 1 M Tris–HAc, pH 7.5 with 1 mL of 1 M MgAc, 1.25 mL of 4 M KAc, and 10 μL of 50 mg/mL bovine serum albumin. Add 0.077 g of DTT and make up to 10 mL with water. Store in 1-mL aliquots at −20 °C (*see* **Note 4**).
7. 2× STEX buffer: Mix 40 mL of 5 M NaCl with 2 mL of Tris–HCl (pH 8.0), 400 μL of 0.5 M EDTA (pH 8.0), and 2 mL of Triton X-100. Make up to 100 mL with water.
8. $T_{10}E_{0.1}$ buffer: Add 1 mL of 1 M Tris–HCl (pH 8.0) and 20 μL of 0.5 M EDTA (pH 8.0) to 80 mL of water. Make up to 100 mL with water and autoclave.
9. 10 mM ATP solution.
10. T4 DNA Ligase (Invitrogen).

2.2.3 Preamplification and Re-amplification

1. AmpliTaq DNA polymerase, supplied with 10× PCR buffer and 25 mM $MgCl_2$ solution (Applied Biosystems, Foster City, CA, USA).
2. SilverStar™ DNA polymerase, supplied with 10× PCR buffer and 50 mM $MgCl_2$ solution (Eurogentec, Seraing, Belgium).

2.2.4 Selective Amplification

1. 1 M Tris–HCl (pH 7.5): Dissolve 12.1 g of Tris in approximately 80 mL of water. Mix well and adjust the pH to 7.5 with hydrochloric acid. Make up to 100 mL with water and autoclave.
2. 1 M $MgCl_2$: Dissolve 2.033 g of magnesium chloride hexahydrate in water. Make up to 10 mL with water.
3. 10× T4 buffer: Mix 2.5 mL of 1 M Tris–HCl, pH 7.5 with 1 mL of 1 M $MgCl_2$. Add 0.077 g of DTT and 0.013 g of spermidine trihydrochloride. Make up to 10 mL with water. Store in 1-mL aliquots at −20 °C (*see* **Note 4**).
4. T4 Polynucleotide Kinase (New England BioLabs, Ipswich, MA, USA).
5. ATP, [γ-^{33}P] 3,000 Ci/mmol (Perkin Elmer, Waltham, MA, USA).
6. AmpliTaq Gold DNA polymerase, supplied with 10× PCR buffer and 25 mM $MgCl_2$ solution (Applied Biosystems).
7. Formamide loading dye: In a 50-mL Falcon tube, mix 49 mL of formamide with 1 mL of 0.5 M EDTA (pH 8.0) and add 0.03 g of bromophenol blue and 0.03 g of xylene cyanol. Store this solution at 4 °C.

2.2.5 SequaMark™ 10 Base Ladder

1. SequaMark™ DNA template (Research Genetics, Huntsville, AL, USA).
2. PCR cleanup kit.
3. $Vent_R$™ (exo-) DNA polymerase, supplied with 10× ThermoPol reaction buffer (New England Biolabs).
4. dNTP/ddTTP mix: Mix 0.6 μL of 10 mM dATP, 2 μL of 10 mM dCTP, 2 μL of 10 mM dGTP, 0.66 μL of 10 mM dTTP, and 14.4 μL of 10 mM ddTTP. Make up to 200 μL with water.

2.2.6 Gel Electrophoresis and Detection

1. 10× Maxam buffer: Dissolve 121 g of Tris and 61.8 g of boric acid in water and make up to 1 L with water.
2. 4.5 % (w/v) denaturing polyacrylamide gel solution: Add 450 g of urea and 112.5 mL of acrylamide/bis-acrylamide (19:1, 40 % stock solution) to a 2-L beaker. Add water to

700 mL and stir for 1 h while heating at 60 °C. When the urea is dissolved, add 100 mL of 10× Maxam buffer and 4 mL of 0.5 M EDTA (pH 8.0). Filter the resulting solution through a 0.45-μm filter with a vacuum pump, and make up to 1 L with water. Store the gel solution at 4 °C in the dark for up to 1 month.

3. 10 % ammonium persulfate (APS): Dissolve 1 g of APS in water and add up to 10 mL with water. Store this solution at 4 °C in the dark for up to 1 month.
4. *N*,*N*,*N*′,*N*′-Tetramethylethylenediamine (TEMED). Store this product at 4 °C in the dark.
5. Sodium acetate (NaAc).

2.3 Restriction Enzymes and Primers

2.3.1 PCR Template Preparation

1. *Bst*YI restriction enzyme (New England BioLabs).
2. *Mse*I restriction enzyme (New England Biolabs).
3. Oligonucleotide *Mse*I-Forward: 5′-GACGATGAGTCCTGAG-3′.
4. Oligonucleotide *Mse*I-Reverse: 5′-TACTCAGGACTCAT-3′.
5. Oligonucleotide *Bst*YI-Forward: 5′-CTCGTAGACTGCGTAGT-3′.
6. Oligonucleotide *Bst*YI-Reverse: 5′-GATCACTACGCAGTCTAC-3′.
7. *Bst*YI adapter (5 μM): Add 5 μL *Bst*YI-Forward (100 μM) and 5 μL *Bst*YI-Reverse (100 μM) to 90 μL of water.
8. *Mse*I adapter (50 μM): Mix 50 μL *Mse*I-Forward (100 μM) and 50 μL *Mse*I-Reverse (100 μM).

2.3.2 Preamplification, Selective Amplification, SequaMark™ 10 Base Ladder, and Re-amplification

1. Preamplification primers: *Bst*YI-T+0 or *Bst*YI-C+0 primer (5′-GACTGCGTAGTGATC(C/T)-3′) and *Mse*I+0 primer (5′-GATGAGTCCTGAGTAA-3′).
2. Selective amplification primers: *Bst*YI-T/C+N primers (5′-GACTGCGTAGTGATC(C/T)N-3′) and *Mse*I+NN primers (5′-GATGAGTCCTGAGTAANN-3′). N represents the selective nucleotides.
3. SequaMark™ primers: Forward: 5′-ACCAGAAGCTGGACGCAG-3′; reverse: 5′-ACACAGGAAACAGCTATGACCA-3′.
4. Re-amplification primers: Forward 1: 5′-AAAAAGCAGGCTGACTGCGTAGTG-3′; reverse 1: 5′-AGAAAGCTGGGTGATGAGTCCTGA-3′; forward 2: 5′-GGGGACAAGTTTGTACAAAAAAGCAGGCT-3′; reverse 2: 5′-GGGGACCACTTTGTACAAGAAAGCTGGGT-3′.

3 Methods

3.1 Double-Stranded cDNA Synthesis

1. For each sample, dilute 2 μg of total RNA into a total volume of 20 μL using DEPC-treated water (*see* **Note 5**).
2. For the first-strand cDNA synthesis mix, combine 8 μL of 5× first-strand buffer, 4 μL of DEPC-treated water, 4 μL of 100 mM DTT, 2 μL of 10 mM dNTP mix, 1 μL of biotin-labeled oligo-dT25 primer (700 ng/μL), and 1 μL of SuperScript™ II Reverse Transcriptase (200 U/μL) for each sample.
3. Add 20 μL of the first-strand cDNA synthesis mix to each sample.
4. Mix well and incubate for 2 h at 42 °C.
5. For the second-strand cDNA synthesis mix, combine 87.4 μL of water, 16 μL of 10× *E. coli* DNA ligase reaction buffer, 6 μL of 100 mM DTT, 3 μL of 10 mM dNTP mix, 5.0 μL of DNA polymerase I (10 U/μL), 1.5 μL of *E. coli* DNA ligase (10 U/μL), and 1.1 μL RNase H (1.5 U/μL) for each sample.
6. Add 120 μL of the second-strand cDNA synthesis mix to the 40 μL of the first-strand reaction cocktail.
7. Mix well and incubate for 1 h at 12 °C, followed by 1 h at 22 °C.
8. Purify the double-stranded cDNA with a cDNA purification kit according to the manufacturer's instructions. Elute the cDNA in 30 μL of elution buffer.
9. Run 8 μL of each cDNA sample on a 1.2 % agarose gel in 0.5× TAE running buffer at 100 V for 15–20 min (*see* **Note 6** and Fig. 1b).

3.2 PCR Template Preparation

This subheading describes the first digestion, 3′-end capturing, second digestion, and adapter ligation.

1. Prepare the first digestion mix by mixing 15 μL of water, 4 μL of 10× RL buffer, and 1 μL of *Bst*YI restriction enzyme (10 U/μL) for each sample.
2. Add 20 μL of the first digestion mix to 20 μL of each cDNA sample.
3. Incubate for 2 h at 60 °C (*see* **Note 7**).
4. For each sample, wash 10 μL Dynabeads with 100 μL 2× STEX. Resuspend the Dynabeads in a final volume of 40 μL 2× STEX per sample (*see* **Note 8**).
5. Mix 40 μL of the resuspended Dynabeads with each digested cDNA sample to give a final volume of 80 μL.

6. Incubate the samples for 30 min at room temperature, with gentle agitation (1,000 RPM) to ensure that the beads remain suspended.
7. Collect the beads with the magnet and remove the supernatants.
8. Remove the tubes from the magnet.
9. Add 100 μL 1× STEX and resuspend the beads (*see* **Note 9**).
10. Transfer the resuspended beads to a fresh tube.
11. Repeat **steps 7–10** four additional times.
12. Collect the beads with the magnet and remove the supernatants.
13. Remove the tubes from the magnet.
14. Add 30 μL of $T_{10}E_{0.1}$ buffer and resuspend the beads.
15. Transfer the resuspended beads to a fresh tube.
16. Prepare the second digestion mix by mixing 5 μL of water, 4 μL of 10× RL buffer, and 1 μL of *Mse*I restriction enzyme (10 U/μL) for each sample.
17. Add 10 μL of the second digestion mix to the 30 μL of resuspended beads.
18. Incubate for 2 h at 37 °C with gentle agitation (1,000 RPM) to ensure that the beads remain suspended.
19. Collect the beads with the magnet.
20. Transfer the supernatant containing the released fragments to a new tube.
21. Prepare the adapter ligation mix by mixing 4 μL of water, 1 μL of *Bst*YI adapter (5 μM), 1 μL of *Mse*I adapter (50 μM), 1 μL of 10 mM ATP, 1 μL of 10× RL buffer, 1 μL of T4 DNA ligase (1 U/μL), and 1 μL of *Bst*YI restriction enzyme (10 U/μL) for each sample.
22. Add 10 μL of the adapter ligation mix to the 40 μL of supernatant.
23. Incubate for 3 h at 37 °C (*see* **Note 10**).
24. After adapter ligation, dilute the samples twofold by adding 50 μL of $T_{10}E_{0.1}$ buffer to each sample (*see* **Note 11**).

3.3 Preamplification

1. Prepare the preamplification mix by mixing 30.8 μL of water, 5.0 μL of 10× PCR buffer, 5.0 μL of 25 mM $MgCl_2$, 1.5 μL of *Bst*YI-T/C+0 primer (50 ng/μL), 1.5 μL of *Mse*I+0 primer (50 ng/μL), 1.0 μL of 10 mM dNTP mix, and 0.2 μL of AmpliTaq DNA polymerase (5 U/μL) for each sample (*see* **Note 12**).
2. Add 45 μL of the preamplification mix to 5 μL of the PCR template.

3. Subject the samples to the following PCR program: Initial denaturation for 1 min at 94 °C, followed by 25 cycles of denaturation at 94 °C for 30 s, annealing at 56 °C for 1 min and elongation at 72 °C for 1 min.
4. Analyze 10 μL of the PCR reaction on a 1.2 % agarose gel in 0.5× TAE running buffer at 100 V for 15–20 min (*see* **Note 13** and Fig. 1c).
5. Dilute the preamplification mix 600-fold with $T_{10}E_{0.1}$ buffer (*see* **Note 14**).

3.4 Selective Amplification

1. Prepare the primer radiolabeling mix by mixing 0.23 μL of water, 0.10 μL of *Bst*YI-T/C+N primer (50 ng/μL), 0.10 μL of γ-^{33}P-ATP (3,000 Ci/mmol), 0.05 μL of 10× T4 buffer, and 0.02 μL of T4 polynucleotide kinase (10 U/μL) for each sample.
2. Incubate the reaction mixture at 37 °C for 45 min.
3. Stop the reaction by incubating the mixture for 10 min at 80 °C.
4. Prepare the selective amplification mix by mixing 6.0 μL of *Mse*I+NN primer (5 ng/μL), 3.9 μL of water, 2.0 μL of 10× PCR buffer, 2.0 μL of 25 mM $MgCl_2$, 0.5 μL of γ-^{33}P-labeled *Bst*YI-T/C+N primer, 0.4 μL of 10 mM dNTP mix, and 0.2 μL of AmpliTaq Gold DNA polymerase (5 U/μL) for each sample.
5. Add 15 μL of the selective amplification mix to 5 μL of the diluted preamplification mixture.
6. Subject the samples to the following PCR program: Initial denaturation for 10 min at 94 °C, followed by 13 cycles touchdown (denaturation at 94 °C for 30 s, annealing for 30 s at an initial temperature of 65 °C, reduced with 0.7 °C per PCR cycle, elongation at 72 °C for 1 min) and 23 additional cycles of denaturation at 94 °C for 30 s, annealing at 56 °C for 30 s, and elongation at 72 °C for 1 min.
7. Add 20 μL of formamide loading dye to each sample.
8. Incubate the samples overnight at −20 °C (*see* **Note 15**).

3.5 Preparation of the SequaMark™ 10 Base Ladder

1. Prepare the SequaMark™ PCR mix by mixing 33.8 μL of water, 5 μL of 10× PCR buffer, 3 μL of 25 mM $MgCl_2$, 2.5 μL of 10 μM forward primer, 2.5 μL of 10 μM reverse primer, 2 μL of SequaMark™ DNA template, 1 μL of 10 mM dNTP mix, and 0.2 μL of AmpliTaq DNA polymerase (5 U/μL).
2. Subject the PCR mix to the following PCR program: 30 cycles of denaturation at 95 °C for 30 s, annealing at 56 °C for 30 s, and elongation at 72 °C for 1 min.

3. Clean the PCR reaction with a PCR cleanup kit according to the manufacturer's instructions.
4. Quantify the product on a nanodrop (*see* **Note 16**).
5. Prepare the SequaMark™ primer radiolabeling mix by mixing 0.23 μL of water, 0.1 μL of 10 μM forward primer, 0.1 μL of γ-^{33}P-ATP (3,000 Ci/mmol), 0.05 μL of 10× T4 buffer, and 0.02 μL of T4 polynucleotide kinase (10 U/μL).
6. Incubate the reaction mixture at 37 °C for 45 min.
7. Stop the reaction by incubating the mixture for 10 min at 80 °C.
8. Prepare the radioactive SequaMark™ PCR mix by mixing 24 μL of dNTP/ddTTP mix, 14.5 μL of water, 5 μL of 10× ThermoPol reaction buffer, 2.5 μL of labeled forward primer, 2.5 μL Vent$_R$™ (exo-) DNA polymerase (2 U/μL), and 1.5 μL (self-amplified) template.
9. Subject the samples to the following PCR program: Initial denaturation for 5 min at 94 °C, 25 cycles of denaturation at 94 °C for 30 s, annealing at 56 °C for 30 s, and elongation at 72 °C for 1 min, followed by a final elongation step at 72 °C for 7 min.
10. Add 50 μL of formamide loading dye to the PCR reaction.
11. Mix carefully and close the tubes.
12. Keep the mixture overnight at −20 °C (*see* **Note 15**).

3.6 Gel Electrophoresis and Detection

1. Clean the glass plate and buffer tank with water and soap (*see* **Note 17**).
2. Clean the surface of the glass plate twice with ethanol and once with acetone.
3. Treat the surface of the buffer tank with Rain-X (*see* **Note 18**).
4. Assemble the gel system.
5. Prepare the gel solution by adding 500 μL of 10 % APS and 100 μL of TEMED to 100 mL of 4.5 % (w/v) denaturing polyacrylamide gel solution. Mix gently.
6. Immediately cast the gel by injecting the gel solution into the gel system and insert the sharktooth comb between the two glass plates with the teeth upwards. Align the holes in the comb with the edge of the glass plate of the buffer tank, and fix the comb with clamps. During this process, carefully avoid introducing air bubbles, because they will damage the front and disturb the gel image.
7. Allow the gel to polymerize for at least 1 h before use (*see* **Note 19**).

8. Prepare the running buffer by diluting 200 mL of 10× Maxam buffer to 2 L with water.
9. Add 8 mL of 0.5 M EDTA (pH 8.0) to the 2 L of 1× Maxam buffer.
10. Dissolve 8.8 g of NaAc in 400 mL of the prepared buffer and pour the resulting solution in the lower buffer tank (*see* **Note 20**).
11. Warm the remaining 1,600 mL of buffer to 50–55 °C (6–7 min at 1,000 W in the microwave oven).
12. Fill the upper buffer tank with the warm buffer.
13. Prerun the gel for 15 min at 100 W to heat up the gel to approximately 50–55 °C.
14. During the prerun, denature the samples and the SequaMark™ 10 base ladder at 95 °C for 5 min.
15. After the prerun, remove the comb and clean the front with a 50-mL syringe containing the running buffer to remove all gel pieces and bubbles from the well.
16. Insert the sharktooth comb into the well with the teeth approximately 0.5 mm into the gel (*see* **Note 21**).
17. Load 3 μL of the PCR product (or ladder) per well for a comb of 72 teeth (*see* **Note 22**).
18. Once all samples are loaded, perform the electrophoresis at a constant power of 100 W for approximately 2 h 45 min or until the dye front reaches the bottom of the gel.
19. After electrophoresis, discard the running buffer and disassemble the gel system.
20. Carefully lift the buffer tank and transfer the gel to a blotting paper.
21. Cover the gel with Saran wrap and dry at 75 °C on a vacuum drier for at least 1 h.
22. To visualize the results, place the dried gel on a phosphorImager screen for 12–16 h or on an X-ray film for 2–3 days (*see* **Note 23**).

3.7 Re-amplifications

1. Place the developed X-ray film back on the gel and align correctly (*see* **Note 23**).
2. Cut the fragments of interest from the gel with a razor blade and transfer the gel pieces to Eppendorf tubes.
3. Add 100 μL of $T_{10}E_{0.1}$ buffer to the gel pieces and crush the pieces to a fine pulp (*see* **Note 24**).
4. Incubate the samples for 1 h at room temperature to allow complete resuspension of the DNA.

5. Centrifuge for 5 min at 11,000 × *g* to separate the blotting paper from the DNA solution. Use the resulting supernatant as template for the re-amplification PCR.
6. Prepare the PCR 1 master mix by mixing 34.8 μL of water, 5.0 μL of 10× PCR buffer, 2.0 μL of 50 mM $MgCl_2$, 1.0 μL of forward 1 primer (10 μM), 1.0 μL of reverse 1 primer (10 μM), 1.0 μL of 10 mM dNTP mix, and 0.2 μL of Silverstar Taq DNA polymerase (5 U/μL) for each sample.
7. Transfer 45 μL of the reaction mixture to 5 μL of the resuspended DNA templates.
8. Subject the samples to the following PCR program: Initial denaturation for 1 min at 95 °C, followed by ten cycles of denaturation at 95 °C for 30 s, annealing at 54 °C for 30 s, and elongation at 72 °C for 1 min. Perform a final elongation at 72 °C for 2 min.
9. Prepare the PCR 2 master mix solution by mixing 28.7 μL of water, 4.0 μL of 10× PCR buffer, 2.0 μL of 50 mM $MgCl_2$, 2.4 μL of forward 2 primer (10 μM), 1.7 μL of reverse 2 primer (10 μM), 1.0 μL of 10 mM dNTP mix, and 0.2 μL of Silverstar Taq DNA polymerase (5 U/μL) for each sample.
10. Transfer 40 μL of the mix to 10 μL of the first PCR reaction and mix well.
11. Subject the samples to the following PCR program: Initial denaturation for 1 min at 95 °C, followed by five cycles of denaturation at 95 °C for 30 s, annealing at 45 °C for 30 s, and elongation at 72 °C for 1 min. This is followed by 20 cycles of denaturation at 95 °C for 30 s, annealing at 55 °C for 30 s, and elongation at 72 °C for 1 min and a final elongation step at 72 °C for 2 min.
12. Analyze 5 μL of the PCR 2 reaction on a 1.2 % agarose gel in 0.5× TAE running buffer at 100 V for 20–25 min (*see* **Note 25** and Fig. 1d).
13. Use the obtained re-amplified fragments for sequencing, either by direct sequencing of the PCR products or after cloning of the fragments into a plasmid vector (*see* **Note 26**).

4 Notes

1. The biotin-labeled oligo-dT25 primer is an oligo-dT primer consisting of a string of 25 deoxythymidine nucleotides. The primer is labeled at its 5′ end with biotin. Use DEPC-treated water to dissolve the lyophilized primer, and store the resuspended primer at −20 °C.
2. EDTA will not go into solution until the pH approaches 8.0. Addition of NaOH pellets will allow the EDTA to dissolve.

3. To prepare the 0.5× TAE buffer, dilute 50 mL of 10× TAE buffer to 1 L with water.
4. Because DTT may precipitate, incubate the buffer briefly at 37 °C to dissolve all DTT prior to use.
5. For cDNA-AFLP analysis, it is essential to have similar quantities of pure and intact RNA for all the samples during the preparation of the total RNA. The integrity of the RNA can be determined by checking an aliquot of the RNA sample on a 1.2 % agarose gel. For any eukaryotic sample, intact total RNA will show two clear bands, the 28S and 18S rRNA bands, with the intensity of the 28S band about twice that of the 18S band (*see* Fig. 1a). The occurrence of a low-molecular-weight smear is indicative of degraded RNA. The purity of the RNA can be assessed spectrophotometrically by determining the absorbance of the samples at 230 nm (A_{230}), 260 nm (A_{260}), and 280 nm (A_{280}). The A_{260}/A_{280} ratio is indicative of the purity of the sample and should be between 1.80 and 2.00. A lower value implies contamination of the RNA sample with proteins or phenolics. The A_{260}/A_{230} ratio is a second measure of RNA purity and should be above 2.00. Lower values indicate contaminants, such as phenolics, carbohydrates, and/or EDTA.
6. The cDNA should appear as a 0.1–4 kbp smear on the gel. The purified cDNA is stable and can be stored for up to 1 year at −20 °C.
7. After 1 h of incubation, briefly centrifuge the tubes because water evaporates from the mixture and condensates in the tube lids.
8. Thoroughly resuspend the Dynabeads before the washing step by pipetting until a uniform brown suspension is obtained. Wash the Dynabeads for maximum 7–8 samples in one tube.
9. The STEX buffer contains the detergent Triton X-100. Care should be taken to avoid the production of foam while washing or resuspending the Dynabeads.
10. Alternatively, the second digestion and adapter ligation can be performed simultaneously by adding the second digestion and adapter ligation mixes at the same time to the 30 μL of resuspended beads. Incubate for 4 h at 37 °C, collect the beads with the magnet, transfer the supernatant to a new tube, and proceed with **step 24** of Subheading 3.2.
11. The diluted samples will serve as template for the preamplifications and can be stored for several years at −20 °C.
12. For the preamplifications, run two separate PCRs. For the first PCR, use the *Bst*YI-C+0 primer and for the second the *Bst*YI-T+0 primer, or vice versa.

13. The preamplification should appear as a 50–500 bp smear on the gel.
14. To avoid large pipetting errors, prepare the 600-fold dilution in two steps. First, make a 20-fold dilution of the PCR product, and subsequently make a 30-fold dilution of the diluted PCR product. If the preamplification product on the gel has a low intensity, prepare a less diluted template for the selective amplifications (for instance, a 100-fold dilution).
15. Keeping the samples at −20 °C improves the quality of the gel image after polyacrylamide gel electrophoresis.
16. The template should have a concentration of 30 ng/μL. To this end, several tubes with PCR product should be combined before the PCR cleanup.
17. All equipment (glass plate, buffer tank, sharktooth comb, and spacers) should be thoroughly cleaned with soap and water prior to gel preparation to remove any residual gel pieces. Remaining pieces of gel or dirt will result in bubbles in the gel that will interfere with the sample running through the gel. After cleaning, rinse the equipment with purified water and dry with tissue paper.
18. Rain-X is a hydrophobic silicone polymer that will prevent the gel from sticking to the surface of the buffer tank.
19. Gels can be prepared 1 day in advance. After polymerization of the gel, remove the clamps and insert water-soaked tissue paper in the gel front. Cover with Saran Wrap and fix with the clamps to prevent the gel from drying out. Store the gel overnight at room temperature.
20. Addition of NaAc in the lower buffer tank will generate an electrolyte gradient, thereby preventing that small cDNA-AFLP fragments run off the gel.
21. To prevent leaking of the sample, do not move the comb once it is pushed into the gel.
22. The volume will vary depending on the comb used for sample loading. For combs of 48 samples, load 4–5 μL; for combs of 72 samples, load 2–3 μL; and for combs of 96 samples, load maximum 2 μL of sample per well. Load a ladder between the different primer combinations.
23. While working in the dark room, align the X-ray film with the gel blot and staple the corners together. Punch holes through the gel blot and the attached X-ray film by using a paper puncher. Place the gel blot and the attached X-ray film in a cassette for exposure. After development of the films, realign the film and gel blots with the paper punch-made holes. Staple the film and the gel back together. Precise alignment is very important for cutting out the correct bands.

24. The gel pieces can be crushed to pulp with a pipette tip. Alternatively, 0.1-mm Zirconia/Silica beads (Biospec Products, Bartlesville, OK, USA) can be used. Add a few mg of beads to the gel piece and add 100 μL of $T_{10}E_{0.1}$ buffer. Mix with a mixer mill for 30 s at 30 Hz.
25. The re-amplified fragments should appear as single bands on the gel.
26. The applied re-amplification method ligates an adapter to the cDNA-AFLP fragments, making the fragments 56 nucleotides longer. This enhances the sequencing quality of short fragments and results in the sequence of the complete tag in one single sequencing reaction. Furthermore, the ligated adapters (*attB1* and *attB2*) allow cloning of the re-amplified fragments in Gateway™ vectors (Invitrogen).

Acknowledgments

This work was supported by the European Framework Programme 7 project SMARTCELL (FP7-KBBE-222716). J.C. is indebted to the Ghent University for a "Bijzondere Onderzoeksfonds" pre-doctoral fellowship.

References

1. Vos P, Hogers R, Bleeker M, Reijans M, van de Lee T, Hornes M, Frijters A, Pot J, Peleman J, Kuiper M, Zabeau M (1995) AFLP: a new technique for DNA fingerprinting. Nucleic Acids Res 23:4407–4414
2. Zabeau M, Vos P (1993) Selective restriction fragment amplification: a general method for DNA fingerprinting. European Patent Application EP 534858A1
3. Breyne P, Dreesen R, Cannoot B, Rombaut D, Vandepoele K, Rombauts S, Vanderhaeghen R, Inzé D, Zabeau M (2003) Quantitative cDNA-AFLP analysis for genome-wide expression studies. Mol Genet Genomics 269:173–179
4. Pollier J, González-Guzmán M, Ardiles-Diaz W, Geelen D, Goossens A (2011) An integrated PCR colony hybridization approach to screen cDNA libraries for full-length coding sequences. PLoS One 6:e24978
5. Vuylsteke M, Peleman JD, van Eijk MJT (2007) AFLP-based transcript profiling (cDNA-AFLP) for genome-wide expression analysis. Nat Protoc 2:1399–1413
6. Bachem CWB, van der Hoeven RS, de Bruijn SM, Vreugdenhil D, Zabeau M, Visser RGF (1996) Visualization of differential gene expression using a novel method of RNA fingerprinting based on AFLP: analysis of gene expression during potato tuber development. Plant J 9:745–753
7. Bachem CWB, Oomen RJFJ, Visser RGF (1998) Transcript imaging with cDNA-AFLP: a step-by-step protocol. Plant Mol Biol Rep 16:157–173
8. Breyne P, Zabeau M (2001) Genome-wide expression analysis of plant cell cycle modulated genes. Curr Opin Plant Biol 4: 136–142
9. Goossens A, Häkkinen ST, Laakso I, Seppänen-Laakso T, Biondi S, De Sutter V, Lammertyn F, Nuutila AM, Söderlund H, Zabeau M, Inzé D, Oksman-Caldentey KM (2003) A functional

genomics approach toward the understanding of secondary metabolism in plant cells. Proc Natl Acad Sci USA 100:8595–8600

10. Rischer H, Orešič M, Seppänen-Laakso T, Katajamaa M, Lammertyn F, Ardiles-Diaz W, Van Montagu MCE, Inzé D, Oksman-Caldentey K-M, Goossens A (2006) Gene-to-metabolite networks for terpenoid indole alkaloid biosynthesis in *Catharanthus roseus* cells. Proc Natl Acad Sci USA 103: 5614–5619
11. Pauwels L, Morreel K, De Witte E, Lammertyn F, Van Montagu M, Boerjan W, Inzé D, Goossens A (2008) Mapping methyl jasmonate-mediated transcriptional reprogramming of metabolism and cell cycle progression in cultured *Arabidopsis* cells. Proc Natl Acad Sci USA 105:1380–1385

Chapter 24

Analysis of RNA-Seq Data with TopHat and Cufflinks for Genome-Wide Expression Analysis of Jasmonate-Treated Plants and Plant Cultures

Jacob Pollier, Stephane Rombauts, and Alain Goossens

Abstract

The recent development of various deep sequencing techniques has led to the most powerful transcript profiling method available to date, RNA sequencing or RNA-Seq. Besides the identification of new genes and new splice variants of known genes, RNA-Seq allows to compare the whole transcriptome of any organism under two or more experimental conditions, such as before and after jasmonate treatment. However, the vast amounts of data generated during RNA-Seq experiments require complex computational methods for read mapping and expression quantification. Here, we describe a detailed protocol for the analysis of deep sequencing data, starting from the raw RNA-Seq reads. First, a quality check is performed on the raw reads to assess the quality of the sequencing. Subsequently, adapters and low-quality sequences are trimmed off the raw reads. The resulting processed reads are mapped to the reference genome, and the mapped reads are counted to generate expression data for the annotated genes for each sample. This method can be used for the analysis of RNA-Seq data of any organism for which a reference genome is available.

Key words Transcript profiling, Gene expression, Transcriptome, RNA sequencing, RNA-Seq, FastQC, TopHat, Cufflinks

1 Introduction

The treatment of plants or plant cell cultures with jasmonates triggers an extensive transcriptional reprogramming of the cells, leading to transcriptional activation or repression of entire metabolic pathways [1, 2]. Since many of these pathways lead to the production of secondary metabolites, comparing the transcriptome before and after jasmonate treatment may allow to identify candidate genes for the biosynthesis of secondary metabolites [2, 3]. In tobacco Bright Yellow 2 (BY-2) cells, for instance, the biosynthesis of nicotine is elicited by jasmonate treatment. Genome-wide transcript profiling of jasmonate-elicited BY-2 cultures has led to a set of tobacco genes [4]

Alain Goossens and Laurens Pauwels (eds.), *Jasmonate Signaling: Methods and Protocols*, Methods in Molecular Biology, vol. 1011, DOI 10.1007/978-1-62703-414-2_24,

from which several new regulators, transporters, and enzymes involved in nicotine biosynthesis were identified in subsequent functional screens [5–9].

Various techniques can be used for genome-wide transcript profiling, including hybridization-based approaches like microarrays, and tag-based sequencing approaches such as cDNA-AFLP [10] (*see* Chapter 23), serial analysis of gene expression (SAGE) [11], and massively parallel signature sequencing (MPSS) [12]. However, the development of deep sequencing technologies has led to a method that is undoubtedly the most powerful transcript profiling technique available to date, RNA sequencing or RNA-Seq [13–15]. RNA-Seq not only allows the identification of new genes and new splice variants of known genes, but it also allows to compare the whole transcriptome under two or more conditions [16], such as before and after jasmonate treatment. For RNA-Seq, RNA is isolated and converted to a set of cDNAs sheared into fragments to which adapters are attached. Subsequently, the cDNA fragments are subject to deep sequencing, resulting in millions of short sequence fragments or reads (typically 30–400 nucleotides long, depending on the sequencing technology used) from one end (single-end) or both ends (paired-end) of the cDNA fragments. To obtain genome-wide quantitative transcript data, the reads are mapped on a reference genome or a de novo assembled set of transcripts [13].

To analyze the vast amounts of data generated during an RNA-Seq experiment, complex computational methods for read mapping, transcriptome reconstruction, and expression quantification are required [16, 17]. Several methods and software exist, but here we use the pipeline relying on TopHat [15] for read mapping and Cufflinks [18] for expression quantification. The method presented here aims to compare the transcriptome under two different conditions (e.g., before and after jasmonate treatment), starting from the raw RNA-Seq reads. First, a quality check is performed on the raw reads, and the adapters and low-quality sequences are trimmed. Subsequently, the processed reads are mapped to the reference genome with TopHat, of which the resulting alignment files are used as input for Cufflinks, which generates normalized expression data for each of the analyzed raw sequencing files.

2 Materials

The software used for analysis of deep sequencing data in this protocol need a 64-bit CPU/computer running on Linux, with a minimal amount of 16 GB of RAM. During processing of RNA-Seq data, several hundred GB of disk space may be required.

2.1 Quality Control

The various RNA-Seq protocols available to date still suffer from biases and sequencing artifacts, such as GC bias, read errors, primer and adapter contaminations, and PCR bias. To assure meaningful downstream processing of the obtained RNA-Seq data, a quality check should be performed on the raw sequencing data [19, 20]. In this protocol, the quality control is done with FastQC, a commonly used program that provides an overview of whether the raw RNA-Seq data have any problems or biases to consider before further analysis. FastQC is a freely available program and can be downloaded from http://www.bioinformatics.babraham.ac.uk/projects/fastqc/.

2.2 Adapter Trimming

Various programs are available to trim adapters from the RNA-Seq data. In this protocol, adapter trimming is performed with fastx_clipper. The fastx_clipper program is part of the FASTX-Toolkit, which is freely available and can be downloaded from http://hannonlab.cshl.edu/fastx_toolkit/.

2.3 Quality Trimming

The quality trimming step will trim ambiguous (N) and low-quality residues from the ends of the reads. In this protocol, the quality trimming is performed with fastq_quality_trimmer. Like the fastx_clipper program, the fastq_quality_trimmer program is part of the FASTX-Toolkit.

2.4 Read Mapping

In the read mapping step, the processed sequencing reads are aligned to the reference genome. In this protocol, read mapping is performed with TopHat [15], which uses the widely used Bowtie program [21, 22] as alignment engine. In addition to the reference genome, TopHat (freely available at http://tophat.cbcb.umd.edu/) needs Bowtie [21, 22] (freely available at http://bowtie-bio.sourceforge.net/index.shtml/) and SAM Tools [23] (freely available at http://samtools.sourceforge.net/) to be installed.

2.5 Read Counting

The counting of the mapped reads is performed with Cufflinks [18], a software program freely available at http://cufflinks.cbcb.umd.edu/. Cufflinks counts the expression of each gene and reports it in "fragments per kilobase of transcript per million fragments mapped" or FPKM [18]. The FPKM value is a measure of the expression of a transcript, normalized by transcript length and the total number of fragments. As such, the FPKM value can be used to compare the expression of the genes in the analyzed samples. However, one should be aware that Cufflinks (but other software too) uses an annotation of the reference genome described in GFF (or GTF) format. This means that the results depend on the quality of the provided annotation. The reported FPKM values relate to the genes described and genes missing in the annotation description file (even though reads map to it) will not be reported. Wrong gene models will report altered FPKM values.

3 Methods

All programs are operated through the UNIX shell. The working directory in which the commands (prefix $) given in this protocol are executed should contain the (zipped) raw sequencing data in FASTQ format. This protocol was designed with the following versions of the above-described programs:

- FastQC version 0.9.1.
- FastX version 0.0.13.
- Bowtie version 2.0.0b6 [22].
- TopHat version 2.0.3 [15].
- SAM Tools 0.1.18 [23].
- Cufflinks 1.3.0 [18].

3.1 Quality Control

1. Unzip the first raw sequence file:

```
$ gunzip name.fastq.gz
```

2. Run FastQC on the unzipped file:

```
$ fastqc name.fastq
```

3. Rezip the raw sequencing file:

```
$ gzip name.fastq
```

4. Repeat **steps 1–3** for all the raw sequence files (*see* **Note 1**).

As output, for each of the analyzed raw sequencing files, the FastQC program will generate a folder containing an HTML-based permanent report that provides an overview of whether the raw RNA-Seq data have problems or biases to consider before further analysis.

3.2 Adapter Trimming

1. Unzip the first raw sequence file (*see* **Note 2**):

```
$ gunzip -c name.fastq.gz >name.fastq
```

2. Trim an adapter from the unzipped data file (*see* **Note 3**):

```
$ fastx_clipper -i name.fastq -o newname.
fastq -l 20 -v -a ADAPTERSEQUENCE
```

3. Remove the original fastq file:

```
$ rm name.fastq
```

4. Zip the trimmed file:

```
$ gzip newname.fastq
```

5. Repeat **steps 1–4** for all the raw sequence files and all used adapters (*see* **Note 4**).

The output of the fastx_clipper program is a new fastq file containing the adapter-trimmed sequences. The original raw sequencing files are unchanged and will not be used any more in the downstream processing.

3.3 Quality Trimming

1. Unzip the first adapter-trimmed sequence file:

```
$ gunzip -c name.fastq.gz >name.fastq
```

2. Perform the quality trimming (*see* **Note 5**):

```
$ fastq_quality_trimmer -i name.fastq -o newname.fastq -v -t 20 -l 65
```

3. Zip the quality-trimmed file:

```
$ gzip newname.fastq
```

4. Remove the original fastq file:

```
$ rm name.fastq
```

5. Repeat **steps 1–4** for all the adapter-trimmed sequence files (*see* **Note 6**).

The output of the quality trimming step is a new (zipped) fastq file containing the quality-trimmed sequences that will be used for downstream processing. The file containing the adapter-trimmed sequences is unchanged, and is not needed any more after this step. To assess the effects of the adapter and quality trimming steps, a new quality control can be performed on the processed reads (Fig. 1).

3.4 Read Mapping

1. Build the bowtie index files from the reference genome (*see* **Note 7**):

```
$ bowtie2-build genomename.fasta genomename
```

2. Unzip the first quality-trimmed sequence file:

```
$ gunzip -c name.fastq.gz >name.fastq
```

3. Make an output directory for the first sequence file:

```
$ mkdir dirname
```

4. Map the reads of the first sample to the reference genome (*see* **Note 8**):

```
$ tophat2 -o ./dirname genomename name.fastq
```

5. Remove the quality-trimmed file:

```
$ rm name.fastq
```

6. Repeat **steps 2–5** to map the reads of the other samples to the reference genome (*see* **Note 9**).

TopHat will write its output into the defined folder. Next to a set of intermediate files, the output consists of a file called `accepted_hits.bam`, which contains a list of read alignments, and which will be used for the read counting. The file containing the quality-trimmed sequences is unchanged, and is not needed any more after this step.

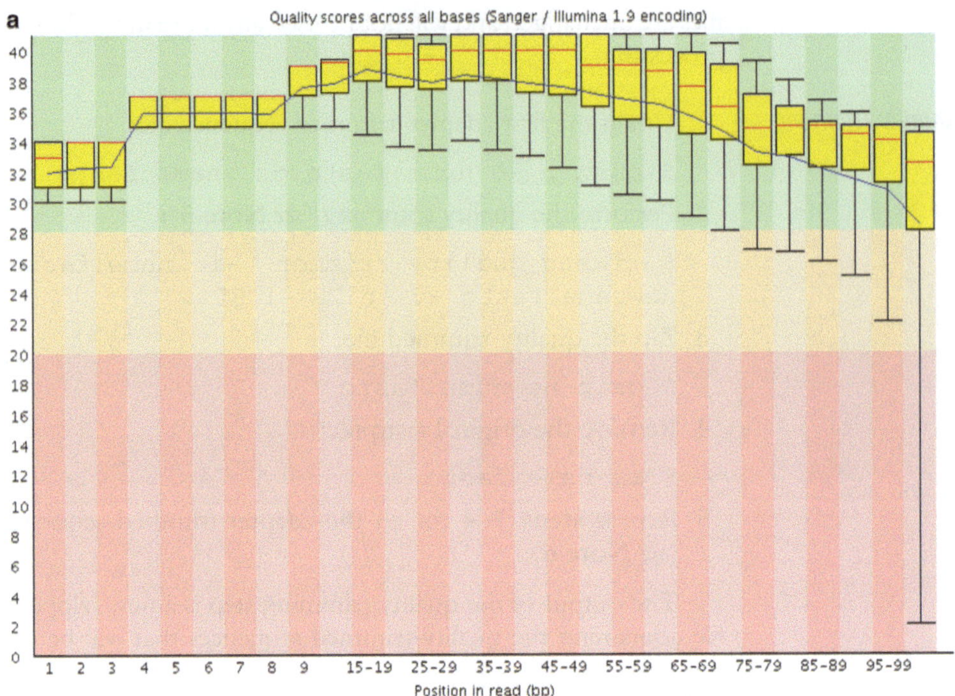

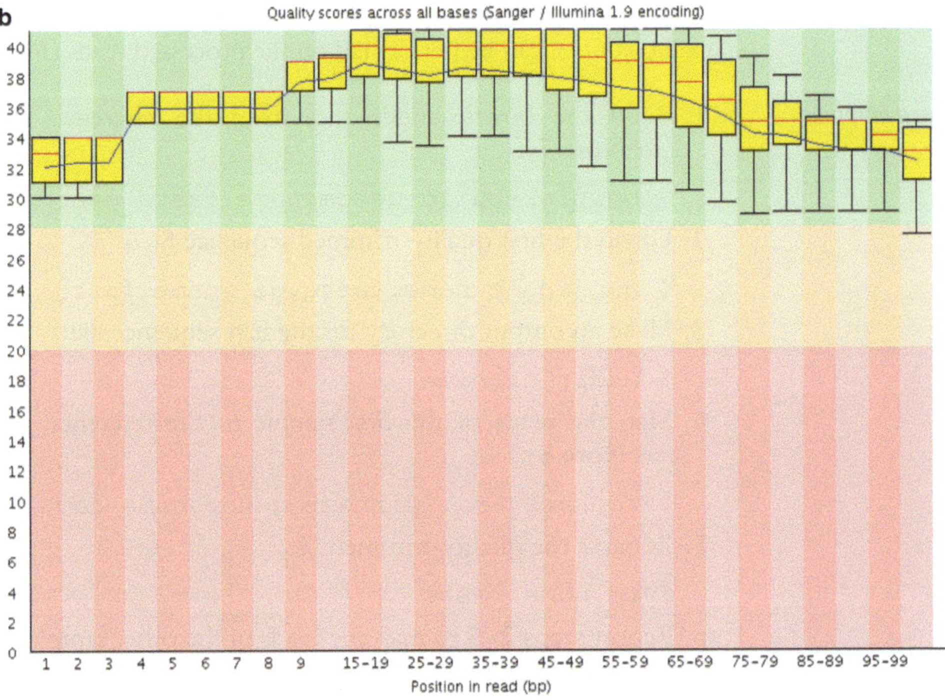

Fig. 1 Box and whisker plot of the per base sequence quality generated by the FastQC quality control program before (**a**) and after (**b**) adapter and quality trimming of the raw RNA-Seq reads. For each of the base positions (*X*-axis), the quality scores are plotted (*Y*-axis), with higher scores representing better base calls. The background *green*, *orange*, and *red* colors represent base calls of good, reasonable, and poor quality, respectively. In most RNA-Seq platforms, it is normal to see the base call quality degrading with the base position (**a**), which is improved after quality trimming of the reads (**b**)

3.5 Read Counting

1. Count the reads of the first sample with Cufflinks (*see* **Note 10**):

```
$ cufflinks -v --compatible-hits-norm -u -o ./
dirname -G genome.gff3 dirname/accepted_hits.
bam
```

2. Repeat **step 1** to count the reads of the other samples (*see* **Note 11**).

As output of the read counting, for each sequencing file, a file (`genes.fpkm_tracking`) is generated containing the FPKM values of all the genes present on the reference genome. The FPKM values can be copied in an Excel table, and used to compare the transcriptome in the different conditions.

4 Notes

1. Running of the quality control can be automated for all the raw RNA-Seq files by making use of the UNIX foreach command in **step 1**:

```
$ foreach i(*.fastq.gz)
      gunzip $i
      foreach j(*.fastq)
         fastqc $j
         gzip $j
         end
      end
```

2. Unlike the quality control, trimming of the adapters will modify the original files. In order not to lose the original data, the raw sequencing data are unzipped whilst keeping the original files unchanged.
3. With this command, the specified adapter sequence will be trimmed, and sequences shorter than 20 nucleotides, or sequences with unknown (N) nucleotides, will be discarded. As output, a new fastq file is generated containing the adapter-trimmed sequences. The –v (verbose) parameter will create a short summary with information on the amount of reads that were processed and trimmed or discarded. When using sequence data generated by the Illumina/Solexa platform, an invalid quality score value error may occur depending on the CASAVA software version that generated the original fastq files. Depending on the case, add the –Q 33 parameter in the command line:

```
$ fastx_clipper -i name.fastq -o newname.
fastq -Q 33 -l 20 -v -a ADAPTERSEQUENCE
```

4. Trimming of the adapters can be automated for all the raw RNA-Seq files by making use of the foreach command in **step 1**:

```
$ foreach i(*.fastq.gz)
     echo $i
```

```
set name = `basename $i .fastq.gz`
echo $name
gunzip -c ${name}.fastq.gz >${name}.fastq
fastx_clipper -i ${name}.fastq -o
new_${name}.fastq -l 20 -v -a
ADAPTERSEQUENCE
gzip new_${name}.fastq
rm ${name}.fastq
end
```

5. When using sequence data generated by the Illumina/Solexa platform, an invalid quality score value error may occur. In this case, add the –Q 33 parameter in the command line:

```
$ fastq_quality_trimmer -i name.fastq -o
newname.fastq -v -t 20 -l 65 -Q 33
```

   The –v (verbose) parameter will create a short summary with information on the amount of reads that were processed and trimmed or discarded. The –t parameter defines the minimum acceptable quality of the base calling. In this case, the quality threshold (*Q*) is 20, meaning that the probability (*P*) of an incorrect base call is 1 %, according to the formula $Q=-10\log_{10}(P)$. The Q20 base call accuracy of 99 % means that a read of 100 bp will likely contain one error. Given the reads will be trimmed from the ends until the quality reaches the minimum required value of 20, it is good to add a minimum length (–l option) for the reads that should be reported in the output. This parameter depends on the length of the input reads and what will be done after trimming. But in any case, one should keep in mind that the shorter a read, the less specific it becomes.

6. Quality trimming can be automated for all the adapter-trimmed RNA-Seq files by making use of the foreach command:

```
$ foreach i(*.fastq.gz)
    echo $i
    set name = `basename $i .fastq.gz`
    echo $name
    gunzip -c ${name}.fastq.gz >${name}.fastq
    fastq_quality_trimmer -i ${name}.fastq
    -o new_${name}.fastq -v -t 20 -l 65
    gzip new_${name}.fastq
    rm ${name}.fastq
    end
```

7. The `bowtie2-build` algorithm builds a Bowtie index from the FASTA file of the reference genome. The Bowtie index is used to align the reads to the genome and consists of a set of six files with suffixes .1.bt2, .2.bt2, .3.bt2, .4.bt2, .rev.1.bt2, and .rev.2.bt2. As input file, a FASTA-file of the complete genome or a comma-separated list of FASTA files containing the reference

sequence (e.g., FASTA-files of the chromosomes) is used. The defined genome name of the index files to write will be used as base name of the set of six files.

8. The command given in the protocol is to run the TopHat script using the default parameters. However, these default parameters are set to process mammalian RNA-Seq reads, and hence, when working with other organisms, such as plants, a more strict setting of certain parameters will keep the number of false positives low. For instance, the command given below restricts the maximum intron size to 6,000 bp (in *Arabidopsis*, over 99.9 % of the introns are shorter than 6,000 bp):

   ```
   $ tophat2 -I 6000 -o ./dirname genomename
   name.fastq
   ```

   For more detailed information about the options available in TopHat, use the help-command:

   ```
   $ tophat2 -h
   ```

9. Read mapping can be automated by making use of the foreach command:

   ```
   $ foreach i(*.fastq.gz)
         echo $i
         set name = `basename $i fastq.gz`
         mkdir ${name}
         gunzip -c ${name}.fastq.gz >${name}.fastq
         tophat2 -o ./${name} genomename ${name}.
         fastq
         rm ${name}.fastq
         end
   ```

10. By adding the `--compatible-hits-norm` option, Cufflinks will normalize the gene expression according to the number of hits within the reference genome, and not the total amount of reads, as is the default. For more detailed information about the options available in Cufflinks, use the help-command:

    ```
    $ cufflinks -h
    ```

    Furthermore, it is important that the headers of the genome annotation file are the same as the headers in the `accepted_hits.bam` file. If they are not the same, Cufflinks will give the expression of all genes as 0 FPKM.

11. Read counting can be automated for all the TopHat output files by making use of the foreach command:

    ```
    $ foreach i(dirname)
          set dirname = `basename $i`
          echo ${dirname}
          cufflinks -v --compatible-hits-norm -u
          -o ./${dirname} -G genome.gff3
          ${dirname}/accepted_hits.bam
          end
    ```

Acknowledgements

This work was supported by the European Framework Programme 7 project SMARTCELL (FP7 KBBE 222716).

References

1. De Geyter N, Gholami A, Goormachtig S, Goossens A (2012) Transcriptional machineries in jasmonate-elicited plant secondary metabolism. Trends Plant Sci 17:349–359
2. Pauwels L, Inzé D, Goossens A (2009) Jasmonate-inducible gene: what does it mean? Trends Plant Sci 14:87–91
3. Pollier J, Moses T, Goossens A (2011) Combinatorial biosynthesis in plants: a (p) review on its potential and future exploitation. Nat Prod Rep 28:1897–1916
4. Goossens A, Häkkinen ST, Laakso I, Seppänen-Laakso T, Biondi S, De Sutter V, Lammertyn F, Nuutila AM, Söderlund H, Zabeau M, Inzé D, Oksman-Caldentey KM (2003) A functional genomics approach toward the understanding of secondary metabolism in plant cells. Proc Natl Acad Sci USA 100:8595–8600
5. De Boer K, Tilleman S, Pauwels L, Vanden Bossche R, De Sutter V, Vanderhaeghen R, Hilson P, Hamill JD, Goossens A (2011) APETALA2/ETHYLENE RESPONSE FACTOR and basic helix-loop-helix tobacco transcription factors cooperatively mediate jasmonate-elicited nicotine biosynthesis. Plant J 66:1053–1065
6. De Sutter V, Vanderhaeghen R, Tilleman S, Lammertyn F, Vanhoutte I, Karimi M, Inzé D, Goossens A, Hilson P (2005) Exploration of jasmonate signalling via automated and standardized transient expression assays in tobacco cells. Plant J 44:1065–1076
7. Häkkinen ST, Tilleman S, Świątek A, De Sutter V, Rischer H, Vanhoutte I, Van Onckelen H, Hilson P, Inzé D, Oksman-Caldentey KM, Goossens A (2007) Functional characterisation of genes involved in pyridine alkaloid biosynthesis in tobacco. Phytochemistry 68: 2773–2785
8. Lackman P, González-Guzmán M, Tilleman S, Carqueijeiro I, Cuéllar Pérez A, Moses T, Seo M, Kanno Y, Häkkinen ST, Van Montagu MCE, Thevelein JM, Maaheimo H, Oksman-Caldentey KM, Rodriguez PL, Rischer H, Goossens A (2011) Jasmonate signaling involves the abscisic acid receptor PYL4 to regulate metabolic reprogramming in *Arabidopsis* and tobacco. Proc Natl Acad Sci USA 108:5891–5896
9. Morita M, Shitan N, Sawada K, Van Montagu MCE, Inzé D, Rischer H, Goossens A, Oksman-Caldentey KM, Moriyama Y, Yazaki K (2009) Vacuolar transport of nicotine is mediated by a multidrug and toxic compound extrusion (MATE) transporter in *Nicotiana tabacum*. Proc Natl Acad Sci USA 106: 2447–2452
10. Bachem CWB, van der Hoeven RS, de Bruijn SM, Vreugdenhil D, Zabeau M, Visser RGF (1996) Visualization of differential gene expression using a novel method of RNA fingerprinting based on AFLP: analysis of gene expression during potato tuber development. Plant J 9: 745–753
11. Velculescu VE, Zhang L, Vogelstien B, Kinzler KW (1995) Serial analysis of gene expression. Science 270:484–487
12. Brenner S, Johnson M, Bridgham J, Golda G, Lloyd DH, Johnson D, Luo S, McCurdy S, Foy M, Ewan M, Roth R, George D, Eletr S, Albrecht G, Vermaas E, Williams SR, Moon K, Burcham T, Pallas M, DuBridge RB, Kirchner J, Fearon K, Mao J, Corcoran K (2000) Gene expression analysis by massively parallel signature sequencing (MPSS) on microbead arrays. Nat Biotechnol 18: 630–634
13. Wang Z, Gerstein M, Snyder M (2009) RNA-Seq: a revolutionary tool for transcriptomics. Nat Rev Genet 10:57–63
14. Mortazavi A, Williams BA, McCue K, Schaeffer L, Wold B (2008) Mapping and quantifying mammalian transcriptomes by RNA-Seq. Nat Methods 5:621–628
15. Trapnell C, Pachter L, Salzberg SL (2009) TopHat: discovering splice junctions with RNA-Seq. Bioinformatics 25:1105–1111
16. Trapnell C, Roberts A, Goff L, Pertea G, Kim D, Kelley DR, Pimentel H, Salzberg SL, Rinn JL, Pachter L (2012) Differential gene and transcript expression analysis of RNA-seq experiments with TopHat and Cufflinks. Nat Protoc 7:562–578
17. Garber M, Grabherr MG, Guttman M, Trapnell C (2011) Computational methods for transcriptome annotation and quantification using RNA-seq. Nat Methods 8:469–477

18. Trapnell C, Williams BA, Pertea G, Mortazavi A, Kwan G, van Baren MJ, Salzberg SL, Wold BJ, Pachter L (2010) Transcript assembly and quantification by RNA-Seq reveals unannotated transcripts and isoform switching during cell differentiation. Nat Biotechnol 28:511–515
19. Wang L, Wang S, Li W (2012) RSeQC: quality control of RNA-seq experiments. Bioinformatics 28:2184–2185
20. Patel RK, Jain M (2012) NGS QC Toolkit: a toolkit for quality control of next generation sequencing data. PLoS One 7:e30619
21. Langmead B, Trapnell C, Pop M, Salzberg SL (2009) Ultrafast and memory-efficient alignment of short DNA sequences to the human genome. Genome Biol 10:R25
22. Langmead B, Salzberg SL (2012) Fast gapped-read alignment with Bowtie 2. Nat Methods 9:357–359
23. Li H, Handsaker B, Wysoker A, Fennell T, Ruan J, Homer N, Marth G, Abecasis G, Durbin R, Subgroup GPDP (2009) The sequence alignment/map format and SAMtools. Bioinformatics 25:2078–2079

Chapter 25

Transcriptome Coexpression Analysis Using ATTED-II for Integrated Transcriptomic/Metabolomic Analysis

Keiko Yonekura-Sakakibara and Kazuki Saito

Abstract

Transcriptome coexpression analysis is an excellent tool for predicting the physiological functions of genes. It is based on the "guilt-by-association" principle. Generally, genes involved in certain metabolic processes are coordinately regulated. In other words, coexpressed genes tend to be involved in common or closely related biological processes. Genes of which the metabolic functions have been identified are preselected as "guide" genes and are used to check the transcriptome coexpression fidelity to the pathway and to determine the threshold value of correlation coefficients to be used for subsequent analysis. The coexpression analysis provides a network of the relationships between "guide" and candidate genes that serves to create the criteria by which gene functions can be predicted. Here we describe a procedure to narrow down the number of candidate genes by means of the publicly available database, designated *Arabidopsis thaliana trans*-factor and *cis*-element prediction database (ATTED-II).

Key words Coexpression analysis, ATTED-II, Correlation coefficients, Transcriptomics, Metabolomics, *Arabidopsis thaliana*, Plant

1 Introduction

The proliferation of plant genome sequencing projects and the subsequent development of high-throughput technologies, including DNA microarrays, have generated a massive amount of biological data sets. In an effort to make these rapidly increasing data sets publicly available, functional genomics data repositories were established, such as Gene Expression Omnibus (GEO, http://www.ncbi.nlm.nih.gov/geo) and ArrayExpress (http://www.ebi.ac.uk/arrayexpress). As of April 2012, over 700,000 microarray data are available through GEO. As a secondary analytical tool, data-mining informatics provides a manner to develop gene coexpression databases.

Gene coexpression databases have been established on publicly available microarray data sets. They provide a list of coexpressed

Alain Goossens and Laurens Pauwels (eds.), *Jasmonate Signaling: Methods and Protocols*, Methods in Molecular Biology, vol. 1011, DOI 10.1007/978-1-62703-414-2_25,

genes and the degree of similarity between gene expression patterns, generally by Pearson's correlation coefficients, Spearman's correlation coefficients, or unique calculated scores between any two genes. Various gene coexpression databases for plants are accessible publicly, such as *Arabidopsis* co-expression tool (ACT) [1], *Arabidopsis* Systems Interaction Database (ASIDB) [2], ATTED-II, Botany Array Resource (BAR) Expression Angler [3], Co-Expression analysis for *Arabidopsis* (CressExpress) [4], Comprehensive Systems-Biology Database (CSB.DB) [5], Gene Co-Expression Analysis Toolbox (GeneCAT) [6], and Plant Gene Expression Database (PED) [7]. Details about these databases, including data sources, calculation methods, and data retrieval tools, are available [8].

Transcriptome coexpression analysis is based on the hypothesis that genes in the same and/or nearby pathway are regulated in a coordinated manner. In other words, coexpressed genes tend to contribute to common or closely related biological processes. Genes of which the expression patterns are highly similar to that of genes with determined functions, the so-called guide genes, can be selected as potential target genes because the probability of their involvement in a common or related pathway is high (Fig. 1). The degree of gene coexpression similarity is measured by correlation coefficients (Table 1). So far, genes encoding enzymes, transcription factors, and complex-forming proteins have been identified based on transcriptome coexpression analysis [8]. Coordinate expression is especially pronounced in plant secondary metabolism. By means of ATTED-II, flavonoid modification enzymes have been determined functionally from among 107 candidates and

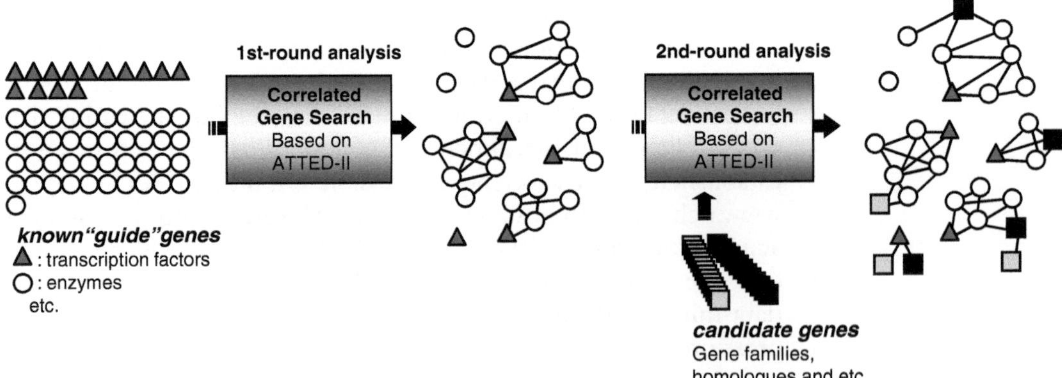

Fig. 1 Basic concept of transcriptome coexpression analysis. Genes that have been identified functionally are used as known "guide" genes. In a first-round analysis, the threshold values are determined for the second-round analysis that is conducted to search candidate target genes. Positive correlations ($r >$ defined value) are indicated by connecting lines

Table 1
Example of a general interpretation of Pearson's correlation coefficients [15]

Pearson's correlation coefficients, r	Degree of correlation[a]
$\|r\|=1.0$	Perfect
$1>\|r\|\geq 0.7$	Strong
$0.7>\|r\|\geq 0.3$	Moderate
$0.3>\|r\|>0$	Weak
$\|r\|=0$	None

[a]The range may vary slightly according to the references

MYB and biosynthetic enzymes in the glucosinolate pathway as well [9–11]. Many examples of coexpression-based identification with ATTED-II are shown on the ATTED-II homepage (http://atted.jp/top_publication.shtml). Here, we describe how to assess the effectiveness of transcriptome coexpression analysis for a given metabolic pathway and to narrow a field of candidate genes by using ATTED-II [12–14].

2 Materials

2.1 Software for Transcriptome Coexpression Analysis

1. ATTED-II (http://atted.jp/).
2. PRIMe, Correlated Gene Search (http://prime.psc.riken.jp/?action=coexpression_index).

2.2 Software for Analysis and Visualization of Networks

1. Pajek (http://vlado.fmf.uni-lj.si/pub/networks/pajek/).
2. BioLayout (http://www.biolayout.org/).
3. Cytoscape (http://www.cytoscape.org/).

3 Methods

3.1 Coexpression Analysis to Generate Hypothesis

ATTED-II can be used to obtain a quick overview of gene coexpression for a given biological process.

1. Open ATTED-II (http://atted.jp/).
2. Enter keyword, GO ID, gene alias, or gene ID and press the search button (*see* **Note 1**). Functional categories and/or loci matching the search are shown.
3. In the Gene Ontology (GO) term results, click the hyperlinks "list" (*see* **Note 2**) and/or "network" (*see* **Note 3**, Fig. 2) for further information.

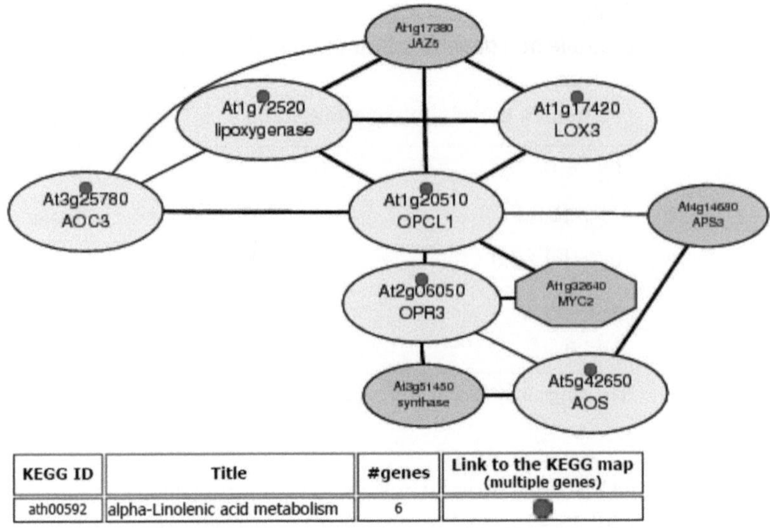

KEGG ID	Title	#genes	Link to the KEGG map (multiple genes)
ath00592	alpha-Linolenic acid metabolism	6	●

The gene list in the network

search coexpressed genes from the selected genes below			
	Locus	**Function**	**Link to the KEGG map (single gene)**
☐	At1g17380	JAZ5 (JASMONATE-ZIM-DOMAIN PROTEIN 5)	
☐	At1g17420	LOX3	●
☐	At1g20510	OPCL1 (OPC-8:0 COA LIGASE1)	●
☐	At1g32640	MYC2	
☐	At1g72520	lipoxygenase, putative	●
☐	At2g06050	OPR3 (OPDA-REDUCTASE 3)	●
☐	At3g25780	AOC3 (ALLENE OXIDE CYCLASE 3)	●
☐	At3g51450	strictosidine synthase family protein	
☐	At4g14680	APS3	
☐	At5g42650	AOS (ALLENE OXIDE SYNTHASE)	●

Fig. 2 ATTED-II Network and gene list output of the GO term jasmonic acid metabolic process (GO:0009694) as of August 2012

4. Within the obtained Locus search results, click the "locus" hyperlink to get more details regarding functional annotations, gene coexpression, gene expression, and predicted *cis*-elements for genes of interest. If there are coexpressed gene networks, transcriptome coexpression analysis may be applicable (*see* Subheading 3.2).

3.2 Transcriptome Coexpression Analysis to Narrow Down Candidate Genes Involved in the Pathway of Interest

1. Collect "guide" genes. These are functionally identified genes known to be in the pathway of interest. Having more guide genes may increase the reliability of the pool of candidate genes. As examples, genes involved in flavonol or jasmonate biosynthesis are listed in Table 2.
2. Check (1) the presence of probes corresponding to guide genes on the microarray chip and (2) the possibility of cross hybridization with these guide genes (*see* **Note 4**).

Table 2
"Guide" genes in flavonol and jasmonate metabolism

Function	Abbreviation	AGI
Flavonol metabolism		
4-Coumarate:CoA ligase	*4CL3*	At1g65060
Chalcone synthase	*CHS*	At5g13930
Chalcone isomerase	*CHI*	At3g55120
Flavanone 3-hydroxylase	*F3H*	At3g51240
Flavonoid 3′-hydroxylase	*F3 H*	At5g07990
Flavonol synthase	*FLS*	At5g08640
Jasmonate metabolism		
Lipoxygenase[a]	*LOX3*	At1g17420
Lipoxygenase[a]	*LOX4*	At1g72520
Allene oxide synthase[a]	*AOS*	At5g42650
Allene oxide cyclase[a]	*AOC3*	At3g25780
Oxophytodienoic acid reductase[a]	*OPR3*	At2g06050
OPC-8:0 CoA ligase1[a]	*OPCL1*	At1g20510

[a]Genes linked to the Kyoto Encyclopedia of Genes and Genomes (KEGG) map in the gene list of GO:0009694 (jasmonic acid metabolic process) in ATTED-II (Fig. 2) were used as "guide" genes for jasmonate metabolism

3. Remove the genes from the list if either the probes corresponding to the guide genes are absent on the microarray chip or cross hybridization to the guide genes is highly possible.
4. Open "Correlated Gene Search" in PRIMe (http://prime.psc.riken.jp/?action=coexpression_index).
5. Enter the *Arabidopsis* Genome Initiative (AGI) codes of guide genes in the Locus ID column.
6. Set parameters, such as Matrix, Methods, and Threshold value, and press Search button (*see* **Note 5**).
7. To check all correlation coefficient values between guide genes, set Matrix at "data sets v.3," Methods at "interconnection of sets," and Threshold value to −1 (*see* **Note 6**).
8. Set Format to HTML.
9. An Order and Display limit can be selected.
10. Check all correlation coefficient values between guide genes and determine the threshold value used for further analysis (*see* **Notes 7–9**).

11. Prepare a list of the candidate genes. Gene families, paralogs, homologs, and/or pathway-related genes are frequently used as candidate genes (*see* **Notes 10** and **11**).
12. Enter the combined list of AGI codes of the guide and candidate genes in the Locus ID column of the Correlated Gene Search. Set the threshold value of correlation coefficients determined above using the guide genes (*see* Subheading 3.2, **step 10**).
13. Set other parameters and press the Search button. "Interconnection of sets" should be chosen in Methods, and the same Matrix data sets should be used as when the threshold value was chosen (*see* **Note 8**). As examples, the networks using guide genes in flavonol metabolism (Table 2) and 120 family-1 glycosyltransferase genes as candidate genes (Matrix: all data set v.3 (1388 data), Methods: interconnection of sets, threshold value 0.667) and guide genes in jasmonate metabolism (Table 2) and all Arabidopsis genes (Matrix: all data set v.3 (1388 data), Methods: union of sets, threshold value 0.669) are shown in Fig. 3.
14. Look at an alternative perspective of candidate genes, i.e., gene annotations, primary sequences, gene expression profiles, etc. (*see* **Note 12**).
15. Select target genes for further analysis (*see* **Note 13**).

4 Notes

1. If there is no hit with a GO term search, check the word used in GO released by The Arabidopsis Information Resource (TAIR) that was used in ATTED. For example, input of "jasmonic acid" should yield nine functional categories and 34 loci and input of "jasmonate" should yield one functional category and 39 loci.
2. "List" shows the list of all the genes with the GO term. In addition, coexpressed genes among the selected genes can be searched by selecting them and pressing "search coexpressed genes." The identical network shown in "network" (*see* **Note 3**) and unconnected genes are shown.
3. "Network" shows the network of all coexpressed genes in the GO term and the list of genes in a coexpressed gene network. Unconnected genes are omitted from the network. As example, network of "GO:0009694" (jasmonic acid metabolic process) is shown in Fig. 2.
4. Not all genes are represented on the ATH1 microarray. The possibility of cross hybridization of guide genes can be checked in the Affymetrix NetAffx™ Analysis Center (http://www.affymetrix.com/analysis/index.affx). If the guide genes cross-hybridize with other genes, their expression patterns are not

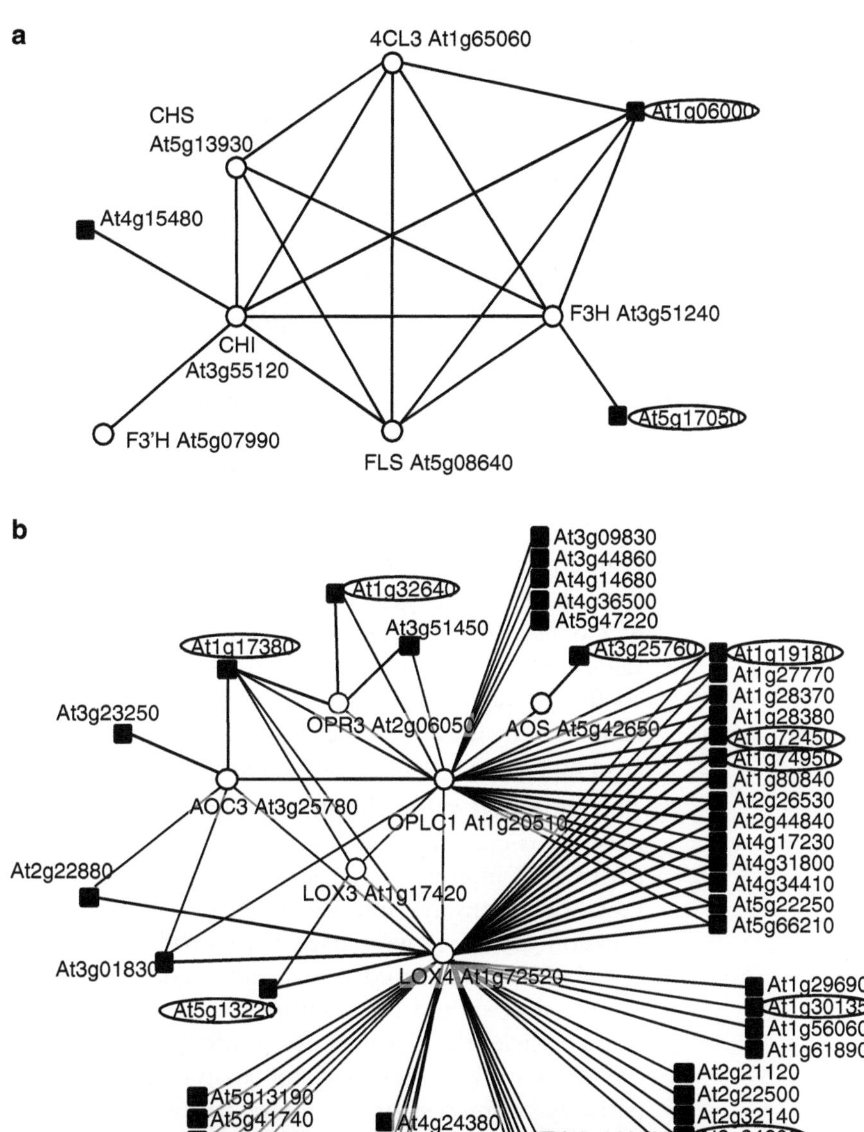

Fig. 3 Coexpression relationships of genes in flavonol and jasmonate synthesis pathways. White and black circles indicate "guide" and candidate genes, respectively. (**a**) Genes in flavonol metabolism (Table 2) and 120 family-1 glycosyltransferase genes [19] are used for analysis with Matrix ("all data set v.3 (1388 data)"), Methods ("interconnection of sets"), and threshold value (0.667). The genes surrounded by circles were identified as flavonoid 3-*O*-glucosyltransferase (At5g17050) and flavonol 7-*O*-rhamnosyltransferase (At1g06000) based on transcriptomics and transcriptome coexpression analyses [4, 20]. (**b**) Genes in jasmonate metabolism (Table 2) are used with Matrix ("all data set v.3 (1388 data)"), Methods ("union of sets"), and threshold value (0.669). The genes surrounded by circles were allene oxide cyclase (*AOC1*, At3g25760), *MYC2* (At1g32640), and jasmonate ZIM-domain (JAZ) proteins (*JAZ1*, At1g19180; *JAZ2*, At1g74950; *JAZ5*, At1g17380; *JAZ6*, At1g72450; *JAZ7*, At2g34600; *JAZ8*, At1g30135; and *JAZ10*, At5g13220)

suitable for analysis. If needed, the contribution of each gene can be estimated by tiling data through TileViz (http://jsp.weigelworld.org/tileviz/tileviz.jsp).

5. A matrix can be chosen among "All data sets v.3 (1388 data)," "All data sets v.1 (771 data)," "Hormone treatments v.1 (236 data)," "Tissue and development v.1 (237 data)," and "Stress treatments v.1 (298data)." At this point, the best Matrix is determined by trial and error.
6. "Interconnection of sets" reports on correlated gene pairs amongst the queried genes only.
7. Normally, the first choice is the lowest correlation coefficient value that is enough to minimally connect all guide genes. General statistical descriptions about relationship and correlation coefficient values are shown in Table 1 [15]. Biologically significant relationships are expected to be above a threshold value from 0.55 to 0.66 [16]. The lowest values that minimally connect all guide genes involved in flavonol and jasmonate metabolism (Table 2) are 0.667 and 0.669, respectively. These values will be used for later analyses. If the lowest correlation coefficient value that is enough to minimally connect all guide genes is quite lower that the above values, omit some of the causative guide genes that are not as likely to be coregulated.
8. The threshold value should be adjusted as the coexpression network is consistent with a known regulatory system for the pathway of interest. The threshold value is dependent on the metabolic pathways and Matrix (data set) used. If gene coexpression databases were used with smaller data sets, the least correlations considered statistically significant, depending on sample size, should be taken into account [17]. If all correlation coefficient values between guide genes are too low, either the Matrix (data set) used for analysis is unsuitable for the pathway of interest or the target pathway is not regulated at the transcriptional level.
9. To understand the relationships between genes at a glance, results can be saved and visualized with network visualization programs, such as Pajek (http://pajek.imfm.si/doku.php) and BioLayout (http://www.biolayout.org/). Instructions on how to use Pajek are given in http://vlado.fmf.uni-lj.si/pub/networks/pajek/howto.htm. Also useful is Cytoscape (http://www.cytoscape.org/).
10. For a nontargeted analysis, the entire *Arabidopsis* genes can be used as the candidate list (*see* **Note 11**).
11. To use all *Arabidopsis* genes as the candidates, input AGI codes of guide genes only in Locus IDs column and set Methods at "union of sets." The latter method searches for all genes correlated with any of the queried genes.

12. In the case of a nontargeted gene search, check the full locus detail information, especially the publication, in TAIR. Annotations released by TAIR are shortened in ATTED-II.
13. For proof of concept of transcriptome coexpression analyses for functional genomics and pitfalls and limitations of this method, *see* Saito et al. [18].

Acknowledgments

We would like to thank Drs. A. Fukushima and Y. Sasaki-Sekimoto for their helpful comments and Dr. T. Obayashi for kind permission of the use of figures on ATTED-II.

References

1. Manfield IW, Jen C-H, Pinney JW, Michalopoulos I, Bradford JR, Gilmartin PM, Westhead DR (2006) *Arabidopsis* Co-expression Tool (ACT): web server tools for microarray-based gene expression analysis. Nucleic Acids Res 34:W504–W509
2. Rawat A, Seifert GJ, Deng Y (2008) Novel implementation of conditional co-regulation by graph theory to derive co-expressed genes from microarray data. BMC Bioinformatics 9:S7
3. Toufighi K, Brady SM, Austin R, Ly E, Provart NJ (2005) The Botany Array Resource: e-Northerns, Expression Angling, and promoter analyses. Plant J 43:153–163
4. Srinivasasainagendra V, Page GP, Mehta T, Coulibaly I, Loraine AE (2008) CressExpress: a tool for large-scale mining of expression data from Arabidopsis. Plant Physiol 147:1004–1016
5. Steinhauser D, Usadel B, Luedemann A, Thimm O, Kopka J (2004) CSB.DB: a comprehensive systems-biology database. Bioinformatics 20:3647–3651
6. Mutwil M, Øbro J, Willats WGT, Persson S (2008) GeneCAT–novel webtools that combine BLAST and co-expression analyses. Nucleic Acids Res 36:W320–W326
7. Horan K, Jang C, Bailey-Serres J, Mittler R, Shelton C, Harper JF, Zhu J-K, Cushman JC, Gollery M, Girke T (2008) Annotating genes of known and unknown function by large-scale coexpression analysis. Plant Physiol 147:41–57
8. Usadel B, Obayashi T, Mutwil M, Giorgi FM, Bassel GW, Tanimoto M, Chow A, Steinhauser D, Persson S, Provart NJ (2009) Co-expression tools for plant biology: opportunities for hypothesis generation and caveats. Plant Cell Environ 32:1633–1651
9. Hirai MY, Sugiyama K, Sawada Y, Tohge T, Obayashi T, Suzuki A, Araki R, Sakurai N, Suzuki H, Aoki K, Goda H, Nishizawa OI, Shibata D, Saito K (2007) Omics-based identification of *Arabidopsis* Myb transcription factors regulating aliphatic glucosinolate biosynthesis. Proc Natl Acad Sci USA 104:6478–6483
10. Yonekura-Sakakibara K, Tohge T, Matsuda F, Nakabayashi R, Takayama H, Niida R, Watanabe-Takahashi A, Inoue E, Saito K (2008) Comprehensive flavonol profiling and transcriptome coexpression analysis leading to decoding gene-metabolite correlations in *Arabidopsis*. Plant Cell 20:2160–2176
11. Yonekura-Sakakibara K, Tohge T, Niida R, Saito K (2007) Identification of a flavonol 7-*O*-rhamnosyltransferase gene determining flavonoid pattern in *Arabidopsis* by transcriptome coexpression analysis and reverse genetics. J Biol Chem 282:14932–14941
12. Obayashi T, Hayashi S, Saeki M, Ohta H, Kinoshita K (2009) ATTED-II provides coexpressed gene networks for Arabidopsis. Nucleic Acids Res 37:D987–D991
13. Obayashi T, Kinoshita K, Nakai K, Shibaoka M, Hayashi S, Saeki M, Shibata D, Saito K, Ohta H (2007) ATTED-II: a database of co-expressed genes and *cis* elements for identifying co-regulated gene groups in *Arabidopsis*. Nucleic Acids Res 35:D863–D869
14. Obayashi T, Nishida K, Kasahara K, Kinoshita K (2011) ATTED-II updates: condition-specific gene coexpression to extend coexpression analyses and applications to a broad range of flowering plants. Plant Cell Physiol 52:213–219

15. Jackson SL (2011) Correlational methods and statistics. In: Jackson SL (ed) *Research Methods and Statistics: A Critical Thinking Approach*. Belmont, CA, Wadsworth, pp 147–170
16. Aoki K, Ogata Y, Shibata D (2007) Approaches for extracting practical information from gene co-expression networks in plant biology. Plant Cell Physiol 48:381–390
17. Frey B (2006) Discovering relationships. In: Frey B (ed) Statistical hacks: tips and tools for measuring the world and beating the Odds. O'Reilly Media, Sebastopol, CA, pp 41–95
18. Saito K, Hirai MY, Yonekura-Sakakibara K (2008) Decoding genes with coexpression networks and metabolomics—"majority report by precogs". Trends Plant Sci 13:36–43
19. Paquette S, Møller BL, Bak S (2003) On the origin of family 1 plant glycosyltransferases. Phytochemistry 62:399–413
20. Tohge T, Nishiyama Y, Hirai MY, Yano M, J-i N, Awazuhara M, Inoue E, Takahashi H, Goodenowe DB, Kitayama M, Noji M, Yamazaki M, Saito K (2005) Functional genomics by integrated analysis of metabolome and transcriptome of Arabidopsis plants over-expressing an MYB transcription factor. Plant J 42:218–235

Chapter 26

A Guide to CORNET for the Construction of Coexpression and Protein–Protein Interaction Networks

Stefanie De Bodt and Dirk Inzé

Abstract

To enable easy access and interpretation of heterogenous and scattered data, we have developed a user-friendly tool for data mining and integration in *Arabidopsis thaliana*, designated CORrelation NETworks (acronym CORNET), allowing browsing of microarray data, construction of coexpression and protein–protein interactions (PPIs), analysis of gene association and transcription factor (TF) regulatory networks, and exploration of diverse functional annotations. CORNET consists of three tools that can be used individually or in combination, namely, the coexpression tool, the PPI tool, and the TF tool. Different search options are implemented to enable the creation of networks centered around multiple input genes or proteins. Functional annotation resources are included to retrieve relevant literature, phenotypes, localization, gene ontology, plant ontology, and biological pathways. Networks and associated evidence of the majority of the currently available data types are visualized in Cytoscape. CORNET is available at https://bioinformatics.psb.ugent.be/cornet.

Key words Coexpression, Protein–protein interactions, Networks, Plants

1 Introduction

In recent years, plant biology has witnessed a true data explosion. However, these data can only be exploited to their full use through data integration, thereby leading to, for instance, the identification of the temporal and spatial activities of protein complexes and the prediction of putative functions for unknown genes [1–5]. To overcome problems in formatting, quality, and integration of data, we developed a user-friendly tool for data mining and integration, designated CORNET, acronym for CORrelation NETworks [6, 7]. In a central database, we collected data on microarray expression with corresponding metadata, describing sampling of tissues, treatments and time points, protein–protein interactions (PPIs), gene–gene associations, regulatory interactions, localization, and functional information. A user-friendly interface allows

Alain Goossens and Laurens Pauwels (eds.), *Jasmonate Signaling: Methods and Protocols*, Methods in Molecular Biology, vol. 1011, DOI 10.1007/978-1-62703-414-2_26,

us to query the database, enabling network construction through a multitude of search options that address different biological questions. Coexpression networks can be obtained by means of user-defined and multiple predefined expression datasets. PPI, TF, and gene association networks can be constructed with both experimentally identified and computationally predicted data. The search options in CORNET are very extensive and flexible. A comprehensive visualization of the networks is generated in Cytoscape, providing a bird's eye view on the results and on the different degrees of reliability of the extracted information [8].

This chapter describes the use of CORNET that consists of the coexpression, PPI, and TF tools to construct molecular networks. As an example, the reader will be guided through the different steps needed to generate and visualize a coexpression network on the jasmonate (JA) ZIM domain (JAZ) protein targets and a PPI network around JAZ proteins.

2 Materials

2.1 Microarray Data

1. The *Arabidopsis thaliana* microarray data from the GeneChip Arabidopsis ATH1 Genome Array (www.affymetrix.com) retrieved from the Gene Expression Omnibus (GEO) database (www.ncbi.nlm.nih.gov/geo) and processed with the Robust Multi-array Average (RMA) procedure implemented in BioConductor [9–12].
2. An up-to-date configuration data file based on The Arabidopsis Information Resource (TAIR; Stanford, CA, USA) (21,428 TAIR10 genes—v14) and provided by Brainarray (www.brainarray.mbni.med.umich.edu) to define relations between probe sets and genes. Only experiments are included comprising two or more replicates.
3. Metadata described by ontology terms and experimental design types (e.g., development_or_differentiation_design, genetic_modification_design, compound_treatment_design, abiotic_stress_design, biotic_stress_design, time_series_design, hormone_treatment_design).
4. Fourteen predefined expression datasets in the coexpression tool: The global expression datasets (AtGenExpress [http://www.weigelworld.org/resources/microarray/AtGenExpress/], Microarray compendium 1, and Microarray compendium 2) and the specific expression datasets (Abiotic stress, Biotic stress, Development, Flower, Genetic modification, Hormone treatment, Leaf, Root, Seed, Stress (abiotic+biotic), and Whole plant).

2.2 Expression Correlation Data

1. To quantify the similarity in expression profiles, the commonly used correlation coefficients of Pearson (PCCs) and Spearman

are calculated. Pearson's method is parametric and based on actual expression values, whereas Spearman's is nonparametric and based on ranks. Both measures range from −1 (anti-correlation) over 0 (no correlation) to +1 (correlation). Only PCCs can be used when multiple expression datasets are selected. In the CORNET database, PCC>0.4 and PCC <−0.4 are stored.

2.3 Protein–Protein Interaction Data

1. Currently, the available experimentally identified PPIs for *Arabidopsis* are from BIND [13], IntAct [14], BioGRID [15], DIP [16], MINT [17], TAIR [18], MIND [19], G-protein interactome [20], and the Arabidopsis Interactome Mapping Consortium [21].
2. The predicted PPIs are from BAR [22], AtPID [23], and ref. 24.

2.4 Gene–Gene Association Data

1. Gene–gene associations, most often identified through computational approaches, represent possible functional links between genes or proteins that do not necessarily interact physically. The AraNet Arabidopsis probabilistic functional gene network (http://www.functionalnet.org/aranet/) is included representing inferred gene–gene associations based on the integration of diverse functional genomics, proteomics, and comparative genomics datasets [25].

2.5 Regulatory Interaction Data

1. The dataset of the AtRegNet data from The Arabidopsis Gene Regulatory Information server (AGRIS) (http://arabidopsis.med.ohio-state.edu/) that contains direct and indirect as well as confirmed and unconfirmed regulatory interactions between TFs and their target genes [26].
2. To computationally identify possible regulatory interactions, microarray data are extracted from the CORNET database that have been annotated with "genetic_modification_design," i.e., experiments in which transgenic plants were profiled. By means of the metadata stored, transgenic and wild-type expression profiles were compared and differentially expressed genes were identified with the BioConductor package *Limma* [27]. Differentially expressed genes show an absolute fold change >2 and a false discovery rate <0.05.

2.6 Functional Annotation Data

1. Functional data consist of localization (predicted and experimental) data, Gene Ontology (GO) annotation, Plant Ontology (PO) annotation, MapMan pathways and processes, PubMed IDs, TAIR phenotypes, and InterPro protein domains [18, 28–32].

3 Methods

3.1 A Quick Tour Around CORNET

The different functionalities of CORNET can be accessed through the menu bar. From left to right, the home page, the coexpression tool, the PPI tool, the TF tool, the Browse experiments page, the Upload and Edit page, and the FAQ page are displayed (*see* Fig. 1a and **Note 1**). The three tools function similarly because the main interface consists of three steps in which the user needs to specify the input genes or proteins, the data to be retrieved, and the search options. In addition to these three tools, CORNET has the functionality to browse the microarray data that are available in the database by means of the ontology terms assigned to describe the experimental conditions. Moreover, the user can upload and edit personal microarray data that can be used in the coexpression tool. Finally, the FAQ page contains information on the data sources, the methodologies, and the functioning of the different tools.

3.2 Construction of Coexpression Networks Around JAZ-Regulated Genes

Coexpression can be retrieved between a set of genes ("Pairwise correlations"), between one or more input gene(s) with all other *Arabidopsis* genes ("Correlations of query gene(s) with neighbors"), and between these other genes ("Correlations between neighbors"). The nature of the expression dataset used to assess coexpression has been shown to be very important [6] (*see* **Note 2**). CORNET allows the user to choose one or more predefined expression datasets (for particular design types or sets of conditions) or to input a user-defined expression dataset.

As an example, we investigate to which extent genes controlled through JA signaling are coexpressed. The input genes are *GLABRA1* (*GL1*), *GL3*, *ENHANCER OF GLABRA3* (*EGL3*), *TRANSPARENT TESTA8* (*TT8*), and *MYB75*, known to belong to the WD-Repeat/bHLH/MYB complex with well-established roles in JA-mediated trichome development and anthocyanin accumulation and negatively regulated by JAZ proteins (*see* Table 1) [33–36]. With the coexpression analysis in CORNET, we can study whether these genes are tightly transcriptionally regulated to enable complex formation and specificity. We will calculate coexpression using the available predefined expression datasets and study the influence of the different search options and expression datasets.

3.2.1 Coexpression Using Global Expression Datasets

1. In the coexpression tool, first reach the page with a description of the tool and the expression datasets used.
2. On this page, select whether the coexpression study will be done with predefined or user-defined expression datasets.
3. Click on "predefined" to analyze the coexpression of the input genes and other *Arabidopsis* genes by using multiple expression datasets.

Fig. 1 Overview of the CORNET coexpression tool. (**a**) Info page. (**b**) Main interface. (**c**) Output page

Table 1
JAZ-regulated genes used as input for the construction of a CORNET coexpression network

AGI identifier	Gene name
AT3G27920	*GL1, ATMYB0*
AT1G63650	*EGL3*
AT5G41315	*GL3*
AT4G09820	*TT8, bHLH42*
AT1G56650	*MYB75, PAP1*

4. Subsequently, in **step 1** of the main interface of the coexpression tool (*see* Fig. 1b), paste the AGI identifiers of the JAZ-regulated genes (*see* Table 1).
5. In **step 2**, select one or more predefined expression datasets, among which three global (compendium 1, compendium 2, and AtGenExpress) and 11 specific expression datasets represent diverse experimental conditions and particular plant organs, treatments, or biological processes, respectively (*see* FAQ for more details) (*see* Fig. 1b). Here, select first the global expression datasets.
6. In **step 3**, provide default parameters for the PCC threshold (>0.8), the *p*-value for significance of correlation (<0.05), and the number of neighbors (top genes ≤10) to be used in this first run of the coexpression tool (*see* Fig. 1b) (*see* **Notes 3–5**).
7. In addition, indicate that "Correlations with neighbors" and "Correlations between neighbors" will be searched.
8. Start the coexpression tool by clicking the "GO" button.
9. On the next page, check the warning that gene *AT5G41315* does not occur on the microarray (*see* Fig. 1c), possibly due to the fact that the gene was not known at the time of the microarray design and, consequently, that no adequate probes were present or no probes uniquely match this gene (e.g., because of a close homolog).
10. Click on one of the links to the heatmap viewer that displays the expression profiles of the input genes in the selected expression datasets and open a new page with a clustered heatmap. The legend for the visual representation in Cytoscape shows, for instance, that the edge color of the coexpression network corresponds to the degree of coexpression (i.e., the correlation coefficient) and that query genes are presented as diamonds.
11. When the correlation coefficients are calculated (in case one expression dataset is selected) or retrieved from the CORNET database (in case multiple expression datasets are selected), allow a new window to pop up (*see* **Note 1**).

12. Click "OK" by which the cy.jnlp file is opened with the Java Webstart Launcher that will download the network file and start Cytoscape.
13. Subsequently, allow the network to be shown in Cytoscape and the default "Organic Layout" to be applied, generating a small coexpression network of 16 genes and 64 coexpression links. Only two of the original query genes (diamonds) are present in this coexpression network.
14. Functional enrichment and the inclusion of known related genes can be used to evaluate the obtained network (*see* **Note 5**).

3.2.2 Coexpression Using Specific Expression Datasets

Although the above-described network already contains useful information, more specific predefined expression datasets will be used besides the global expression datasets, namely, the leaf, development, and hormone expression datasets that are relevant for the genes under study. The other parameter settings are the same as in the first run (*see* Subheading 3.2.1, **step 6**).

1. Generate a coexpression network of 34 genes and 175 coexpression links. Of the four query genes present in the network, *EGL3* and *ATMYB0* (or *GL1*) are part of the same coexpression cluster, whereas *TT8* is part of a second cluster and *PAP1* (or *MYB75*) is coexpressed only with one unknown gene.
2. In the next run, change the PCC and top gene thresholds to expand the obtained coexpression network. For instance, an increase in the number of top genes from "10" to "20" results in a coexpression network of 60 genes and 383 edges in which *EGL3* shares many coexpressed genes with *ATMYB0* (or *GL1*) as in the first network, but has also a number of specific coexpression partners, whereas the same unknown gene coexpresses with *PAP1* (or *MYB75*). Finally, a decrease in the PCC threshold to 0.75 clearly delineates the three coexpression clusters with *EGL3* as a hub between two clusters. These observations are in agreement with the described partially redundant functions of *TT8* and *EGL3* (and *GL3*) [37]. Unfortunately, *GL3* was not included in our network because this gene is not represented on the Affymetrix ATH1 array. An overview of the different runs is given in Fig. 2 and the final network is shown in Fig. 3.

3.2.3 Functional Interpretation of Coexpression Networks

1. To obtain functional information on the genes in the network, use the Node Attribute Browser of Cytoscape (lower panel) that allows viewing of the different functional annotations (*see* Subheading 2.6 and Fig. 3) (*see* **Note 5**).
2. For information on the coexpression links, such as the number of predefined expression datasets that show coexpression (attribute "matching datasets") and the correlation coefficients

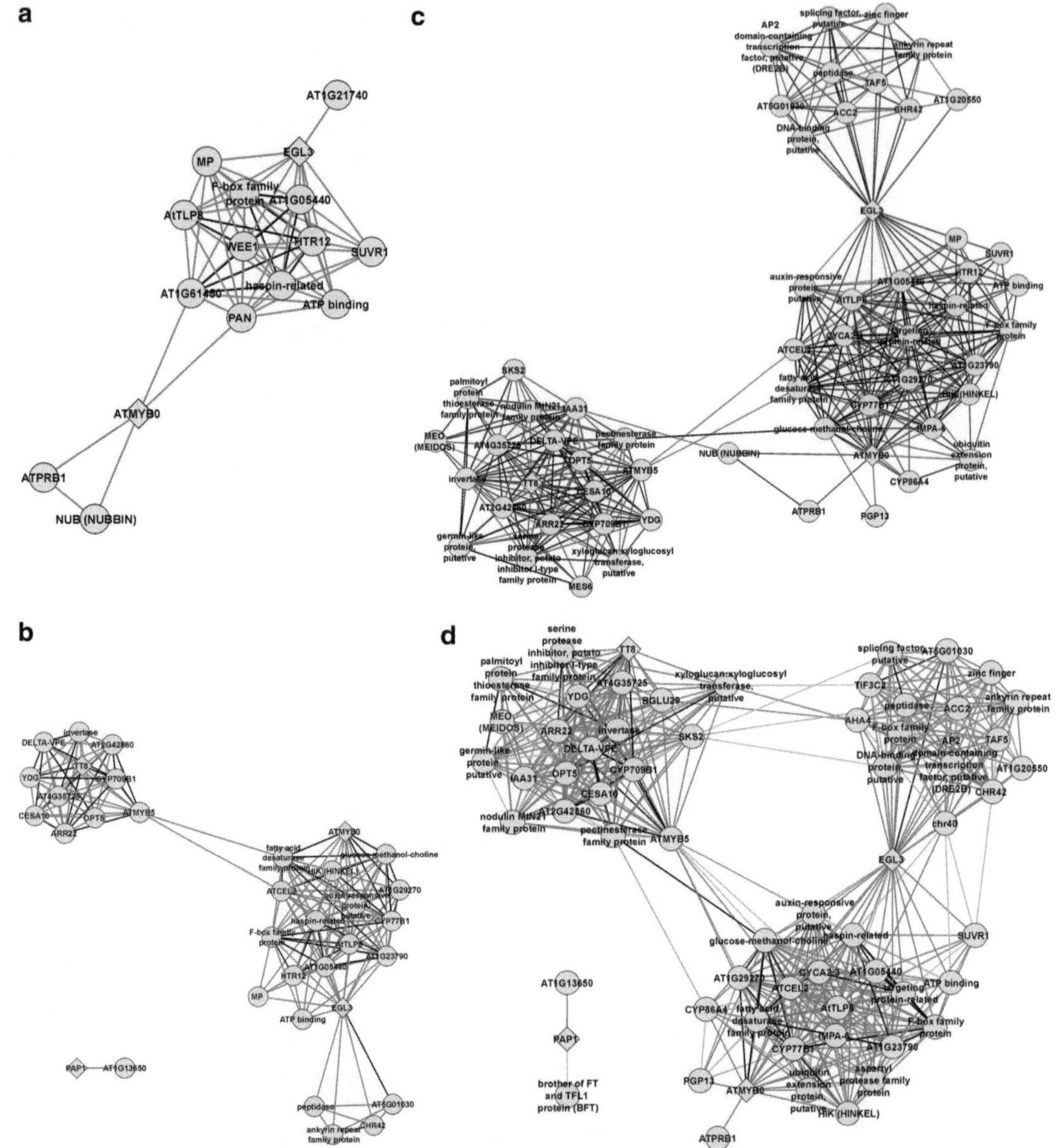

Fig. 2 Different networks resulting from subsequent runs of the coexpression tool. (**a**) Correlations with neighbors + Correlations between neighbors + 10 neighbors + PCC > 0.8 + *p*-value < 0.05 + global expression datasets. (**b**) Correlations with neighbors + Correlations between neighbors + 10 neighbors + PCC > 0.8 + *p*-value < 0.05 + global expression datasets + specific expression datasets. (**c**) Correlations with neighbors + Correlations between neighbors + 20 neighbors + PCC > 0.8 + *p*-value < 0.05 + global expression datasets + specific expression datasets. (**d**) Correlations with neighbors + Correlations between neighbors + 20 neighbors + PCC > 0.75 + *p*-value < 0.05 + global expression datasets + specific expression datasets

for each of the expression datasets (attribute "dataset coefficients"), utilize the Edge Attribute Browser of Cytoscape.

3. In addition, make use of the Cytoscape LinkOut to go back to the original data source by right-clicking on a gene in the network and clicking "Plants_Arabidopsis" and the database of interest.

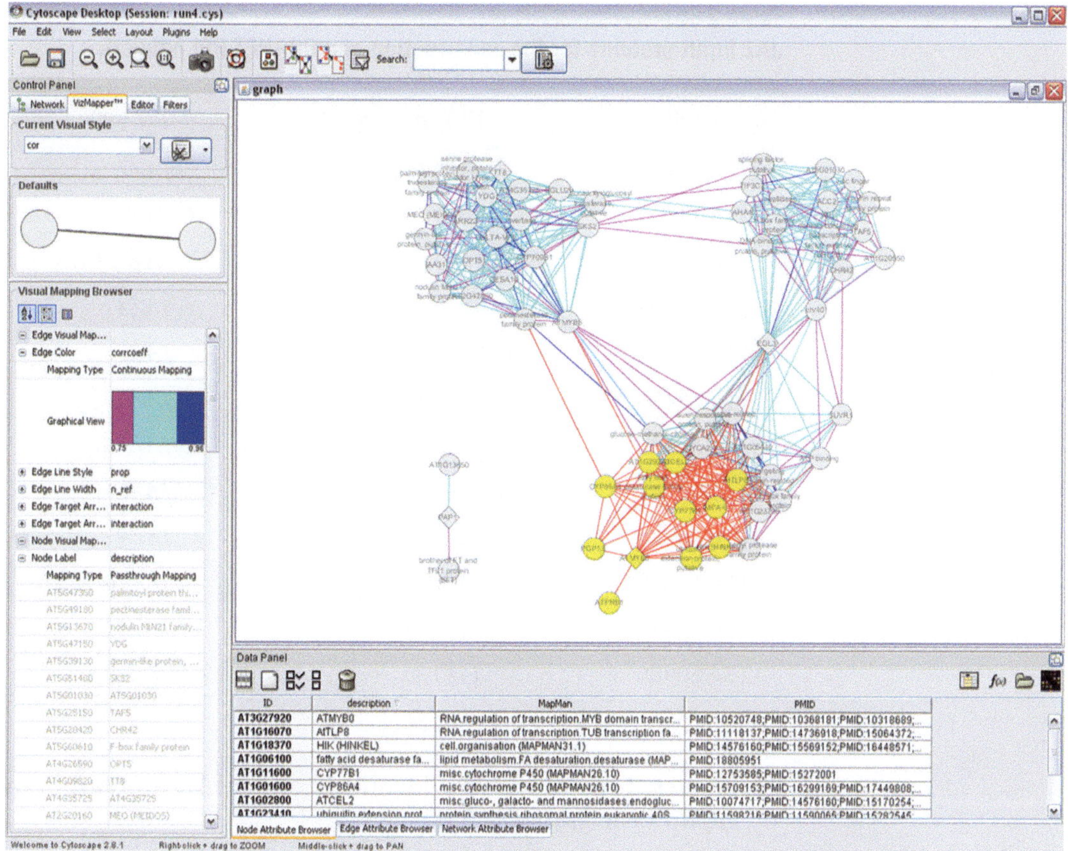

Fig. 3 Cytoscape visualization of the final coexpression network around JAZ-regulated genes

3.3 Construction of the JAZ Protein–Protein Interactome

JAZ proteins are part of a gene family of 12 members that are characterized by the presence of ZIM and Jas protein domains [36, 38–40]. JAZ proteins act as JA coreceptors together with CORONATINE INSENSITIVE1 (COI1), an F-box protein that is part of an Skp1/Cullin/F-box complex (SCF^{COI1}). In addition, these proteins function as transcriptional repressors of gene expression, because JAZ proteins are degraded in an SCF^{COI1}-dependent manner at low JA levels. To unravel the molecular mechanisms by which the different JAZ proteins control JA signaling, numerous interaction studies have identified JAZ targets.

Here, by means of CORNET, we compile known PPI data of JAZ proteins and construct a network of PPIs identified by different experimental procedures. Network visualization in Cytoscape allows us to browse the network and inspect the metadata assembled by CORNET, such as the number of times an interaction is identified, the type of experiment that was used to identify the interaction, and the publication that describes the interaction.

3.3.1 PPI Network Construction

1. Click on "PPI tool," through which the main interface will appear to define the CORNET search in three different steps.

Table 2
JAZ input proteins for the construction of a CORNET PPI network

AGI identifier	TIFY gene name	JAZ gene name
AT1G19180	*TIFY10a*	*JAZ1*
AT1G74950	*TIFY10b*	*JAZ2*
AT3G17860	*TIFY6b*	*JAZ3/JAI3*
AT1G48500	*TIFY6a*	*JAZ4*
AT1G17380	*TIFY11a*	*JAZ5*
AT1G72450	*TIFY11b*	*JAZ6*
AT2G34600	*TIFY5b*	*JAZ7*
AT1G30135	*TIFY5a*	*JAZ8*
AT1G70700	*TIFY7*	*JAZ9*
AT5G13220	*TIFY9*	*JAZ10/JAS1*
AT3G43440	*TIFY3a*	*JAZ11*
AT5G20900	*TIFY3b*	*JAZ12*

2. In **step 1**, paste the AGI identifiers of the JAZ proteins in the input form (*see* Table 2).
3. In **step 2**, select the databases and type of interactions to be retrieved. For instance, select all PPI databases by clicking "Select all" and choose to query experimental data only by clicking "All experimental" (*see* **Note 6**).
4. In **step 3**, specify the search options. For example, by selecting "Interactions of query protein(s) with neighbors," not only all possible interactions between JAZ proteins are retrieved but also interactions with other *Arabidopsis* proteins.
5. In addition, check whether these other proteins interact as well by selecting "Interactions between neighbors."
6. Finally, by selecting the appropriate fields next to the "GO" button, choose either to continue to the coexpression or the TF tool or to visualize localization information on the generated network.
7. When all parameters are set, click the "GO" button to initialize the CORNET database search. Once all PPI data and metadata for interactions as well as proteins in the network are retrieved, they are reformatted to an XML-based network file readable by Cytoscape (xgmml format) that starts automatically. The network will be displayed in a fully functional Cytoscape session.

3.3.2 Network Integration with Cytoscape

Compared to the retrieval of each of the datasets separately, the advantage of CORNET is compelling because by clicking a few buttons, all PPI data of interest are retrieved and displayed. The completeness of the PPI network depends on whether or not the original PPI data had been submitted to one of the PPI databases compiled by CORNET (*see* Subheading 2.3). In the case of the JAZ protein interactome, a number of PPIs that had been reported previously in the literature are missing, namely, interactions of JAZ proteins with R2R3MYB and DELLA proteins [35, 41], besides some interactions between NINJA and JAZ proteins [42].

1. To integrate these interactions with the CORNET network, use the Cytoscape functionalities that allow importing and merging of networks by means of a generated tab-delimited text file of the missing PPI data extracted manually from the literature (*see* Table 3) [40].
2. Import the data into Cytoscape by clicking the "Import Network from Table" wizard (Cytoscape > File > Import) that uses Text or MS Excel inputs.
3. Select the input file containing the missing PPI data.
4. To merge this network with the generated CORNET network, opt for the columns containing the AGI identifiers by choosing column 3 as "Source interaction" and column 4 as "Target interaction."
5. Click "Text file import options," thereby changing the "Network Import Options."
6. Replace "pp" by "new."
7. Click "Import" to generate a new network, named according to the file name PPI_JAZ_missing_ID.txt (see "Network" tab on the left panel).
8. Subsequently, merge the CORNET network and the new network with missing PPIs with the plug-in "Advanced Network Merge" that links the network "graph" (i.e., the CORNET network) and the "PPI_JAZ_missing_ID.txt" network.
9. Select the network and click the arrow pointing to the right panel.
10. When both networks appear in the "Selected networks" window, click the "Merge" button, generating a new "Union" network.
11. To distinguish the CORNET PPIs from the new PPIs, change the edge style of the interactions with the Vizmapper tool (left panel), in which the attribute "interaction" is assigned to the "Edge Line Style" parameter through discrete mapping in the "Visual Mapping Browser." A network is obtained in which the new PPIs are displayed as dashed edges (Fig. 4).

Table 3
Missing JAZ PPI data

Protein 1	Protein 2	AGI protein 1	AGI protein 2
JAZ1	GL3	AT1G19180	AT5G41315
JAZ1	EGL3	AT1G19180	AT1G63650
JAZ1	TT8	AT1G19180	AT4G09820
JAZ1	PAP1	AT1G19180	AT1G56650
JAZ1	GL1	AT1G19180	AT3G27920
JAZ1	GAI	AT1G19180	AT1G14920
JAZ1	RGA	AT1G19180	AT2G01570
JAZ1	RGL1	AT1G19180	AT1G66350
JAZ3	RGA	AT3G17860	AT2G01570
JAZ8	GL3	AT1G30135	AT5G41315
JAZ8	EGL3	AT1G30135	AT1G63650
JAZ8	TT8	AT1G30135	AT4G09820
JAZ8	PAP1	AT1G30135	AT1G56650
JAZ8	GL1	AT1G30135	AT3G27920
JAZ9	GL3	AT1G70700	AT5G41315
JAZ9	EGL3	AT1G70700	AT1G63650
JAZ9	TT8	AT1G70700	AT4G09820
JAZ9	RGA	AT1G70700	AT2G01570
JAZ10	GL3	AT5G13220	AT5G41315
JAZ10	EGL3	AT5G13220	AT1G63650
JAZ10	TT8	AT5G13220	AT4G09820
JAZ10	GL1	AT5G13220	AT3G27920
JAZ11	GL3	AT3G43440	AT5G41315
JAZ11	EGL3	AT3G43440	AT1G63650
JAZ11	TT8	AT3G43440	AT4G09820
JAZ11	GL1	AT3G43440	AT3G27920
JAZ11	PAP1	AT3G43440	AT1G56650
NINJA	JAZ2	AT4G28910	AT1G74950
NINJA	JAZ3	AT4G28910	AT3G17860
NINJA	JAZ4	AT4G28910	AT1G48500

(continued)

Table 3 (continued)

Protein 1	Protein 2	AGI protein 1	AGI protein 2
NINJA	JAZ6	AT4G28910	AT1G72450
NINJA	JAZ9	AT4G28910	AT1G70700
NINJA	JAZ10	AT4G28910	AT5G13220
NINJA	JAZ11	AT4G28910	AT3G43440
NINJA	JAZ12	AT4G28910	AT5G20900

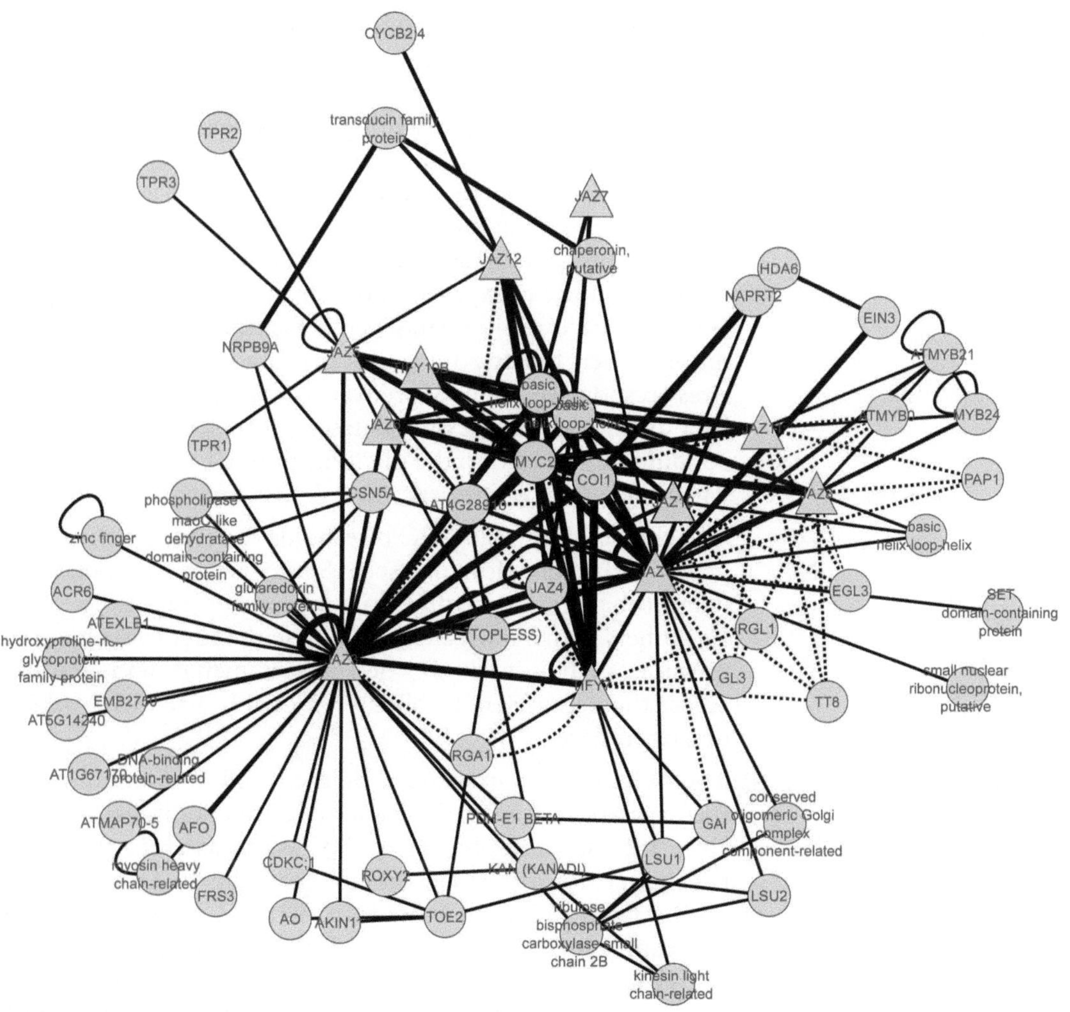

Fig. 4 Protein–protein interaction network around JAZ proteins. *Solid and dashed edges* represent experimental PPIs retrieved from CORNET and from the literature, respectively

4 Notes

1. CORNET can be accessed through the URL https://bioinformatics.psb.ugent.be/cornet. The tool is fully functional in Firefox and Safari browsers. First-time users might need to accept a certificate before accessing the Web site. The site is ideally viewed at a 1,280 × 1,024 resolution. Pop-ups need to be allowed in the browser before the "GO" button is pressed. After calculations and database queries, Cytoscape will start automatically from the Web. In other words, Cytoscape does not have to be installed, but an up-to-date version of Java is required to enable the Cytoscape WebStart. Please refer to the FAQ page for further details.
2. The choice to use one or more predefined expression datasets is coupled to the biological question to be addressed. Several predefined expression datasets are provided, such as global expression datasets representing diverse experimental conditions and tissue-specific or treatment-specific expression datasets. Depending on the nature of the studied genes and the interest, different input expression datasets can be imagined. Global expression datasets will be used when a general view is required on the coexpression of, for instance, unknown genes and specific expression datasets representing abiotic stress conditions when looking for genes that are similar to a drought stress-responsive gene. Moreover, coexpression can be calculated by multiple expression datasets, corresponding to diverse conditions, and lead to the identification of those conditions in which the genes of interest show similar expression patterns.
3. The choice of suitable PCC thresholds depends on the type of genes and expression datasets. Some genes coexpress with many other genes (tightly transcriptionally regulated genes) and others with a few genes. We advise the user to start a coexpression analysis with the default parameters and gradually adjust the parameters to obtain reasonably sized networks (as described here).
4. The PCC value is influenced by the number of conditions in the expression dataset. As the number of conditions considerably varies between predefined expressions datasets, the *p*-value is a better coexpression measure than the PCC to compare results from different expression datasets. When more than one expression dataset is selected, the PCCs and *p*-values will be retrieved from the database (PCC >0.4 or PCC <-0.4, *p*-value <0.05) and displayed as Cytoscape attributes. When only one predefined expression dataset is used, either Pearson's or Spearman's correlation coefficients are calculated at the moment allowing to retrieve correlations between PCC $= 0.4$ and PCC $= -0.4$ as well.

5. Complementary to a gene-by-gene functional analysis of the network as described in Subheading 3.2.3, functional enrichment can be analyzed by using, for instance, the Cytoscape plug-in BiNGO [43] that allows the user to verify whether the genes in the obtained network have similar molecular functions or are involved in similar biological processes. In addition, plug-ins are interesting for downstream analysis of the CORNET networks, such as Agilent Literature Search (with network representation of the text mining results), MCODE (connectivity-based clustering), jActiveModules (visualizing network dynamics), Network Analysis (calculating network metrics such as node degree and shortest path length), Network Modifications (union, intersection, difference of networks), and CytoSQL (importing data from local mySQL databases).
6. Although some selected databases contain only computationally predicted interactions, data will not be retrieved from these databases thanks to the "All experimental" selection.

Acknowledgments

We would like to thank Frederik Delaere, Michiel Van Bel, and Lieven Sterck for assistance and helpful suggestions. This work was supported by grants from Ghent University ("Bijzonder Onderzoeksfonds Methusalem project" no. BOF08/01M00408), the Interuniversity Attraction Poles Programme (IUAP VI/25 and VI/33), initiated by the Belgian State, Science Policy Office, the European Union 6th Framework Programme ("AGRONOMICS," LSHG-CT-2006-037704), and the Research Foundation—Flanders (postdoctoral fellowship to S.D.B.).

References

1. Brown DM, Zeef LAH, Ellis J, Goodacre R, Turner SR (2005) Identification of novel genes in Arabidopsis involved in secondary cell wall formation using expression profiling and reverse genetics. Plant Cell 17:2281–2295
2. Gachon CMM, Langlois-Meurinne M, Henry Y, Saindrenan P (2005) Transcriptional co-regulation of secondary metabolism enzymes in *Arabidopsis*: functional and evolutionary implications. Plant Mol Biol 58:229–245
3. Lisso J, Steinhauser D, Altmann T, Kopka J, Müssig C (2005) Identification of brassinosteroid-related genes by means of transcript co-response analyses. Nucleic Acids Res 33:2685–2696
4. Rautengarten C, Steinhauser D, Büssis D, Stintzi A, Schaller A, Kopka J, Altmann T (2005) Inferring hypotheses on functional relationships of genes: analysis of the *Arabidopsis thaliana* subtilase gene family. PLoS Comput Biol 1(e40):0297–0312
5. Usadel B, Obayashi T, Mutwil M, Giorgi FM, Bassel GW, Tanimoto M, Chow A, Steinhauser D, Persson S, Provart NJ (2009) Co-expression tools for plant biology: opportunities for hypothesis generation and caveats. Plant Cell Environ 32:1633–1651
6. De Bodt S, Carvajal D, Hollunder J, Van den Cruyce J, Movahedi S, Inzé D (2010) CORNET: a user-friendly tool for data mining and integration. Plant Physiol 152: 1167–1179
7. De Bodt S, Hollunder J, Meulemeester N, Nelissen H, Inzé D (2012) CORNET 2.0: Integrating plant co-expression, protein–protein interactions, regulatory interactions, gene associations and functional annotations. New Phytol 195(3):707–720

8. Shannon P, Markiel A, Ozier O, Baliga NS, Wang JT, Ramage D, Amin N, Schwikowski B, Ideker T (2003) Cytoscape: a software environment for integrated models of biomolecular interaction networks. Genome Res 13:2498–2504
9. Irizarry RA, Bolstad BM, Collin F, Cope LM, Hobbs B, Speed TP (2003) Summaries of Affymetrix GeneChip probe level data. Nucleic Acids Res 31:e15
10. Irizarry RA, Hobbs B, Collin F, Beazer-Barclay YD, Antonellis KJ, Scherf U, Speed TP (2003) Exploration, normalization, and summaries of high density oligonucleotide array probe level data. Biostatistics 4:249–264
11. Gautier L, Cope L, Bolstad BM, Irizarry RA (2004) affy–analysis of Affymetrix GeneChip data at the probe level. Bioinformatics 20:307–315
12. Gentleman RC, Carey VJ, Bates DM, Bolstad B, Dettling M, Dudoit S, Ellis B, Gautier L, Ge Y, Gentry J, Hornik K, Hothorn T, Huber W, Iacus S, Irizarry R, Leisch F, Li C, Maechler M, Rossini AJ, Sawitzki G, Smith C, Smyth G, Tierney L, Yang JYH, Zhang J (2004) Bioconductor: open software development for computational biology and bioinformatics. Genome Biol 5:R80
13. Bader GD, Betel D, Hogue CWV (2003) BIND: the Biomolecular Interaction Network Database. Nucleic Acids Res 31:248–250
14. Hermjakob H, Montecchi-Palazzi L, Lewington C, Mudali S, Kerrien S, Orchard S, Vingron M, Roechert B, Roepstorff P, Valencia A, Margalit H, Armstrong J, Bairoch A, Cesareni G, Sherman D, Apweiler R (2004) IntAct: an open source molecular interaction database. Nucleic Acids Res 32:D452–D455
15. Stark C, Breitkreutz B-J, Reguly T, Boucher L, Breitkreutz A, Tyers M (2006) BioGRID: a general repository for interaction datasets. Nucleic Acids Res 34:D535–D539
16. Salwinski L, Miller CS, Smith AJ, Pettit FK, Bowie JU, Eisenberg D (2004) The database of interacting proteins: 2004 update. Nucleic Acids Res 32:D449–D451
17. Chatr-aryamontri A, Ceol A, Montecchi Palazzi L, Nardelli G, Schneider MV, Castagnoli L, Cesareni G (2007) MINT: the Molecular INTeraction database. Nucleic Acids Res 35:D572–D574
18. Rhee SY, Beavis W, Berardini TZ, Chen G, Dixon D, Doyle A, Garcia-Hernandez M, Huala E, Lander G, Montoya M, Miller N, Mueller LA, Mundodi S, Reiser L, Tacklind J, Weems DC, Wu Y, Xu I, Yoo D, Yoon J, Zhang P (2003) The *Arabidopsis* Information Resource (TAIR): a model organism database providing a centralized, curated gateway to *Arabidopsis* biology, research materials and community. Nucleic Acids Res 31:224–228
19. Lalonde S, Sero A, Pratelli R, Pilot G, Chen J, Sardi MI, Parsa SA, Kim D-Y, Acharya BR, Stein EV, Hu H-C, Villiers F, Takeda K, Yang Y, Han YS, Schwacke R, Chiang W, Kato N, Loqué D, Assmann SM, Kwak JM, Schroeder JI, Rhee SY, Frommer WB (2010) A membrane protein/signaling protein interaction network for *Arabidopsis* version AMPv2. Front Physiol 1:24
20. Klopffleisch K, Phan N, Augustin K, Bayne RS, Booker KS, Botella JR, Carpita NC, Carr T, Chen J-G, Cooke TR, Frick-Cheng A, Friedman EJ, Fulk B, Hahn MG, Jiang K, Jorda L, Kruppe L, Liu C, Lorek J, McCann MC, Molina A, Moriyama EN, Mukhtar MS, Mudgil Y, Pattathil S, Schwarz J, Seta S, Tan M, Temp U, Trusov Y, Urano D, Welter B, Yang J, Panstruga R, Uhrig JF, Jones AM (2011) Arabidopsis G-protein interactome reveals connections to cell wall carbohydrates and morphogenesis. Mol Syst Biol 7:532
21. Braun P, Carvunis A-R, Charloteaux B, Dreze M, Ecker JR, Hill DE, Roth FP, Vidal M, Galli M, Balumuri P, Bautista V, Chesnut JD, Kim RC, de los Reyes C, Gilles P, Kim CJ, Matrubutham U, Mirchandani J, Olivares E, Patnaik S, Quan R, Ramaswamy G, Shinn P, Swamilingiah GM, Wu S, Byrdsong D, Dricot A, Duarte M, Gebreab F, Gutierrez BJ, MacWilliams A, Monachello D, Mukhtar MS, Poulin MM, Reichert P, Romero V, Tam S, Waaijers S, Weiner EM, Cusick ME, Tasan M, Yazaki J, Ahn Y-Y, Barabási A-L, Chen H, Dangl JL, Fan C, Gai L, Ghoshal G, Hao T, Lurin C, Milenkovic T, Moore J, Pevzner SJ, Przulj N, Rabello S, Rietman EA, Rolland T, Santhanam B, Schmitz RJ, Spooner W, Stein J, Vandenhaute J, Ware D (2011) Evidence for network evolution in an *Arabidopsis* interactome map. Science 333:601–607
22. Geisler-Lee J, O'Toole N, Ammar R, Provart NJ, Millar AH, Geisler M (2007) A predicted interactome for Arabidopsis. Plant Physiol 145:317–329
23. Cui J, Li P, Li G, Xu F, Zhao C, Li Y, Yang Z, Wang G, Yu Q, Li Y, Shi T (2008) AtPID: Arabidopsis thaliana protein interactome database–an integrative platform for plant systems biology. Nucleic Acids Res 36:D999–D1008
24. De Bodt S, Proost S, Vandepoele K, Rouzé P, Van de Peer Y (2009) Predicting protein–protein interactions in *Arabidopsis thaliana* through integration of orthology, gene ontology and co-expression. BMC Genomics 10:288

25. Lee I, Ambaru B, Thakkar P, Marcotte EM, Rhee SY (2010) Rational association of genes with traits using a genome-scale gene network for *Arabidopsis thaliana*. Nat Biotechnol 28:149–156
26. Yilmaz A, Mejia-Guerra MK, Kurz K, Liang X, Welch L, Grotewold E (2011) AGRIS: the Arabidopsis Gene Regulatory Information Server an update. Nucleic Acids Res 39:D1118–D1122
27. Smyth GK (2004) Linear models and empirical Bayes methods for assessing differential expression in microarray experiments. *Stat Appl Genet Mol Biol* 3, Article 3
28. Harris MA, Clark J, Ireland A, Lomax J, Ashburner M, Foulger R, Eilbeck K, Lewis S, Marshall B, Mungall C, Richter J, Rubin GM, Blake JA, Bult C, Dolan M, Drabkin H, Eppig JT, Hill DP, Ni L, Ringwald M, Balakrishnan R, Cherry JM, Christie KR, Costanzo MC, Dwight SS, Engel S, Fisk DG, Hirschman JE, Hong EL, Nash RS, Sethuraman A, Theesfeld CL, Botstein D, Dolinski K, Feierbach B, Berardini T, Mundodi S, Rhee SY, Apweiler R, Barrell D, Camon E, Dimmer E, Lee V, Chisholm R, Gaudet P, Kibbe W, Kishore R, Schwarz EM, Sternberg P, Gwinn M, Hannick L, Wortman J, Berriman M, Wood V, de la Cruz N, Tonellato P, Jaiswal P, Seigfried T, White R (2004) The Gene Ontology (GO) database and informatics resource. Nucleic Acids Res 32:D258–D261
29. Usadel B, Nagel A, Thimm O, Redestig H, Blaesing OE, Palacios-Rojas N, Selbig J, Hannemann J, Piques MC, Steinhauser D, Scheible W-R, Gibon Y, Morcuende R, Weicht D, Meyer S, Stitt M (2005) Extension of the visualization tool MapMan to allow statistical analysis of arrays, display of corresponding genes, and comparison with known responses. Plant Physiol 138:1195–1204
30. Heazlewood JL, Verboom RE, Tonti-Filippini J, Small I, Millar AH (2007) SUBA: the Arabidopsis Subcellular Database. Nucleic Acids Res 35:D213–D218
31. Avraham S, Tung C-W, Ilic K, Jaiswal P, Kellogg EA, McCouch S, Pujar A, Reiser L, Rhee SY, Sachs MM, Schaeffer M, Stein L, Stevens P, Vincent L, Zapata F, Ware D (2008) The plant ontology database: a community resource for plant structure and developmental stages controlled vocabulary and annotations. Nucleic Acids Res 36:D449–D454
32. Hunter S, Apweiler R, Attwood TK, Bairoch A, Bateman A, Binns D, Bork P, Das U, Daugherty L, Duquenne L, Finn RD, Gough J, Haft D, Hulo N, Kahn D, Kelly E, Laugraud A, Letunic I, Lonsdale D, Lopez R, Madera M, Maslen J, McAnulla C, McDowall J, Mistry J, Mitchell A, Mulder N, Natale D, Orengo C, Quinn AF, Selengut JD, Sigrist CJA, Thimma M, Thomas PD, Valentin F, Wilson D, Wu CH, Yeats C (2009) InterPro: the integrative protein signature database. Nucleic Acids Res 37:D211–D215
33. Ishida T, Kurata T, Okada K, Wada T (2008) A genetic regulatory network in the development of trichomes and root hairs. Annu Rev Plant Biol 59:365–386
34. Balkunde R, Pesch M, Hülskamp M (2010) Trichome patterning in *Arabidopsis thaliana*: from genetic to molecular models. Curr Topics Dev Biol 91:299–321
35. Qi T, Song S, Ren Q, Wu D, Huang H, Chen Y, Fan M, Peng W, Ren C, Xie D (2011) The jasmonate-ZIM-domain proteins interact with the WD-repeat/bHLH/MYB complexes to regulate jasmonate-mediated anthocyanin accumulation and trichome initiation in *Arabidopsis thaliana*. Plant Cell 23:1795–1814
36. Kazan K, Manners JM (2012) JAZ repressors and the orchestration of phytohormone crosstalk. Trends Plant Sci 17:22–31
37. Zhang F, Gonzalez A, Zhao M, Payne CT, Lloyd A (2003) A network of redundant bHLH proteins functions in all TTG1-dependent pathways of *Arabidopsis*. Development 130:4859–4869
38. Chico JM, Chini A, Fonseca S, Solano R (2008) JAZ repressors set the rhythm in jasmonate signaling. Curr Opin Plant Biol 11:486–494
39. Staswick PE (2008) JAZing up jasmonate signaling. Trends Plant Sci 13:66–71
40. Pauwels L, Goossens A (2011) The JAZ proteins: a crucial interface in the jasmonate signaling cascade. Plant Cell 23:3089–3100
41. Hou X, Lee LYC, Xia K, Yan Y, Yu H (2010) DELLAs modulate jasmonate signaling via competitive binding to JAZs. Dev Cell 19:884–894
42. Pauwels L, Fernández Barbero G, Geerinck J, Tilleman S, Grunewald W, Cuéllar Pérez A, Chico JM, Vanden Bossche R, Sewell J, Gil E, García-Casado G, Witters E, Inzé D, Long JA, De Jaeger G, Solano R, Goossens A (2010) NINJA connects the co-repressor TOPLESS to jasmonate signalling. Nature 464:788–791
43. Maere S, Heymans K, Kuiper M (2005) *BiNGO*: a Cytoscape plugin to assess overrepresentation of Gene Ontology categories in Biological Networks. Bioinformatics 21:3448–3449

Index

Alain Goossens and Laurens Pauwels (eds.), *Jasmonate Signaling: Methods and Protocols*, Methods in Molecular Biology, vol. 1011, DOI 10.1007/978-1-62703-414-2,

O

P

Q

R

S

T

U

W

X

Y

MIX
Papier aus verantwortungsvollen Quellen
Paper from responsible sources
FSC® C105338

If you have any concerns about our products,
you can contact us on
ProductSafety@springernature.com

In case Publisher is established outside the EU,
the EU authorized representative is:
Springer Nature Customer Service Center GmbH
Europaplatz 3, 69115 Heidelberg, Germany

Printed by Libri Plureos GmbH
in Hamburg, Germany